Soft Robotics

Alexander Verl • Alin Albu-Schäffer • Oliver Brock
Annika Raatz (Eds.)

Soft Robotics

Transferring Theory to Application

Springer

Editors
Alexander Verl
Fraunhofer IPA
Stuttgart, Germany

Oliver Brock
TU Berlin
Berlin, Germany

Alin Albu-Schäffer
DLR Institute of Robotics and Mechatronics
Oberpfaffenhofen, Germany

Annika Raatz
Leibniz Universität Hannover
Hannover, Germany

ISBN 978-3-662-51549-5 ISBN 978-3-662-44506-8 (eBook)
DOI 10.1007/978-3-662-44506-8

Springer

Softcover reprint of the hardcover 1st edition 2015

Printed on acid-free paper

Springer is a brand of Springer Berlin Heidelberg
Springer Berlin Heidelberg is part of Springer Science+Business Media
(www.springer.com)

Preface

For about the last ten years, the scientific world of robotics has been influenced by worldwide innovative stimuli from research inspired by biological processes. Instead of rigid structures, the trend is now towards the use of soft, pliable organic structures, materials, and surfaces. In doing so, scientists obtain their inspiration from diverse biological organisms such as humans, vertebrates, caterpillars, snakes, octopuses, starfish and plant roots. They try to understand natural mechanisms and use this information to develop a new generation of robots – "soft robots". This new class of robot may be utilized in unsafe, dynamic task environments, to grip and manipulate unknown objects, move around in rough terrain, interact with humans in top security situations and even be capable of the visionary research topic of self-repair.

In numerous initiatives, especially in the USA, Japan, Italy and Switzerland, but also in Germany, some of these technologies have already been transferred to initial applications. In view of this already advanced pioneering work, in the spring of 2014 we came to the conclusion that there is an urgent need to gain an overview of the state of research in selected European institutions. To do this, we organized a symposium that was held at Fraunhofer IPA in Stuttgart in Germany on 23rd and 24th June 2014. There, over the course of three plenary sessions and four forums, 30 scientists gave their assessment of the situation and reported on their first successes in the promising research field of »soft robotics«.

We took the opportunity, offered by such a positive response to the event, to ask speakers to contribute towards this book by submitting details of their projects. Among other things, the geographic locations of the authors also showed that quite a few universities are taking different approaches as they explore this new field of research.

We were always aware of the fact that this has only given us a glimpse of the situation and highlighted merely a handful of representative research concepts. The 22 contributions to the book are intended not only to intensify people's individual research efforts but also to encourage more interdisciplinarity. That's because we're only just beginning to develop the in-depth cooperations that are needed between engineers, biologists, material scientists, medical doctors, chemists and mathematicians to turn promising new ideas into radical innovations.

As well as the 77 authors, a few scientists were also particularly involved in completing the book. The editors would like to take this opportunity to thank them for their commitment. We would also like to express our thanks to he German Academic Society for Assembly, Handling and Industrial Robotics (MHI) and its president Bernd Kuhlenkötter as well as the directors Jörg Franke and Thorsten Schüppstuhl for their kind support in realizing the symposium and in planning this book.

Fig. 1 Geographic locations of the authors

We would especially like to thank Mr. Hans-Friedrich Jacobi for his input and cooperation and for managing the overall coordination of the various contributions to the book, as well as Mr. Landherr for preparing the book for print.

We would also like to thank Mrs. Hestermann-Beyerle and Mrs. Kollmar-Thoni from Springer Verlag for their excellent advice and encouraging support.

Alexander Verl   Alin Albu-Schäffer   Oliver Brock   Annika Raatz

Contents

Part I Outline

1 Introduction

There's no doubt that robots render valuable services in our day-to-day lives. The technological developments that have taken place over recent decades in the field of mechatronics form the basis of this, especially with the development of more refined sensors and advances in computing and software technologies. In general, product innovations and markets develop incrementally: the cost-benefit ratio improves from one generation to the next with regard to products as well as services. Thanks to the use of technologies inspired by biology, robotics is currently receiving unconventional impetus: instead of implementing the rigid mechanical structures of the past, a new robotics paradigm is now starting to focus on soft, pliable, sensitive, organic representations – on "soft robotics".

The term "soft robotics" has been deliberately coined by this emerging field of research to describe soft, organic embodiments with biologically-inspired sensors and knowledge processing combined with intuitive, safe and more sensitive interaction. To achieve this, materials in particular have to feature high levels of functional integration: electronics, sensors, and drive technology need to be linked more effectively with one another. Compact sensors and new manufacturing processes will be required, with microsystem sensors enabling robots to react reliably to the environment. These new robot systems demand different manufacturing methods: instead of conventional assembly, a greater emphasis will be placed on continuous processes, for example it will be possible to construct robots and their respective subassemblies additively using a "growing" technique. New product opportunities will also arise in the manufacturing environment through the use of "soft robotics" as a third helping hand, in the medical field with invasive micro robots used in diagnosis and treatment, in logistics with lifting aids worn on the body, and even in the home.

The aim of this book is to outline a framework for the above-mentioned lines of research concerning "soft robotics" that are still open, thus giving researchers and practitioners a better understanding of the fundamental aspects involved, the general requirements to be observed, and the concepts, methods, and tools they can use. Due to the intrinsic differences between traditional and soft robots, the concepts of both actuation and sensation need to be revisited, especially when it comes to interaction with the environment; for example soft robotic sensors may present at new levels such as internal, external and on the surface of robots. In order to reflect both the biologically-inspired approach and the development and construction of a "soft robot" or subassembly, new modeling and simulation methods and control processes need to be conceived and tested. In addition to the application of these principles, material issues, design methods and manufacturing requirements also play a decisive role in the process of: Transferring Theory to Applications. As far as potential fields of use are concerned, this process can be already outlined by means of initial presentable specimens of "soft robot" subassemblies. This allows the orientation framework to be broken down into the four following sections:

1. Sensors and actuators
2. Modeling, simulation and control
3. Materials, design and manufacturing
4. Soft robotics applications

On categorizing all of the contributions made into these sections, it was impossible to avoid some of the topics discussed from overlapping. This is due to the fact that the "soft robotics" field of research is very young and some areas need to be better defined. In this regard, there is still plenty of room for more definitions, models and calculations as well as for further experiments to be carried out. For example, four questions concerning the design of a "soft robot" still need answering:

- Does a differentiation have to be made between the features of "soft" and "pliable" within the context of a soft robot (definition)?
- What should a "soft robot" look like (morphology)?
- What should a "soft robot" be made from (material)?
- How should a "soft robot" move (locomotion)?

Of course, in the future many other questions will arise, be solved, or dismissed while continuing research in the field of soft robotics. The topics selected for the book are intended to set the ball rolling in order to gain more detailed knowledge and also aim to stimulate scientific discussion.

2 Abstracts

New Concepts for Distributed Actuators and Their Control

Welf-Guntram Drossel, Holger Schlegel, Michael Walther, Philipp Zimmermann, André Bucht

Fraunhofer Institute for Machine Tools and Forming Technology IWU, Chemnitz

Abstract Recently, decreasing costs for robots and control components have led to a broader acceptance of different kinds of robots. Hence, various fields of application start to flourish. As this is especially true for the field of service robotics it is typically implying an increasing physical human-machine-interaction. In this case a soft appearance yields major benefits, as it prevents injuries corresponding to an inherent safety of the system and, in theory, enables the robot to obtain virtually unlimited degrees of freedom. In this chapter the possibilities of the use of shape memory alloys for distributed actuators will be discussed by reference to application examples and implications for the control of such systems will be pointed out in detail.

Artificial Muscles, Made of Dielectric Elastomer Actuators - A Promising Solution for Inherently Compliant Future Robots

In Seong Yoo, Sebastian Reitelshöfer, Maximilian Landgraf, Jörg Franke

Friedrich-Alexander-Universität Erlangen-Nürnberg

Abstract The cutting-edge robotic technology can deal with a lot of complex tasks. However, one of the most challenging technological obstacles in robotics is the development of soft actuators. Remaining challenges in the field of drive technology can be overcome with innovative actuator concepts, for example dielectric elastomer actuators (DEAs). DEAs show numerous advantages in comparison to prevailing robotic actuators that are based on geared servomotors: They are form-flexible, inherently compliant, can store and recuperate kinetic energy, feature high power-to-weight ratio and high energy density that is comparable to human skeletal muscles, and finally can be designed to perform natural motion patterns other than rotation. In this article, after a review on disadvantages of state-of-art robotic drives, which are stimulus for a research on the promising drive solution, benefits of DEAs will be presented with regard to the possibility of applications in soft robotics. Finally, the article will conclude with a brief report on the ongoing research effort at the Institute for Factory Automation and Production Systems (FAPS) with two major foci – the development of an automated manufacturing process for stacked DEAs and a lightweight control hardware.

Musculoskeletal Robots and Wearable Devices on the Basis of Cable-driven Actuators

Martin Haegele, Christophe Maufroy, Werner Kraus, Maik Siee, Jannis Breuninger

Fraunhofer Institute of Manufacturing Engineering and Automation IPA, Stuttgart

Abstract Cable-driven actuators are a promising alternative for future kinematic designs, particularly when the combination of lightweight, high strength, compact designs and dynamic motions are required. Powered exoskeletons or wearable robots are typical candidates of these novel actuators as has been demonstrated by previous research. This chapter focusses on current work in cable-driven actuators, introduces the Myorobotics toolkit for supporting the engineer to build up prototypes from cable-actuates modules and gives an outlook to using cable-driven actuation for advanced wearable robots.

Capacitive Tactile Proximity Sensing: From Signal Processing to Applications in Manipulation and Safe Human-Robot Interaction

Stefan Escaida Navarro, Björn Hein, and Heinz Wörn

Karlsruhe Institute of Technology (KIT)

Abstract Recently we have shown developments on capacitive tactile proximity sensors (CTPS) and their applications. In this work we give an overview of these developments and put them into a more general perspective, emphasizing what the common grounds are for the different applications, i.e., preshaping and grasping, haptic exploration as well as collision avoidance and safe human-robot interaction. We discuss issues related to signal processing and the design of a smart skin for the robot arm and its end-effector. On a higher level we discuss the concept of proximity servoing and its use for the above mentioned applications.

Perception of Deformable Objects and Compliant Manipulation for Service Robots

Jörg Stückler and Sven Behnke

University of Bonn

Abstract We identified softness in robot control as well as robot perception as key enabling technologies for future service robots. Compliance in motion control compensates for small errors in model acquisition and estimation and enables safe physical interaction with humans. The perception of shape similarities and deformations allows a robot to adapt its skills to the object at hand, given a description of the skill that generalizes between different objects. In this chapter, we present our approaches to compliant control and object manipulation skill transfer for service robots. We report on evaluation results and public demonstrations of our approaches.

Soft Robot Control with a Behaviour-Based Architecture

Christopher Armbrust, Lisa Kiekbusch, Thorsten Ropertz, Karsten Berns

University of Kaiserslautern

Abstract In this chapter, we explain how behaviour-based approaches can be used to control soft robots. Soft robotics is a strongly growing field generating innovative concepts and novel systems. The term “soft” can refer to the basic structure, the actuators, or the sensors of these systems. The soft aspect results in a number of challenges that can only be solved with new modelling, control, and analysis methods whose novelty matches those of the hardware. We will present prior achievements in the area of behaviour-based systems and suggest their application in soft robots with the aim to increase the fault tolerance while improving the reaction to unexpected disturbances.

Optimal Exploitation of Soft-Robot Dynamics

Sami Haddadin

Leibniz University Hannover

Abstract Inspired by the elasticity contained in human muscles, elastic soft robots are designed with the aim of imitating motions as observed in humans or animals. Especially reaching peak velocities using stored energy in the springs is a task of significant interest. In this chapter, general results on maximizing a soft-robot's end-point velocity by using elastic joint energy are presented and discussed.

Simulation Technology for Soft Robotics Applications

Jürgen Roßmann, Michael Schluse, Malte Rast, Eric Guiffo Kaigom, Torben Cichon

RWTH Aachen University

Abstract Soft robots are implied to be inherently safe, and thus "compatible", not only with human coworkers in a production environment, but also with the "family around the house". Such soft robots today still hold numerous new challenges for their design and control, for their commanding and supervision approaches as well as for human-robot interaction concepts. The research field of eRobotics is currently underway to provide a modern basis for efficient soft robotic developments. The objective is to effectively use electronic media - hence the "e" at the beginning of the term - to achieve the best possible advance in the research field. A key feature of eRobotics is its capability to join multiple process simulation components under one "software roof" to build "Virtual Testbeds", i.e. to alleviate the dependancy on physical prototypes and to provide a comprehensive tool chain support for the analysis, development, testing, optimization, deployment and commanding of soft robots.

Concepts of Softness for Legged Locomotion and Their Assessment

Andre Seyfarth, Katayon Radkhah, Oskar von Stryk

Technische Universität Darmstadt

Abstract In human and animal locomotion, compliant structures play an essential role in the body and actuator design. Recently, researchers have started to exploit these compliant mechanisms in robotic systems with the goal to achieve the yet superior motions and performances of the biological counterpart. For instance, compliant actuators such as series elastic actuators (SEA) can help to improve the energy efficiency and the required peak power in powered prostheses and exoskeletons. However, muscle function is also associated with damping-like characteristics complementing the elastic function of the tendons operating in series to the muscle fibers. Carefully designed conceptual as well as detailed motion dynamics models are key to understanding the purposes of softness, i.e. elasticity and damping, in human and animal locomotion and to transfer these insights to the design and control of novel legged robots. Results for the design of compliant legged systems based on a series of conceptual biomechanical models are summarized. We discuss how these models compare to experimental observations of human locomotion and how these models could be used to guide the design of legged robots and also how to systematically evaluate and compare natural and robotic legged motions.

Mechanics and Thermodynamics of Biological Muscle - A Simple Model Approach

Syn Schmitt and Daniel Haeufle

University of Stuttgart

Abstract Macroscopic muscle models allow for a detailed analysis of the mechanic and thermodynamic function of biological muscles. Here we summarize results from various simulation studies which emphasize the extraordinary design features of biological muscles. Discussed are the benefits resulting from (2) wobbling masses and the muscles soft-tissue inertia effects, (2) biological damping,

(3) internal mass distribution, (4) stabilising properties of active muscles in upright stance and periodic hopping, (5) reduced control effort due to these stabilising effects. We present approaches to systematically transfer these results to technical actuators and exploit these properties in the next generation of functional artificial muscles.

Nanostructured Materials for Soft Robotics – Sensors and Actuators

Raphael Addinall[1], Thomas Ackermann[1,2] and Ivica Kolaric[1]

[1] Fraunhofer Institute for Manufacturing Engineering and Automation IPA, Stuttgart

[2] Graduate School of Excellence advanced Manufacturing Engineering (GSaME), University of Stuttgart

Abstract The advances in nanotechnology during the past two decades have led to several breakthroughs in material sciences. Ongoing and future tasks are related to the transfer of the unique properties of nanostructured materials to the macroscopic behaviour of composite structures and the system integration of novel materials for improved mechanical, electronic and optical devices. Nanostructured carbons, especially carbon nanotubes, are promising candidates as novel material for future applications in several fields. One of the big aims is the utilisation of the unique intrinsic mechanical and electronic properties of carbon nanotubes for sensing and actuation devices. The combination of excellent electrical conductivity and mechanical deformation makes carbon nanotubes ideal for applications in sensors and actuators and opens new possibilities in construction design of next generation robotic systems, which can be built with soft, bendable and stretchable materials. This chapter gives a brief overview on the properties of carbon nanotubes and their potential for actuators and sensors in soft robotics.

Fibrous Materials and Textiles for Soft Robotics

M. Milwich[1,2], S.K. Selvarayan[1], G.T. Gresser[1]

[1]Institute of Textile Technology & Process Engineering Denkendorf, [2]Hochschule Reutlingen

Abstract Soft, mechanically compliant robots are developed to safely interact with a "human environment". The use of textiles and fibrous (composite-) materials for the fabrication of robots opens up new possibilities for "Softness/Compliance" and safety in human-robot interaction. Besides external motion monitoring systems, textiles allow on-board monitoring and early prediction, or detection, of robot-human contact. The use of soft fibers and textiles for robot skins can increase the acceptance of robots in human surroundings. Novel topology optimization tools, materials, processing technologies and biomimetic engineering allow developing ultra-light-weight, multifunctional, and adaptive structures.

Opportunities and Challenges for the Design of Inherently Safe Robots

Annika Raatz, Sebastian Blankemeyer, Gundula Runge, Christopher Bruns, Gunnar Borchert

Leibniz University Hannover

Abstract An approach for solving the challenges that arise from the increased complexity of modern assembly tasks is believed to be human robot co-operation. In these hybrid workplaces humans and robots do not only work on the same task or interact during certain assembly steps, but also have overlapping workspaces. Therefore, 'safe robots' should be developed that do not harm workers in case of a collision. In this chapter, an overview of methods for designing a hardware based soft robot that is inherently safe in human-machine interaction is given. Recent projects show that robots could be soft enough for interaction but they are not able to resist forces that occur in the assembly process. Current solutions show that the designer of such robots must face a trade-off between softness and dexterity on the one hand and rigidity and load carrying capabilities on the other hand. A promising approach is to integrate variable stiffness elements in the robotic system. The chapter classifies two main design rules to achieve stiffness variability, the tuning of material properties and geometric parameters. Existing solutions are described and four concepts are presented to show how different mechanisms and materials could be combined to design safe assembly robots with a variable stiffness structure.

Aspects of Human Engineering – Bio-optimized Design of Wearable Machines

C. Hochberg, O. Schwarz, U. Schneider

Fraunhofer Institute for Manufacturing Engineering and Automation IPA, Stuttgart

Abstract This chapter outlines important factors for the design process of wearable robots. First, the challenges are discussed and possible user groups are detailed and categories of devices given. Then, major differences of classical design methods from the field of robotics are illustrated. This is due to linking between the machine and the user and challenges of user intention detection. Finally, some design approaches, guidelines and best practices for the development of wearable devices are discussed.

3D Printed Objects and Components Enabling Next Generation of True Soft Robotics

Andreas Fischer, Steve Rommel, Alexander Verl

Fraunhofer Institute for Manufacturing Engineering and Automation IPA, Stuttgart

Abstract Soft robotics in the content of true softness, with regards to components, parts, or the complete robot, are the next step in the development of tools for humans, especially when used in close proximity. Considering the fact that robots are a multilevel extension of the human body, and that their main purpose should be to help humans perform tasks, then focusing on the development of soft-materials, and product design options allowing for flexibility and softness by design is necessary, for the next development level of the tool "robot". Using additive manufacturing in combination with new materials, design methods, and biomimicry / biomimetics is a key in that development, but also very challenging due to the multi-level complexity. An understanding of the real world tasks required to be performed, and abstracting this information into new applications and robotic designs in the combination mentioned above, is shown in the chapter, functioning as a basis and overview of the state-of-the-art.

Soft Hands for Reliable Grasping Strategies

Raphael Deimel and Oliver Brock

University of Technology Berlin

Abstract Recent insights into human grasping show that humans exploit constraints to reduce uncertainty and reject disturbances during grasping. We propose to transfer this principle to robots and build robust and reliable grasping strategies from interactions with environmental constraints. To make implementation easy, hand hardware has to provide compliance, low inertia, low reaction delays and robustness to collision. Pneumatic continuum actuators such as PneuFlex actuators provide these properties. Additionally they are easy to customize and cheap to manufacture. We present an anthropomorphic hand built with PneuFlex actuators and demonstrate the ease of implementing a robust multi-stage grasping strategy relying on environmental constraints.

Task-specific Design of Tubular Continuum Robots for Surgical Applications

Jessica Burgner-Kahrs

Leibniz University Hannover

Abstract Tubular continuum robots are the smallest among continuum robots. They are composed of multiple, precurved, superelastic tubes. The design space for tubular continuum robots is infinite: each one of the component tubes can be individually parameterized in terms of its length, segmental curvatures, diameter, and material properties. Ad-hoc selection of those parameters is extremely challenging, since the elastic coupling of concentrically arranged and actuated tubes is hard to predict with common sense, especially under the presence of workspace constraints. In this chapter, an overview of the design process is given and the current state of the art in task-specific design of tubular continuum robots is reviewed.

Soft Robotics with Variable Stiffness Actuators: Tough Robots for Soft Human Robot Interaction

Sebastian Wolf, Thomas Bahls, Maxime Chalon, Werner Friedl, Markus Grebenstein, Hannes Höppner, Markus Kühne*, Dominic Lakatos, Nico Mansfeld, Mehmet Can Özparpucu, Florian Petit, Jens Reinecke, Roman Weitschat and Alin Albu-Schäffer

German Aerospace Center (DLR)

Abstract Robots that are not only robust, dynamic, and gentle in the human robot interaction, but are also able to perform precise and repeatable movements, need accurate dynamics modeling and a high-performance closed-loop control. As a technological basis we propose robots with intrinsically compliant joints, a stiff link structure, and a soft shell. The flexible joints are driven by Variable Stiffness Actuators (VSA) with a mechanical spring coupling between the motor and the actuator output and the ability to change the mechanical stiffness of the spring coupling. Several model based and model free control approaches have been developed for this technology, e.g. Cartesian stiffness control, optimal control, reactions, reflexes, and cyclic motion control.

Soft Robotics Research, Challenges, and Innovation Potential, Through Showcases

Cecilia Laschi

The BioRobotics Institute, Scuola Superiore Sant'Anna, Pisa

Abstract Soft robotics, intended as the use of soft materials in robotics, is a young yet promising and growing research field. The need for soft robots emerged in robotics, for facing unstructured environments, and in artificial intelligence, too, for implementing the embodied intelligence, or morphological computation, paradigm, which attributes a stronger role to the bodyware and its interaction with the environment. Using soft materials for building robots poses new technological challenges: the technologies for actuating soft materials, for embedding sensors into soft robot parts, for controlling soft robots are among the main ones. Though still in its early stages of development, soft robotics is finding its way in a variety of applications, where safe contact is a main issue, in the biomedical field, as well as in exploration tasks and in the manufacturing industry. Literature in soft robotics is increasingly rich, though scattered in many disciplines. The soft robotics community is growing worldwide and initiatives are being taken, at international level, for consolidating this community and strengthening its potential for disruptive innovation.

Flexible Robot for Laser Phonomicrosurgery

Dennis Kundrat, Andreas Schoob, Lüder A. Kahrs, Tobias Ortmaier

Institute of Mechatronic Systems, Leibniz University Hannover

Abstract In this contribution we present a customized flexible robot developed as endoscopic device for laser phonomicrosurgery. Following the idea of soft robotics we describe the conventional clinical setting and adjunct benefits of the proposed assistance device to facilitate gentle surgery and usability in the operating room. Design constraints are obtained from medical image data implementing a mechanical design comprising compliant and flexible sections, actuation unit and multifunctional tip. We present results of a proof of concept experiment using a patient phantom, demonstrating the applicability of our system for laryngeal access.

Soft Components for Soft Robots

Jamie Paik

Ecole Polytechnique Federale de Lausanne

Abstract Typical robot platforms comprise rigid links with fixed degrees-of-freedom, solid blocks of transmission and actuator, and superficial positioning of sensors: they are often optimized for the given design criteria but are unable to execute instantaneous changes to the robot's initial mechanism design. The real-life incidences, however, require robots to face complex situations filled with unprogrammed tasks and unforeseen environmental changes. One of the growing efforts in the field that address such juxtaposing design paradigm is *soft robotics*: augmentations of "softness" in robots to complement, adapt, and reconfigure to the contingent assignments. Although the "softness" invokes and relates to many facets of robot design in both soft and hardware, this manuscript focuses on describing some critical hardware components. I will present several on going research on actuation and sensor solutions for soft robotics application as well as novel methods and materials for sensor and actuation integration.

Soft Robotics for Bio-mimicry of Esophageal Swallowing

Steven Dirven[1], Weiliang Xu[1], Leo Cheng[2]

[1] Department of Mechanical Engineering, The University of Auckland

[2] Auckland Bioengineering Institute, The University of Auckland

Abstract The field of soft robotics is continuing to expand into exploring the possibilities for novel, non-skeletal, transport and locomotion systems inspired by biological phenomena. Application of these techniques toward development of an anthropomorphic esophageal swallowing robot requires overcoming of many soft robotic design and characterization challenges. Additionally, soft-robots require vastly different methods of specification and validation than traditional robots, as they typically exhibit less well-defined degrees of freedom. This chapter reveals a series of novel methods to: establish interdisciplinary specifications for the esophageal swallowing process, develop a soft robotic analogue in the engineering domain, and demonstrate its capability.

Part II Sensors and Actuators

3 New Concepts for Distributed Actuators and Their Control

Welf-Guntram Drossel, Holger Schlegel, Michael Walther, Philipp Zimmermann, André Bucht

Fraunhofer Institute for Machine Tools and Forming Technology IWU, Chemnitz

Abstract Recently, decreasing costs for robots and control components have led to a broader acceptance of different kinds of robots. Hence, various fields of application start to flourish. As this is especially true for the field of service robotics it is typically implying an increasing physical human-machine-interaction. In this case a soft appearance yields major benefits, as it prevents injuries corresponding to an inherent safety of the system and, in theory, enables the robot to obtain virtually unlimited degrees of freedom. In this chapter the possibilities of the use of shape memory alloys for distributed actuators will be discussed by reference to application examples and implications for the control of such systems will be pointed out in detail.

3.1 Introduction

For decades, industrial robots are one of the main drivers for the ongoing automation of production processes. In the beginning, the automotive industry with its high demand for automated processes was the pacesetter for the development of new and improved robots. Meanwhile, the decreased costs for robots and control components led to a much broader acceptance of different kinds of robots in different fields of application. Especially in the last few years, one market growing rapidly is service robotics. Service applications are very often characterized by direct human-machine-interaction. In this context the safety for humans directly in contact with the robot will be essential. Current developments therefore focus on integrated sensor and control solutions aiming at the avoidance of collisions and minimization of the impact in case of unwanted contact between human and machine. However, the safety derived from the use of sensors and control algorithms is not an inherent safety. In case of collisions conventional stiff structures, as commonly used at present, can create a huge force. This is critical, especially in collisions with soft objects like human bodies for instance. Latest research therefore focuses on solutions for so called soft structures which can create an inherent safety. Soft structures are nature-like elastic objects made of polymers or other soft materials. Nevertheless, the use of compliant structures causes drawbacks in precision and obtainable forces. To overcome conventional stiff supporting structures there is a need for actuators which work precisely as well as achieving large forces. Conventional actuator technologies like electric drives are not suitable for elastic bodies with a nearly infinite number of degrees of freedom (DOF). Instead an actuator technology is needed which provides the possibility to design distributed actuators with a high level of structural integration and a soft characteristic similar to the surrounding structure.

Smart materials like dielectric elastomers (EAP) or shape memory alloys (SMA) are able to convert electrical energy into a mechanical reaction without a copper inductor or any kind of magnetic material. In contrast they possess a naturally given elasticity as well as the possibility to integrate actuator elements into the structure. Due to this, smart materials are suitable for the use as actuators in soft robotic structures. The actuating performance of smart materials is given by the stress and the strain which can be created for actuating use. Fig. 3.1 shows a comparison of different smart materials. It is obvious that the obtainable stress and strain varies clearly for the different materials. Hence, the different smart materials are not in a direct competition, in fact they complement one another.

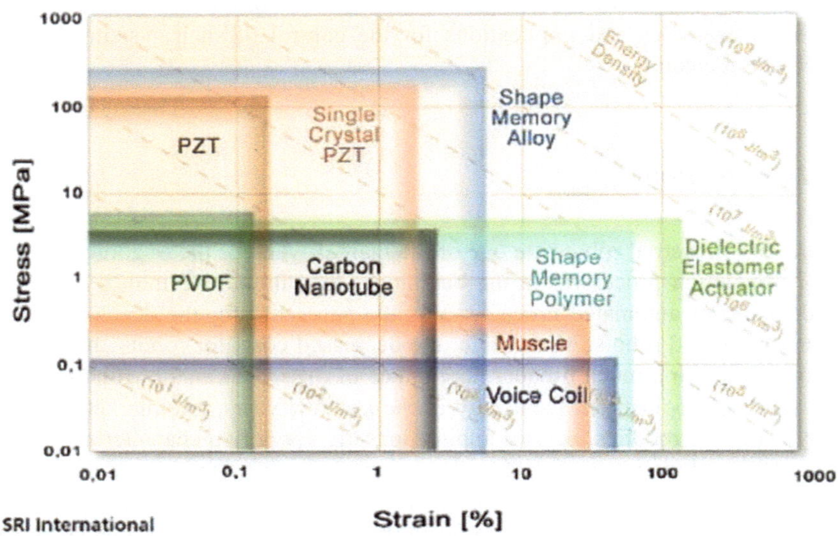

Fig. 3.1 Actuator performance of different smart materials (source: *DARPA* and *SRI International*)

In this chapter, the potential as well as the limitations of the use of shape memory alloys in soft robotic structures will be discussed. A short overview concerning the material basics is given, followed by some mechanical and control design rules which have to be considered for the design of SMA actuator systems. One of the unique features of SMA is the integrability. This chapter shows an approach for producing SMA actuator structures with distributed actuator elements using textile manufacturing processes. Finally, this chapter discusses some aspects regarding control concepts for distributed actuators.

3.2 Shape Memory Alloys as Flexible Actuators

3.2.1 Basics

Thermal shape memory alloys have the special ability to 'remember' and reassume their original shape. After experiencing a permanent plastic distortion below a specific critical temperature the shape can subsequently be reattained by means of heating up above this temperature. A reversible austenite-martensite phase transformation is required for the development of the shape memory effect. Analogous to steel, the high temperature phase is called austenite and the low temperature phase is called martensite. In an ideal situation the austenite phase will be converted into the martensite phase as a result of shear. Due to diffusion-free rearrangement processes on the molecular level, this generates a change in the stacking sequence of the crystal lattice levels and therefore a change in the structure of the crystal lattice.

Heating the SMA actuator, for instance a wire, then causes the described phase transition shown in Fig. 3.2a. The transformation from martensite into austenite starts at the temperature A_s and is finished at A_f. Above the austenite finish-temperature the lattice structure is fully austenitic. During this process, some thermal energy is converted into mechanical energy and can be used for actuation. Cooling down the wire induces the re-transformation from austenite to martensite. It starts at temperature M_s and is finished when the martensite finish temperature M_f is reached. The hysteresis effect within the temperature regime is shown in Fig. 3.2a. The absolute values of the temperatures strongly depend on the alloy composition and the heat treatment process previous to the application. Values for commonly used alloys are given in table 3.1. The material most commonly applied is NiTi, also called *Nitinol.* NiTi is commercially available as wire, rod, tube or sheet. Actuator bodies can be made out of these semi-finished products by various processes.

Due to the different lattice structures, two different stress-strain-curves exist as shown in Fig. 3.2b. In the low-temperature martensitic phase a small linear increase region is followed by a so-called plateau-stress where the wire can easily be deformed, almost without increasing the applied external stress. After resetting the stress to zero, a plastic distortion remains within the wire. In the high temperature austenitic phase the linear increase region is significantly wider, the Young's modulus is two to three times higher. Furthermore, applying a high amount of stress will cause a so called super-elastic behavior.

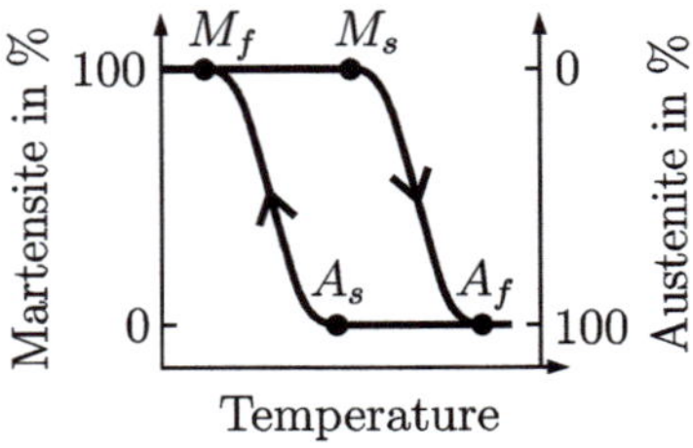

(a) Phase transformation process

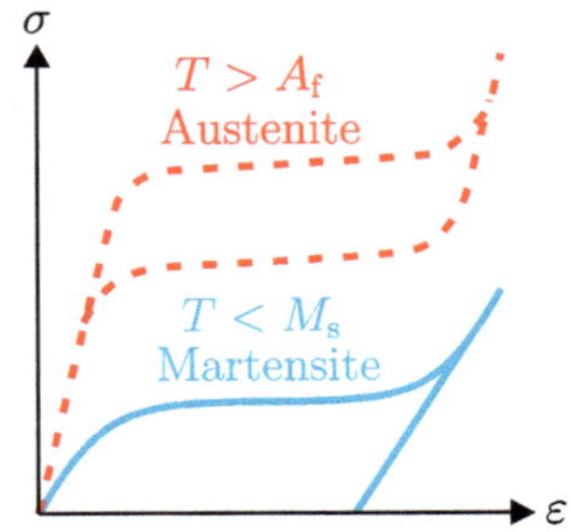

(b) Stress-strain-curve

Fig. 3.2 Characteristics of shape memory alloys

Table 3.1 Transition temperatures of different SMAs [9, 10]

SMA	Transformation temperature A_f in °C	Hysteresis $A_f - M_s$ in K
NiTi	–100...120	15...30
NiTiCu	–100...120	10...20
CuAlNi	–150...200	20...30
CuZnAl	–200...120	10...20

During the phase transition from martensite to austenite, via heating, the wire is able to perform mechanical work. The amount of work depends on the mechanical boundary conditions of the wire. In the case of a free wire the resulting work will be zero, but the actuator deformation would be at the maximum. In contrast, blocking the wire will cause a very high actuation force but no deformation. The resulting work output will also be zero. Using a spring with a defined stiffness as boundary element instead causes a deformation as well as a reaction force and therefore results in a usable workload. The amount of that workload depends on the stress-strain-curves of the used material and the design of the spring. The mechanical design of such actuators has been described previously and details can be found in [1].

Pre-stretched one-way SMA wires contract by heating above a certain austenite start temperature. However, without an applied load these wires do not reattain their original position upon cooling. Systems using one-way SMA wires therefore have to be used in combination with a force creating, pull-back component such as an applied constant load or mass, a spring, or an antagonistically arranged second SMA wire.

3.2.2 Control Design

The material behavior of SMAs is strongly nonlinear, which results in a relatively ambitious request for position control realization. Nevertheless linear control approaches as described in [8] are a suitable choice. This is owed to the simplicity of the algorithms and the possibility to optimize the controller by adjusting only a few parameters. Furthermore, such control algorithms are relatively robust concerning possible variations in actuator parameters. Overall, linear controllers offer the possibility to realize adequate closed loop controls without profound knowledge in control design.

External Sensing

Controlled drives for positioning applications always require a closed loop control of the actual position. Due to the high precision demands of most applications the control path necessitates an accurate external position sensor. This can be achieved by a laser triangulation sensor for example. Under certain circumstances as described in [13] the control loop of an SMA wire can be described by a simple first order lag element that can be controlled by a simple PI-controller. However, the integral part of the controller is always set as a compromise between heating and cooling and would be different for the three previously mentioned actuator principles. Due to the necessary comparability of the actuator principles, this work only focuses on the application of simple proportional controllers. Fig. 3.3 shows the simplified control loop of a single SMA wire that consists of a PI-controller, a power limitation, and the control path as a first order lag element.

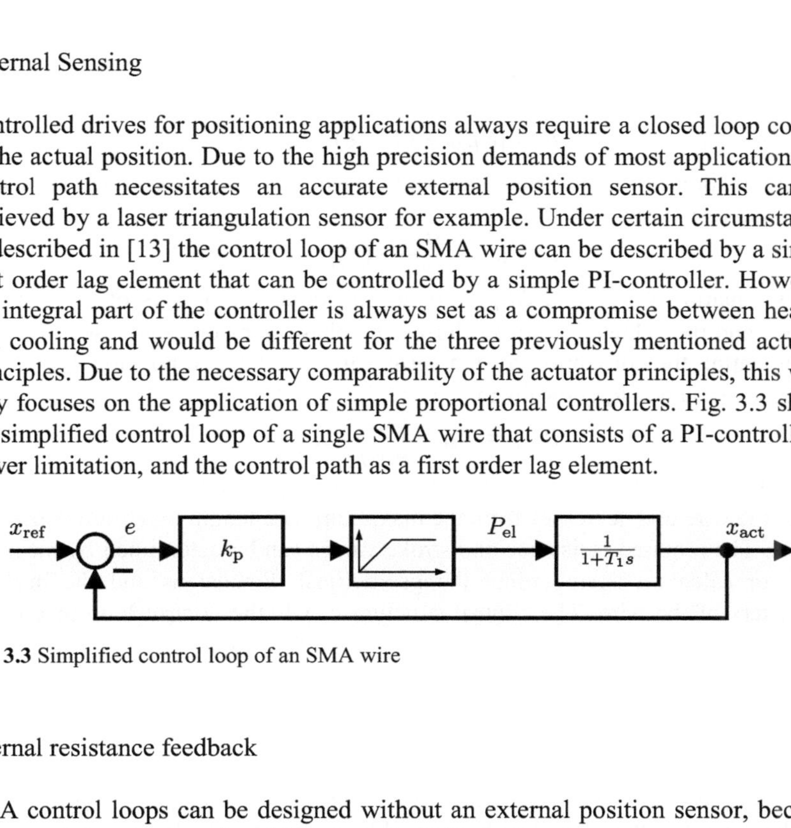

Fig. 3.3 Simplified control loop of an SMA wire

Internal resistance feedback

SMA control loops can be designed without an external position sensor, because the material possesses the ability to gain information about the actual stroke by measuring the resistance [12]. During the phase transformation from martensitic to austenitic lattice, and the involved changes in the structure of the SMA material, the status of transformation correlates with the electrical resistance. The lattice structure in the austenitic state is more regular than in the martensitic state. Therefore, the specific electrical resistance of austenite is significantly lower than the one of martensite. The information about the actual wire stroke can be determined by measuring the electrical resistance during positioning operations. Compared to the implementation of an external position sensor, this method can be achieved by significantly less effort, because an electrical interface is already needed to control the power input of the actuator.

As shown in the references [8] and [17], there are different possibilities to model the length-resistance-correlation of SMA wires. This varies from elementary linear approaches to complex approaches considering the temperature influences on the specific resistance and the changing geometry during the deformation of the wire. According to the intended application an implementation on a rapid prototyping system is necessary. In combination with the requirements regarding the positioning accuracy of the intended positioning devices using a linear approach seems to be adequate. In [1] a linear interpolation of the specific resistance from martensite ρ_M to austenite ρ_A is given with the starting wire length L_0 and the wire cross section A_q by:

$$R = \left[\xi\, \rho_M + (1 - \xi) \cdot \rho_A\right] \cdot \frac{L_0}{A_q} \tag{1}$$

The elimination of the martensite amount ξ results in a linear correlation of wire length and resistance:

$$L = L_M - \frac{\Delta L_{ref}}{\Delta R_{ref}} \cdot (R_M - R) \tag{2}$$

In this equation ΔL_{ref} and ΔR_{ref} are the maximal differences of the wire resistance and the achievable stroke during the phase transformation (see Fig. 3.4). Implementing the correlation to transform a given reference position L_{ref} into a reference resistance R_{ref} results in:

$$R_{ref} = R_M - \frac{\Delta R_{ref}}{\Delta L_{ref}} \cdot L_{ref} \tag{3}$$

It has to be remarked, that the reference position L_{ref} always represents a contraction of the wire and decreases from the maximum wire length in the full martensitic state. The required resistance and stroke values can be determined by measurements or calculations, regarding the geometrical dimensions and the material parameters of the wire. The control structure equals the control loop in Fig. 3.3 with an additional resistance calculation in the feedback loop.

Control concept for antagonistic wires

The existence of two SMA actuators in an antagonistic arrangement necessitates an enlargement of the control path by an additional first order lag element that represents the second SMA wire. The wires are coupled at their mechanical output as shown in Fig. 3.5. As described in [17] the phase transition temperatures of SMAs increase with increasing tension. This correlation possesses the ability to raise these temperatures in the first wire by heating the second [11]. Due to the increasing difference between ambient and phase transformation temperatures in the first wire, the transformation from austenite to martensite will occur faster. The control concept bases on splitting the control value to benefit from this effect. Fig. 3.5 shows that the first actuator only acts in case of positive control values and the second actuator only acts for negative values.

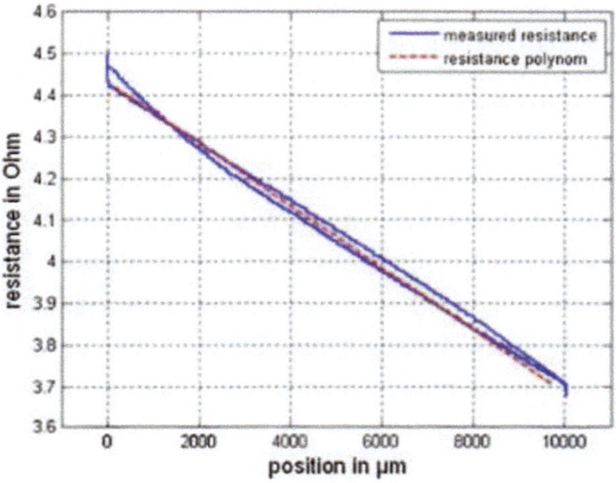

Fig. 3.4 Length-resistance-correlation (*X2* wire from *Memry*)

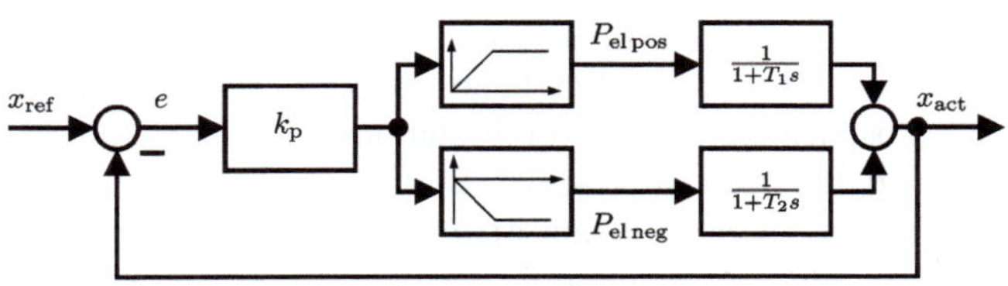

Fig. 3.5 Control concept for an antagonistic arrangement

Due to the coupling between the SMA wires it has to be considered that only one wire is heated at a certain time. This is realized by the limitation of the mechanical tension between the wires. If the tension exceeds a defined constant the control value is set to zero either until the tension decreases, or the control values indicate changes.

3.2.3 Structural Integration

Most SMA applications have a limited number of DOFs and their positions are fixed. The actuating wires are fixed straight to well defined points of force transmission. Bionic approaches, like exemplarily shown in Fig. 3.6, need completely different design processes. In [18] and [3] approaches are given which are each based on woven structures in combination with SMA springs. Reference [18] presents a mesh-worm prototype making use of an antagonistic actuation scheme inspired by the hydro-stat skeleton of oligochaetes. According to the bionic example a mesh structure with two groups of muscle fibers is utilized, consisting of four longitudinal muscle fibers on the inside and a circumferential fiber wrapped around the outside of the mesh tube. The octopus robot shown in [3] comprises both cable-driven actuation systems globally, and SMA spring actuators locally. This hybrid actuation system is yielded by a braided sheath also featuring soft sensors based on *Electrolycra*, which is a textile material whose conductivity depends on how tightly it is stretched.

Fig. 3.6 Soft-bodied robot inspired by the morphology and behaviour of the octopus [3]

The production of bionic applications is a crucial challenge besides their design. Single prototypes are manufactured by hand. Thus, the arrangement of the actuating elements can be shaped freely, with limitations only set by kinematic requirements and the available installation space. With such freely shaped soft structures building the basis for the construction of robots, manual production is not cost-effective. One possibility for a cost-effective production lies within the usage of textile processing technologies. The processing of metallic threads is basically possible, even if the typical pseudo-plastic and pseudo-elastic behavior of the shape memory wires constrain the processing. Fig. 3.7 shows a shape memory alloy sleeve produced by an automatic circular knitting process. The SMA wires are embedded axially into the textile structure and are fixed by the textile knitted fabric. The SMA sleeve is intended to be used for surgical suction. The surgeon can shape the suction sleeve into any form to ease the entering in minimally invasive surgeries. After the surgery the suction sleeve is autoclaved and returns to its original shape due to the heat generated by the autoclave cleaning. At the moment, the surgical aspirator is tested medically. With an active control of single SMA wires, the principle can also be used in reverse. Hence, the sleeve can then be applied as an active controllable structure.

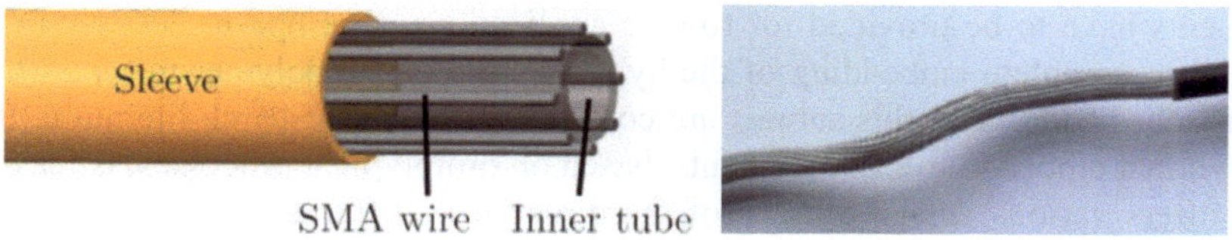

(a) Schematic view (b) Circular knitted sample

Fig. 3.7 SMA sleeve manufactured by a circular knitting process

What's more, the automatic production of textile 2D pre-products with integrated shape memory material is possible. During a joint research project with the *Institute for Special Textiles and Flexible Materials Thüringen-Vogtland* (*TITV*) the *Fraunhofer IWU* was able to show that processes like knitting, weaving and knotting are principally suitable for this approach. Fig. 3.8 shows samples of the hybrid structures mentioned above. By controlling the embedded SMA wires the stiffness and the geometry of the pre-products can be influenced. Subsequent research aims at adaptive bandages and ortheses with adjustable supporting effects, which enhance the healing process. Moreover, it has to be shown to what extent such textile pre-products will be feasible as a basis for soft structures. In contrast to prototypical approaches (cf. Fig. 3.6 and 3.7) these automatically manufactured actuating elements cannot be positioned freely. In fact manufacturing possibilities and limitations have to be considered alongside the kinematic dimensioning of the structure. Embedded actuators can only be arranged within the textile where the automatic feed of the SMA wires is possible.

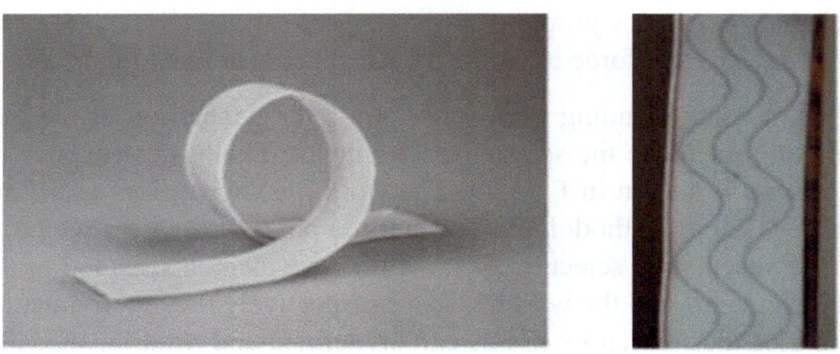

(a) Woven fabric (b) Stitched sample

Fig. 3.8 Hybrid 2D textiles with integrated SMA wires

In general, textile structures show highly oriented stiffness values. On the one hand it is possible to reach high stiffness values in the pulling direction. On the other hand the stiffness will be much lower in lateral or push direction. This characteristic destines textile structures as basic elements for soft structures. To fulfill the requirements regarding precision and bearing capacity additional load bearing elements have to be provided for robotic applications. One option lies within the partial or complete embedding of the hybrid textile into a polymer matrix. Many approaches addressing this subject are covered by current research literature. It is shown that embedding SMA elements, based on prototypical processes, is feasible and offers various design options of the directional stiffnesses of the composite structure. Processing technologies which are suitable for industrial production are not yet available. However, they are moving into the focus of current research.

The use of shape memory actuators allows the design of active structures with a vast amount of degrees of freedom. The simplicity of the single actuating elements results in large creative leeway. A controlled movement will only be achieved by the setup of appropriate powerful control algorithms.

3.3 Control and Feedback Control of Distributed Actuators

Another focus of this chapter is devoted to the control of a specified motion of a structure, e. g. a robot arm, with distributed actuators. The control problem consists of three main tasks: the selection of the mechanical structure, path planning, and the feedback control of the SMA elements [4]. The definition of a mechanical structure includes:

Selection of numbers and degrees of freedom (rotational or translational motion) of SMA actuators

Combination of multiple SMA elements to achieve higher elements of the kinematic chain, for instance joints

Measures to increase the force or work space respectively manipulability

A crucial point when planning a robot system is the generation of the robot path out of the requirements in the selected environment. The basic sequence of a path planning process is shown in Fig. 3.9. The work plan synergizes user knowledge and automated design methodology. The starting point of the iterative process is a model that represents the selection of the kinematic elements, including the combined SMA elements, and the working space respectively the environment.

In this model, desired target points can be defined and used for the calculation of a draft of the TCP path. In a first user interaction, a basic structure of the handling device or robot kinematics can be set. Based on the selected mechanical structure the path of the manipulator is planned in the next step. Methods of dynamic calculation and collision avoidance are included. The final path should be collision free and abide by physical limitations, e. g. hardware or acceleration limits, as well as taking into account deformations caused by the own weight or external loads, for instance, tools. Due to the reasons mentioned before, optimization approaches will be applied to the calculation of the path [14, 16].

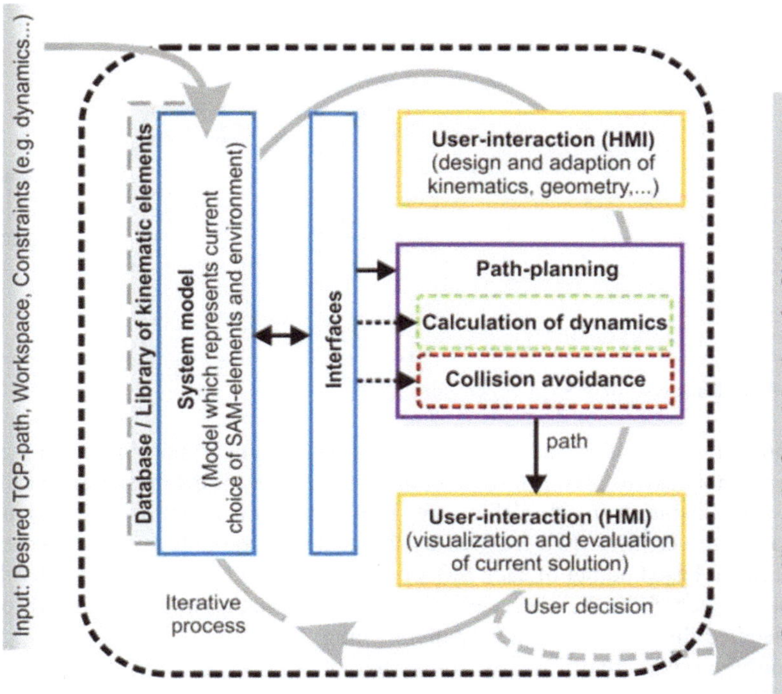

Fig. 3.9 Iterative procedure for path planning and selection of kinematic elements

In the next step, the robot kinematics and the calculated path can be visualized and evaluated concerning collisions and accessibility in form of a second user interaction. Possibly occurring collisions can be detected, presented to the user in a concentrated form, and finally, extracted. At this point it is the decision of the user whether the solution is satisfactory or needs to be re-optimized. The illustrated process is performed by an adaptation of the kinematic structure, building on the optimized TCP paths.

The complexity of the feedback control problem is reduced by the task-oriented subdivision of the SMAs, e. g. as single joint or movement direction. These collocations are controlled individually, as shown in Fig. 3.10. However, a multivariable controller is necessary, because most tasks are not possible to be fulfilled by a single SMA unit.

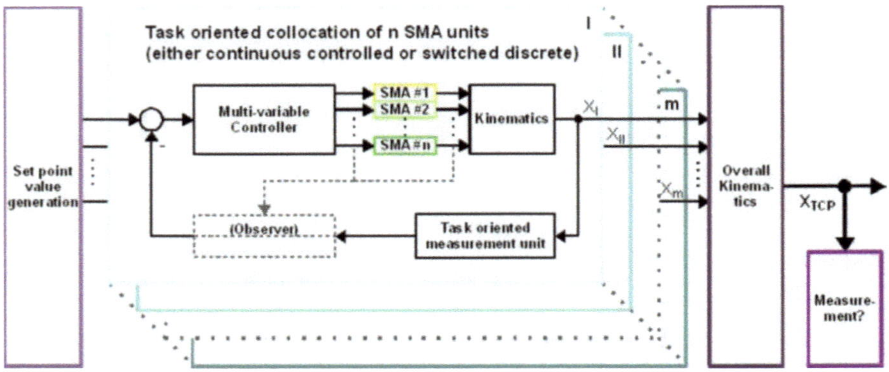

Fig. 3.10 Feedback control of SMA collocations

The collocation should include a task oriented measurement unit. Since the measurement is not easy to perform, or potentially there is no direct measurement to be found or realizable in an affordable way. This is why it is common in literature to apply an observer [5, 6, 15]. It is fed by the actuating variable and the resistance measurement based output of each SMA and the task oriented measurement value. The model needed for the observer needs to include the kinematics but has to be more detailed than the one used for path planning.

A multi-variable control is super-ordinated to the inner resistance control loops of the SMA. The controller should be designed robustly, e. g. according to H_∞. Adaptive control concepts may also be taken into account. Moreover, standard controllers could be used, but then the control allocation problem comes into focus. The control output, in terms of the manipulated variable, has to be distributed among the redundant SMA units taking into account the individual limits and static gains. A static and possibly nonlinear distribution could be considered. Additionally, there are several applicable optimization algorithms shown in literature, mainly used for aircraft control [2, 7].

The collocation of SMA units may also comprise discretely switched SMA units. They are more easy to realize and do not need feedback control. These can be used to gain more force or stroke. Whereas the feedback controlled SMAs may behave differently, resulting from kinematic influences like load changes or from the state of the switched SMA, so that a variable structure control has to be considered to increase the performance.

With the collocations being controlled individually, the position x_{TCP} of the end effector is not fed back. This may lead to a deviation from the desired position due to the uncertainties in the overall kinematics. However, with modern industrial robots it is common to teach-in crucial positions for the process manually. Thus, a less accurate absolute positioning is tolerable.

3.4 Conclusions and Outlook

Existing approaches already address the integration of shape memory actuators into single flexible structures and their successful control. Nevertheless, to fully take advantage of the inherent possibilities of soft robotic systems, a complete integration approach is to be examined in future research. An advanced control circuit has to be set up to efficiently operate the adaptive system as a whole. A control loop for each single wire, and a supervisory loop for each group of wires within one actuation element have to be designed and optimized. Finally an optimized overall control has to be set up for the complete set of actuating elements to operate the entire flexible structure safely and efficiently.

Turning to shape memory actuators the use of distributed actuating concepts in particular yields the opportunity to overcome the dynamic limitations of conventional actuating approaches. As can be expected, the usually critical value of the cooling time of the SMA wire limits the dynamics of the system to a certain extent. Obviously, the use of numerous actuating elements easily overcomes this obstacle by operating the different elements with a defined time delay.

3.5 References

[1] Besselink PA (1996) Procedure for the calculation of the geometry of a resistance heated NiTi-actuator. In: Actuator Proceedings of the 5th International Conference on New Actuators, pp 421–424

[2] Bodson M (2002) Evaluation of optimization methods for control allocation. Journal of Guidance, Control, and Dynamics 25(4):703–711

[3] Cianchetti M, Follador M, Mazzolai B, Dario P, Laschi C (2012) Design and development of a soft robotic octopus arm exploiting embodied intelligence. In: IEEE International Conference on Robotics and Automation (ICRA), pp 5271–5276, DOI 10.1109/ICRA.2012.6224696

[4] Ghasemi A (2012) Specified motion and feedback control of engineering structures with distributed sensors and actuators. Dissertation, University of Kentucky

[5] Hädrich J, Walther M, Schlegel H, Neugebauer R (2011) Sensordatenfusion zur Zustandserfassung nachgiebiger Leichtbauarme. In: 21st International Scientific Conference, Mittweida, pp 45–49

[6] Hädrich J, Walther M, Schlegel H, Neugebauer R (2011) Zustandserfassung für die Lageregelung eines nachgiebigen Roboterarms. In: SPS/IPC/DRIVES, Nürnberg, pp 423–431

[7] Härkegård O (2002) Dynamic control allocation using constrained quadratic programming. In: AIAA Guidance, Navigation, and Control Conference, Monterey, California

[8] Jung J (2009) Aufbau eines Greifmechanismus mit FG-Drahtaktoren. Diploma thesis, Technische Universität Dresden

[9] Kohl M (2002) Entwicklung von Mikroaktoren aus Formgedchtnislegierungen, Wissenschaftliche Berichte, vol 6718. Forschungszentrum Karlsruhe

[10] Lagoudas DC (2008) Shape Memory Alloys - Modeling and Engineering Applications. Springer

[11] Langbein S, Czechowicz A (2012) Adaptive resetting of SMA actuators. Journal of Intelligent Material Systems and Structures 23(2):127–134, DOI 10.1177/1045389X11431741

[12] Meier H, Czechowicz A, Dilthey S (2008) Regeln von FG-Legierungen mit Widerstandsrückkopplung: Sensorik-Arrays überflüssig. Mechatronik (11/12):24–27

[13] Neugebauer R, Bucht A, Pagel K, Jung J (2010) Numerical simulation of the activation behavior of thermal shape memory alloys. In: Proceedings SPIE, vol 7645, pp 76,450J–76,450J–12, DOI 10.1117/12.847594

[14] Neugebauer R, Hipp K, Hofmann S, Schlegel H (2011) Einsatz von Verfahren der Simulationsbasierten Optimierung zur Reglerparametrierung unter Berücksichtigung von definierbaren Nebenbedingungen. In: Mechatronik, Dresden, pp 247–252

[15] Neugebauer R, Walther M, Hädrich J, Schlegel H (2011) Zustandserfassung flexibler Leichtbauarme durch Multisensor Daten Integration. In: Fachtagung Mechatronik, Dresden, pp 285–290

[16] Neugebauer R, Hipp K, Hellmich A, Schlegel H (2012) Increased Performance of a Hybrid Optimizer for Simulation Based Controller Parameterization. JAMRIS - Journal of Automation, Mobile Robotics & Intelligent Systems (1):42–45

[17] Oelschlaeger L (2004) Numerische Modellierung des Aktivierungsverhaltens von Formgedaechtnisaktoren am Beispiel eines Schrittantriebes. Dissertation, Ruhr-Universität Bochum

[18] Sangbae K, Hawkes E, Kyujin C, Joldaz M, Foleyz J, Wood R (2009) Micro artificial muscle fiber using NiTi spring for soft robotics. In: IEEE/RSJ International Conference on Intelligent Robots and Systems, pp 2228–2234, DOI 10.1109/IROS.2009.5354178

4 Artificial Muscles, Made of Dielectric Elastomer Actuators - A Promising Solution for Inherently Compliant Future Robots

In Seong Yoo, Sebastian Reitelshöfer, Maximilian Landgraf, Jörg Franke

Friedrich-Alexander-Universität Erlangen-Nürnberg

Abstract The cutting-edge robotic technology can deal with a lot of complex tasks. However, one of the most challenging technological obstacles in robotics is the development of soft actuators. Remaining challenges in the field of drive technology can be overcome with innovative actuator concepts, for example dielectric elastomer actuators (DEAs). DEAs show numerous advantages in comparison to prevailing robotic actuators that are based on geared servomotors: They are form-flexible, inherently compliant, can store and recuperate kinetic energy, feature high power-to-weight ratio and high energy density that is comparable to human skeletal muscles, and finally can be designed to perform natural motion patterns other than rotation. In this article, after a review on disadvantages of state-of-art robotic drives, which are stimulus for a research on the promising drive solution, benefits of DEAs will be presented with regard to the possibility of applications in soft robotics. Finally, the article will conclude with a brief report on the ongoing research effort at the Institute for Factory Automation and Production Systems (FAPS) with two major foci – the development of an automated manufacturing process for stacked DEAs and a lightweight control hardware.

4.1 Drawbacks of Prevailing Robotic Actuators

Leading-edge high-DOF robotic systems show a broad set of astonishing capabilities, such as walking in rough and randomly structured terrain or grasping objects adaptively by merging multi-modal sensory data. However, most of the robotic systems today are actuated by prevailing geared servomotors, hydraulic or pneumatic actuator systems.

The performance of these robot systems is usually impaired due to technological limitations of these actuators: Their dynamics and agility are severely limited primarily due to their poor power-to-weight ratio and rigid kinematics. Also, nearly none of the established robotic systems can be operated untethered for a sufficient amount of time and range distance. Furthermore, many mechatronic prosthetic devices driven by geared servomotors are affected by lack of compliance due to the rigidly coupled mechanical components, for example finger elements in active prosthetic hands. [1]

Reasons for the widespread application of the established geared servomotors are manifold: straightforwardness, precision and availability. First of all, the physical behavior of the electromechanical drives is well understood to a great extent, which simplifies their modeling, control and implementation in robotic systems. Secondly, high gear ratio enables a precise positioning of immediately coupled kinematic elements and thus of the entire system. Finally, they are available in a variety of performance categories and for any possible applications.

However, the major dilemma regarding this kind of robotic actuators is that the system architecture of a small, lightweight, efficient and fine positionable servomotors often operates optimally at a higher rotational speed range with lower torques. A typical robotic application on the contrary requires smaller displacements at higher torques. In order to gain higher torques at the cost of angular speed, precision gears are inevitable for such driving units. Unfortunately, the mechanical elements for power transmission severely impair the overall dynamics and the backdrivability of servomotors [2]. Some lightweight gears might feature a desirable performance-to-weight ratio, but they are not capable of enduring cyclic, fluctuating mechanical impulses, for example recoil forces that a humanoid robot would experience during a bipedal walking.

This drawback of geared servomotors with regard to dynamics and capability of enduring impulsive loads has been addressed by numerous research efforts utilizing elastic mechanical elements such as springs, chains or cables that are connected in series with other rigid components. However, these serial elastic actuators (SEA) pose an additional mass to a robotic system and therefore worsen the overall specific power of the system.

In order to achieve a higher dynamic capability, hydraulic or pneumatic actuators are often applied in robotic systems. They feature a comparatively higher energy density [3] and can thus contribute to reducing inertia of moving extremities in a robotic system. Furthermore, with hydraulic or pneumatic drives a compliant actuation behavior can be realized. However, the advantages of hydraulic or pneumatic systems are attended by severe disadvantages, such as low overall efficiency that impedes an energy autarkic, untethered operation of robotic systems, unless a mobile combustion engine is used, which entails other drawbacks [4].

In conclusion, future robotic systems urgently require a new generation of innovative drive systems that is dynamic, lightweight, energy efficient and above all, inherently compliant. With regard to this immediate need, dielectric elastomer actuators (DEAs), a kind of electroactive polymeric composite system, depict a considerably promising solution. In following, after a brief description of DEAs and their functional principles, advantageous attributes and technological potentials of DEAs will be discussed, which encourage their application in soft robotic systems.

4.2 Benefits of DEAs in Soft Robotics

Dielectric elastomer actuators belong to a category of electroactive polymers (EAPs), which is a functionalized material that can perform a mechanical work with electric energy. Principally, a single DEA element can be described as a flexible planar capacitor that consists of two form flexible conducting electrodes (usually a few microns thin layers of carbon grease or graphite powder) and an elastomeric dielectric material (e.g. silicone rubbers) in between those electrodes. Despite the complexity of electromechanical effect in a DEA that involves dialectic polarization and electrostriction in the dielectric layer, a basic working principle of a DEA can well be described with a simplified equation shown below, with following physical parameters:

$$p = \varepsilon_0 \varepsilon_r E^2 = \varepsilon_0 \varepsilon_r \frac{U^2}{z^2} \qquad (1)$$

Maxwell pressure for actuation of DEA p
Absolute permittivity of vacuum ε_0
Relative permittivity of dielectrics in DEA ε_r
Electric field E
Actuation voltage U, and
Thickness of the dielectric layer z

When an electric voltage – in the range of a few kilovolts depending on relative permittivity, electric breakdown strength and thickness of the dielectric layer – is applied across the two conductible electrodes, the dielectric layer will be compressed as the differently polarized electrodes are attracted to each other. In addition, electrostriction occurs in the dielectric medium and contributes to the overall deformation of DEA. Under the assumption that the dielectric material is isotropic and incompressible, the electrically induced deformation of DEA in both x- and y-directions will lead to a mechanical stroke in the orthogonal direction, i.e. contractile motion.

In addition to this effect that can be utilized as a linear drive, DEAs have a wide spectrum of advantageous properties that can contribute to next-generation robotic systems. In following, benefits of DEAs that make them a promising drive solution will be discussed in various technological perspectives.

4.2.1 Capability of Energy Recuperation

Due to their inherent elasticity, DEA are form flexible and therefore able to deform under external loads. The mechanical load can be saved as elastic potential energy and used in next movement. Prevailing robotic systems that are equipped with ordinary geared servomotors should overcome the moment of inertia of their rigid kinematic components after every movement by mechanically decelerating the rotational drives in joints, through which the kinetic energy is inevitably dissipated in terms of waste heat. On the contrary, DEAs can recuperate a part of kinetic energy at the end of an actuation, so that the overall energy efficiency of a robotic system can be drastically enhanced. Moreover, the inherent elasticity can be utilized not only to recuperate kinetic energy but also to actively harvest electric energy through a mechanical deformation of a DEA.

4.2.2 Intrinsic Compliance and Adaptability

Another favorable mechanical characteristic of DEAs that can be derived from the material elasticity is their intrinsic compliance, i.e. the ability to yield to an applied force without the help of a sophisticated control algorithm or other mechanical elements with elasticity such as springs. With regard to robotic application, this can significantly contribute to safety of robotic systems that operate in the vicinity of human, e.g. collaborative robots in an assisted assembly process or personal assistant robots. Along with the inherent compliance that is described above, DEAs are also form flexible. This advantageous characteristic of DEAs can especially be exploited in adaptive gripping systems that are capable of grasping unknown objects with random, unstructured geometry and texture ad hoc. In combination with an intelligent sensor technology, such as a stereoscopic vision system, DEAs can build an adaptive gripping system that features a high error tolerance and the capability of in situ adaptation.

4.2.3 Outstanding Power-to-Weight Ratio

In contrast to serial elastic actuators involving rigid mechanical drive elements, e.g. electric servomotors, gears and springs, DEAs can generate mechanical force but still remain compliant and remarkably lightweight. An extensive overview of the specific properties of various actuator materials can be found in [5]. For instance, a silicone based, pre-strained dielectric elastomer can even surpass a human skeletal muscle in terms of the specific elastic energy density [6, 7]. Next to the capability of energy recuperation, the high power-to-weight ratio is also a distinguishing property of DEAs that can enhance the overall efficiency of a robotic system.

4.2.4 Capability of Self-sensing

A DEA can also function as a dielectric elastomer sensor (DES) at the same time, which is a capacitive tactile or strain sensor. A wide spectrum of sensor applications of dielectric elastomer can be found in [8]. If a DES deforms under a mechanical load, the distance between two compliant electrodes will be reduced and the capacity will change accordingly. With the relationship between force, deformation and electric field strength known, a change of the capacity can be interpreted as force and strain. Unlike widespread robotic drives with integrated potentiometers, DEAs are capable of recognizing its current state of itself without additional sensory devices, while the control of such multi-degree of freedom kinematic system will not be trivial [8, 9].

4.2.5 Noiseless Actuation

DEAs can function without emitting noises, as they do not involve any moving mechanical parts. This might be an important requirement for robotic systems that are to be applied next to human, e.g. personal service robots.

4.3 Current Research Efforts

In this section, the current research project at the Institute of Factory Automation and Production Systems (FAPS) on DEAs will be presented. The project deals with two main topics: automatic manufacturing of DEAs and lightweight control hardware.

4.3.1 Manufacturing Artificial Muscles Based on DEA

In order to develop a macroscopic drive system that are capable of driving a complex kinematic system ("artificial muscle" for robotic systems), a vast number of multiple DEAs should be piled over each other to build a "stacked" DEA. The hierarchic structure of human skeletal muscles is considered as a useful analogous model of a macroscopic actuator that consists of numerous smallest contractile elements (Fig. 4.1). One of the important research goals in the development of stacked multilayer DEA is minimizing the actuation voltage. As the equation above already implies, the first option to achieve a required actuation pressure with a low voltage is reducing the thickness of the dielectric material (z) below 100 µm [10]. The main technologic challenge here is producing microns thin homogenous layers from different materials – dielectrics and functionalized dielectrics working as conducting electrodes stacked over each other in alternating sequence – without losing the adhesion between these layers and thus the integrity of the whole materials system.

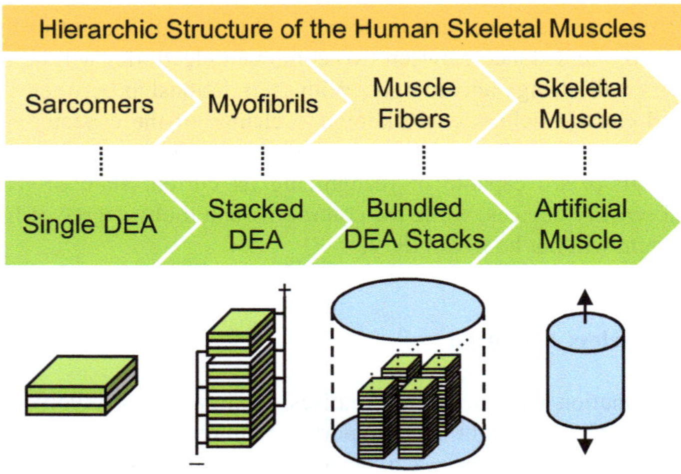

Fig. 4.1 Biomimetically inspired design scheme of DEA based artificial muscle.

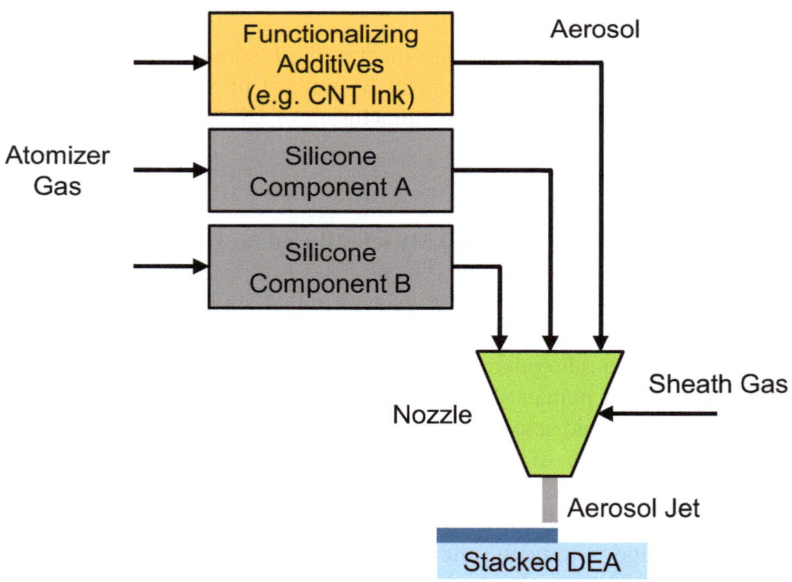

Fig. 4.2 Principal scheme of manufacturing stacked DEAs with multi-component Aerosol Jet Printing.

To exploit the suitability to create homogeneous layers with a broad variety of materials [11], the automated manufacturing of multilayer DEAs using Aerosol Jet printing is being investigated to build up complex structured stack actuators

(Fig. 4.2). First experiments have been carried out using a commercially available silicone based elastomer to achieve a layer thickness below 10 μm. A detailed description of the results can be found in [12].

4.3.2 Lightweight Power Electronics

Another key aspect of this ongoing research is the development of lightweight power electronics with a long-term objective of integrating DEAs in a mobile robotic system. In order to prevent an impairment of the overall specific power and the energy efficiency of a complex kinematic system, a large number of actuators should be driven by smallest possible number of power sources, which pose a considerable additional mass.

Therefore, one single high-voltage source (a DC-DC converter) is implemented, while independent contraction of individual DEA is realized with a control hardware in combination with a pulse width modulated (PWM) signals (Fig. 4.3). Using PWM, DC-DC converters for each DEA can be substituted by a simpler and lightweight control hardware that consists of a single power source and lightweight semiconductor elements like optocouplers [13] or metal oxide-field-semiconductor field-effect transistors (MOSFET) for each actuator. The usage of PWM might also allow a simultaneous monitoring of the DEA activity by comparing the measured capacity of each actuator with the known PWM input signal.

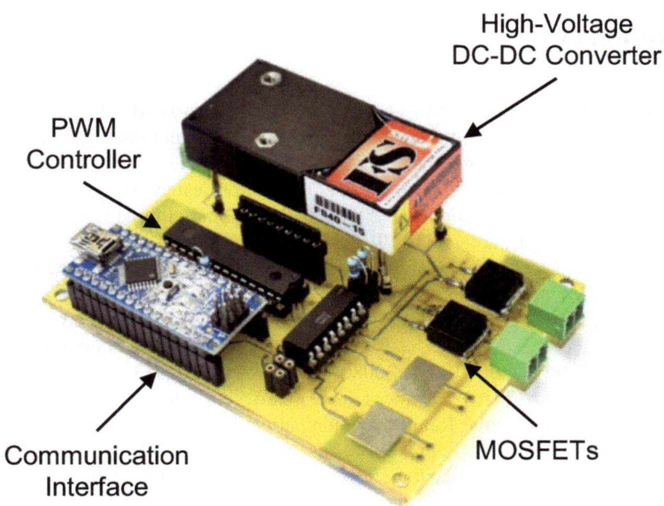

Fig. 4.3 Hardware for pulse width modulated control of multiple DEAs.

4.4 Summary and Future Challenges

This article has given a brief account of motivations and visions of using DEAs as "soft" drive system for "soft" robotic systems. Taking advantage of their favorable properties, such as intrinsic material compliance, adaptability, high energy efficiency and power-to-weight ratio, DEAs show a promising potential of application in soft robotics. The preliminary results, along with other achievements of other research groups cited above, support the idea of building "artificial muscles" of next-generation robotics systems based on DEAs.

The focus of upcoming research efforts will be on further development of an automated manufacturing process using the Aerosol Jet printing and its optimization in respect of the quality and interlayer integrity of the stacked DEAs. The next step will be topologic design and realization of electromechanical interfaces in the stack of DEAs. As regards the other main research area, the lightweight power electronics and control systems will be continuously improved with help of simulation and modeling of the electromechanically coupled mechanism in stacked DEA.

Acknowledgments The authors gratefully acknowledge the research grant of the Bavarian State Government within the framework of the biomimetic initiative "Bionicum Forschung" for the development of biomimetically inspired artificial muscles that is presented in this article.

4.5 References

[1] Franke J (2014) Artificial muscles, made of dielectric elastomer actuators: A promising solution for inherently compliant future robots. 1st International Symposium on Soft Robotics in Germany. Fraunhofer IPA Symposium F291

[2] Seok S, Wang A, Otten D, Kim S (2012) Actuator design for high force proprioceptive control in fast legged locomotion. 2012 IEEE/RSJ International Conference on Intelligent Robots and Systems. doi:10.1109/IROS.2012.6386252

[3] Caldwell DG, Medrano-Cerda GA, Goodwin M (1995) Control of pneumatic muscle actuators. IEEE Control Systems Magazine 15:40-48

[4] Raibert M, Blankespoor K, Nelson G, Playter R (2008) BigDog, the rough terrain quadruped robot. Proceedings of the 17th World Congress of the International Federation of Automatic Control. doi:10.3182/20080706-5-KR-1001.01833

[5] Brochu P, Pei Q (2010) Advances in dielectric elastomers for actuators and artificial muscles. Macromolecular Rapid Communications 31(1):10-36

[6] Kornbluh RD, Pelrine R, Pei Q, Oh S, Joseph J, Bar-Cohen Y (2000) Ultrahigh strain response of field-actuated elastomeric polymers. Proc. SPIE 3987, Smart Structures and Materials 2000: Electroactive Polymer Actuators and Devices (EAPAD). doi:10.1117/12.387763

[7] Pelrine R, Kornbluh R, Joseph J, Heydt R, Pei Q, Chiba S (2000) High-field deformation of elastomeric dielectrics for actuators. Materials Science and Engineering C 11(2):89-100

[8] Bauer S, Bauer-Gogonea S, Graz I, Kaltenbrunner M, Keplinger C, Schwödiauer R (2013) 25th anniversary article: A soft future: From robots and sensor skin to energy harvesters. Adv Mater 2013:1-13. doi:10.1002/adma.201303349

[9] Kim S, Laschi C, Trimmer B (2013) Soft robotics: a bioinspired evolution in robotics. Trends in Biotechnology 31(5):287-294

[10] Carpi F, Bauer S, De Rossi D (2010) Stretching dielectric elastomer performance. Science 330(6012):1759-1761

[11] Goth C, Putzo S, Franke J (2011) Aerosol Jet printing on rapid prototyping materials for fine pitch electronic applications. 61st Electronic Components and Technology Conference (ECTC). doi:10.1109/ECTC.2011.5898664

[12] Landgraf M, Reitelshöfer S, Franke J, Hedges M (2013) Aerosol Jet printing and lightweight power electronics for dielectric elastomer actuator. 3rd International Electric Drives Production Conference (EDPC). doi:10.1109/EDPC.2013.6689733

[13] Gisby TA, Calius EP, Xie S, Anderson IA (2008) An adaptive control method for dielectric elastomer devices. Proceedings of SPIE 6927, Electroactive Polymer Actuators and Devices (EAPAD). doi:10.1117/12.776503

5 Musculoskeletal Robots and Wearable Devices on the Basis of Cable-driven Actuators

Martin Haegele, Christophe Maufroy, Werner Kraus, Maik Siee, Jannis Breuninger

Fraunhofer Institute of Manufacturing Engineering and Automation IPA, Stuttgart

Abstract Cable-driven actuators are a promising alternative for future kinematic designs, particularly when the combination of lightweight, high strength, compact designs and dynamic motions are required. Powered exoskeletons or wearable robots are typical candidates of these novel actuators as has been demonstrated by previous research. This chapter focusses on current work in cable-driven actuators, introduces the Myorobotics toolkit for supporting the engineer to build up prototypes from cable-actuates modules and gives an outlook to using cable-driven actuation for advanced wearable robots.

5.1 Introduction

Powered exoskeletons or wearable robots, handling aids, balancers etc. are referred to as robotic devices which are actuated mechanisms fulfilling the characteristics of (collaborative) industrial robots or service robots, but lacking either number of programmable axes or the required degree of automation or autonomy.

A powered exoskeleton may be defined as an active mechanical device that is essentially anthropomorphic in nature, is "worn" by an operator and fits closely to his or her body, and works in concert with the operator's movements.

These kinematic machines are quite new, but may be considered as breakthrough in various application areas from human augmentation for almost any task, therapeutic aids, and interfaces for controlling or commanding complex machinery up to the implementation of some kind of tele-presence [1]. As devices for human augmentation, exoskeletons could be used for example to ease the working conditions of the ageing workforce by providing support during physically-demanding manual tasks.

In recent years, cable-driven actuation has received significant attention, mainly due to advancements in high-strength cable materials which allow the transmission of high forces resulting in surprisingly high stiffness and power density, therefore pushing lightweight designs of wearable devices.

The goal of this chapter is to characterize cable-driven kinematics in relation to conventional robot designs, and to explore their use in the field of wearable devices. In particular, recent research conducted at Fraunhofer IPA with respect to new kinds of kinematics and transmissions, as well as their applications to robotic devices and exoskeletons is reported.

5.2 Short State of the Art: From Musculoskeletal Robots to Wearable Devices

Musculoskeletal robots take their inspiration from the organization and properties of the musculoskeletal systems of mammals and birds. These systems feature lightweight structures and intrinsically compliant actuators, mimicking the properties of biological skeletons and muscles. Musculoskeletal robots offer several advantages in comparison to traditional robots. For instance, peak forces could be reduced that may result from unexpected collisions of the robot with its environment. This is especially advantageous in situations where humans and robots work in close proximity or need to interact, as it protects both the robot's actuators and the human user [2]. Compliant, spring-like operation of the actuators can also greatly increase the energy efficiency during periodic motions - a crucial issue for legged robots [3]. Finally, musculoskeletal robotic systems are useful as tools for the study of the principles underlying motor control in animals, as technical test beds used to test hypotheses or assumptions derived from neurobiological experiments.

Numerous robots have been built based on musculoskeletal principles, including compliant robotic arms (e.g. the Airic's arm [4] by Festo, using pneumatic muscles), legged robots (including the JenaWalker [5], the BioBiped [6] or the pneumatic walking and jumping robots from Hosoda lab [7]), as well as full humanoid robots (such as Kojiro [8], Kotaro [9] and ECCEROBOT [10, 11]).

Many of the mentioned designs are building on alternative transmissions compared to classical robot designs composed of a high-torque servo motor and a low-backlash reduction gear. These classical designs tend to be relatively expensive, heavy, and bulky. Particularly, high ratio reduction gears have been a continuous limitation to low-cost robot designs. Some selected examples on alternative robot kinematic designs are depicted in Fig. 5.1. The bionic handling assistant with its unique combination of body structure, actuation and generative manufacturing process is activated through bellowed air chambers [12].

Another way of effectively producing motion is through enlacement. As an example the simple Do-Helix mechanism achieves highest ratios of contractive force to weight ratios. The Quad-Helix transmission combines two Do-Helix principles on one shaft therefore allowing a full arm actuation with one motor only. As an implementation example the light-weight Isella experimental robot arm is composed of 4 Quad-Helix actuators producing 4 kinematic degrees of freedoms [13].

Apart from serial kinematics, parallel kinematics are equally subject to cable-driven transmission leading to surprising stiffness, precision and dynamics of the controlled end-effector [14]. Applications of these novel types of scalable, cost-effective parallel kinematics are seen in simulators, shipyards, logistics and general high-speed manipulation.

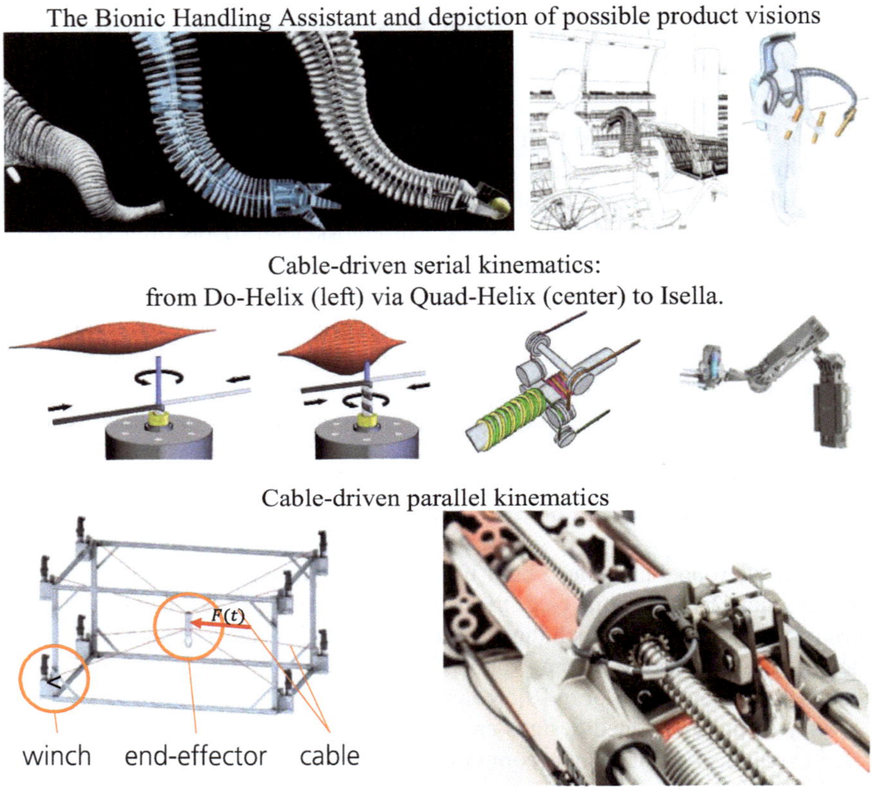

Fig. 5.1 Examples of alternative robot kinematic designs. Principle of the bionic handling assistant and configurations of cable-driven kinematic configurations

Similarly, the described challenges are subject of the Myorobotics project which not only incorporates a new transmission design based on wires, but aims to develop a "commercial-grade", modular and reconfigurable toolkit, coined the "Myorobotics Toolkit" [15, 16], for developing musculoskeletal robotic platforms. It is designed to be used by experimenters of different disciplines and aims to allow them to create, configure and operate their experimental systems based on their individual needs. In addition to academic settings, the toolkit also targets the industrial sector for applications that require the capability to mimic biological structures while maintaining high flexibility and reasonable costs.

5.3 The Myorobotics Toolkit

5.3.1 Overview

The Myorobotics framework, illustrated in , is made up of the *Design Primitives Library* (DPL), which regroups all the modular mechanical and electronic hardware, and the *Controller Library and MYODE*. The latter bundles the software tools that support typical tasks aimed at designing, simulating, operating and optimizing robotic assemblies. Three robots are shown as examples in the lower half of Fig. 5.2, (from left to right: a 2-DoF arm, a monopod hopper and a quadruped under development) to illustrate the kind of robots that can be built with the toolkit.

In the following, we focus on the Design Primitives Library and present in more details its modular components.

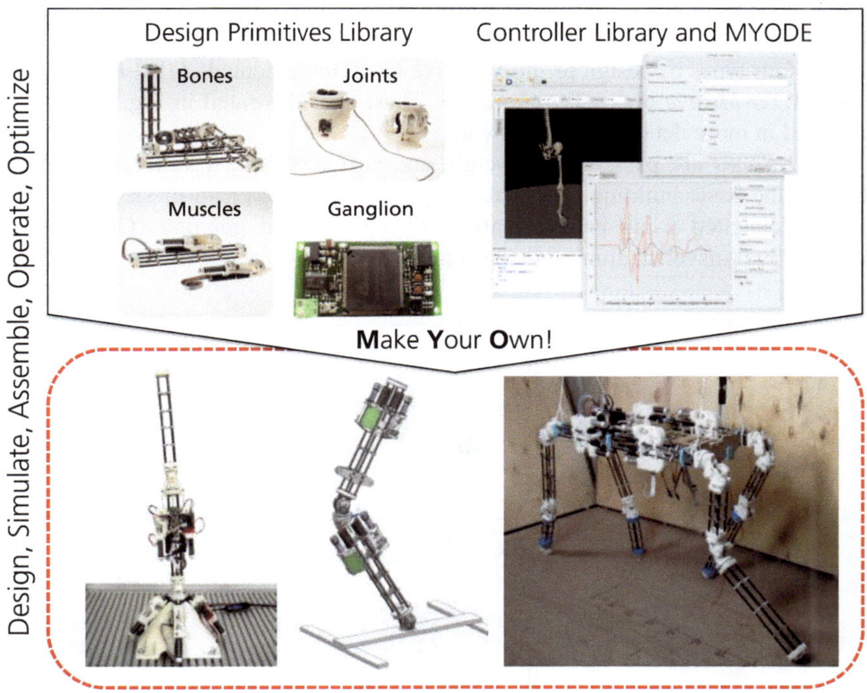

Fig. 5.2 The Myorobotics framework provides a modular toolkit for users to be able to design, assemble, simulate and operate customized musculoskeletal robots.

5.3.2 Design Primitives Library (DPL)

The DPL regroups all the toolkit hardware, organized in engineered modules making up the robot structure, actuation and sensing. In contrast to conventional robots, these modules, called Design Primitives (DPs), are strongly inspired by the human and animal musculoskeletal system, allowing the user to create Myorobots mimicking biological limbs. The toolkit organization aims to maintain as much as possible the capability to design and build engineered bio-inspired robots, while offering the modularity and flexibility offered by construction toolkits. The DPs can be configured and then assembled to custom-made robots, using integrated electromechanical interfaces that reduce cable clutter and simplify assembly. In addition, the mechanical design and production of the DP strongly leverage additive manufacturing techniques, allowing cost effective production, optimized mechanical design and compact integration of the sensors. As these production techniques are becoming widely accessible to the robotic community, we envision that interested users will be able to produce most of the mechanics in-house and be able to quickly develop additional customized modules for their specific application.

So far, four types of design primitives have been implemented: *MYO-Bone*, *MYO-Joint*, *MYO-Muscle* and *MYO-Ganglion*. These are illustrated in Fig. 5.3 and introduced in more details in the following.

MYO-Bones are passive, lightweight and stiff mechanical structural modules that are the basic building-blocks forming the kinematic chain or skeleton. They are implemented using parallel carbon fiber-reinforced polymer (CFRP) tubes bundled by transversal aluminum spacers.

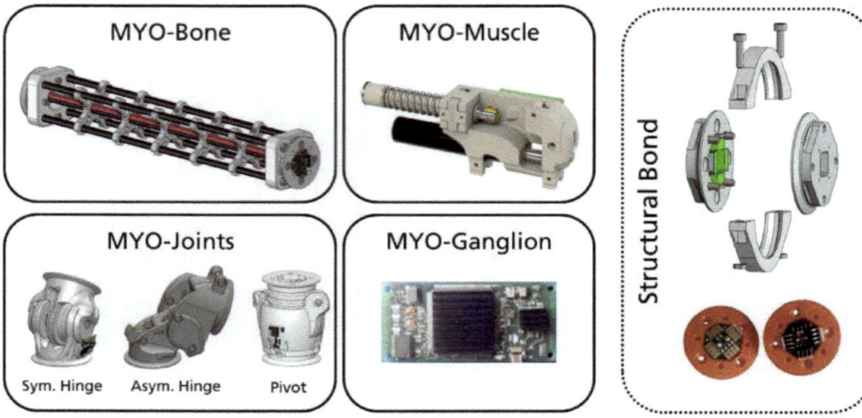

Fig. 5.3 Illustration of the four types of Design Primitives (left boxes with continuous lines) and the Structural Bond, the electromechanical interface used to connect MYO-Bones and MYO-Joints (right box with broken lines).

This allows the easy integration of the electric cabling and offers broad fixation possibilities for MYO-Muscles or MYO-Ganglia. The ends of the MYO-Bone feature an electromechanical interface (coined Structural Bond) to mount MYO-Joints.

MYO-Joints are passive mechanical connector modules, complementing the MYO-Bone to form the skeleton of the robot. Several types of MYO-Joints are envisioned, covering the diversity of articulations found in nature. So far, three kinds of one degree of freedom (1-DoF) MYO-Joints have been developed: two hinge joints, with symmetric (*Sym. Hinge*) and asymmetric (*Asym. Hinge*) angle ranges, and a *Pivot* joint. All MYO-Joints feature ball bearings for low friction operation and are equipped with absolute angle sensors. They also provide fixation points for the attachment of the MYO-Muscle tendon cables.

An electromechanical interface, coined Structural Bond, was designed to connect MYO-Bones and MYO-Joints (Fig. 5.3). Its mechanical interface, composed of two conical flanges, is joined together by two conical clamps tightened with two screws. Printed circuit boards with spring loaded contacts are integrated in the flanges and establish the electrical connection together with the mechanical fixation.

MYO-Muscles are intrinsically-compliant actuators, implemented as Series Elastic Actuators [17], i.e. with an elastic element arranged in series with the muscle contractile element. MYO-Muscles are unidirectional actuators (i.e. they only generate force under tension, not compression) connected to the skeleton using HPPE cables, mimicking the function of the biological tendons. Bidirectional actuation of a joint therefore requires two antagonist muscles. The current implementation is shown in Fig 5.4. It is based on a geared brushless DC servo (combined with a winch to wind up the tendon cable) and an exchangeable die spring as a series elastic element. A sensor measuring the deflection of the spring enables to compute the tension in the tendon cable connecting the MYO-Muscle to the MYO-Joint. The spring deflection is measured with a precision of 15 µm.

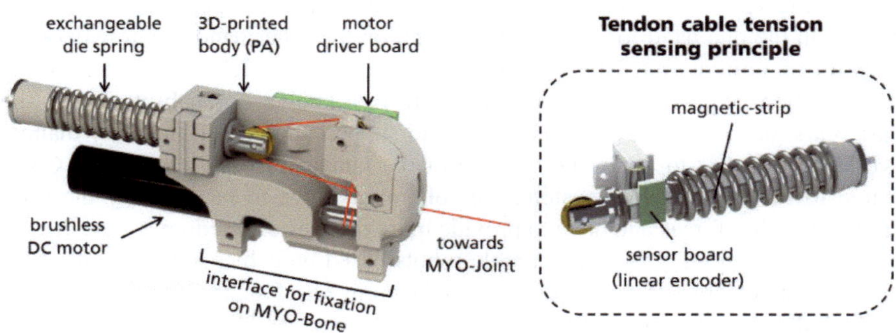

Fig. 5.4 Current implementation of the MYO-Muscle.

Finally, the MYO-Ganglion is a high speed communication and data processing unit able to collect local sensory information from other Design Primitives via CAN bus (1 Mbit/s) and communicate with other MYO-Ganglia using the automotive network communications protocol FlexRay (10 Mbit/s). It can also control up to four MYO-Muscles and, like its biological analogue, allow the execution of local, decentralized control strategies, such as reflexes.

Future visions regarding the use and capabilities of the Myorobotics toolkit comprise the design of lightweight robot arms and wearable robots as is depicted in Fig. 5.5. In the latter case, the Myorobotics toolkit could be especially useful during the R&D and prototyping phase of the wearable robot design.

Fig. 5.5 Visions of lightweight compliant systems, including robot arms and wearable robots, developed using the Myorobotics toolkit.

5.4 Wearable Cable-Driven Robots

Supporting systems for the lower limb were typically developed for two applications. On the one hand for military usage and on the other hand to allow paraplegics for an upright gait. Upper limb support expands the field of use for soft robots. Stationary passive motion machines are well known in rehabilitation. In the last years, body-worn structures appeared to support the human in heavy manual work. The goal is to avoid the damage of the human locomotion system. Compared to the rehabilitation system, the soft robots have to be worn by the user, transmit higher forces and have to provide more degrees of freedom.

In the last decade, several wearable robots based on cables appeared of which four wearable cable-driven robots are displayed in Fig. 5.6. The muscle suit from Koba Lab uses pneumatic artificial muscles in the back and Bowden cables for the force transmission to the joint of the upper body support [18]. With a weight of 9 kg, the muscle suit enabled a person lifting up to 50 kg. The Harvard Biodesign Laboratory already developed several generations of lower limb exoskeletons. The latest system is a so called soft exosuit using textiles and cables as actuators [19].

The ErgoSkeleton (formerly known as strongarm vest) is the product of the US start-up strongarmtech. One cable per arm equipped with a braking mechanism relieves the back in typical industrial handling applications. The system is passive and therefore lightweight [20]. Exhaustive work on cable-driven exoskeletons is performed by Agrawal Lab at Columbia University. The kinematic design of the exoskeleton is inspired by a cable-driven parallel robot: The topology is parallel and the cable tensions have to be taken into account [21].

Muscle Suit (Koba Lab) | Soft Exosuit (Harvard University) | ErgoSkeleton (Strongarmtech) | Cable-driven exoskeleton (Uni. of Delaware)

Fig. 5.6 Examples of state of the art wearable cable-driven robots

The overview over the state of the art shows, that there exist already cable-driven systems for lower limb as well as upper limb. The actuation principles reach from passive system over pneumatic muscles to servo drives. The main application of the research devices is rehabilitation. Strongarmtech focusses with its product on the support of workers.

5.4.1 Requirements and Structure of a Body Worn Lifting Aid

The requirements of a lifting aid can be characterized by the load which has to be manipulated and the performance in picks per hour as shown in Fig. 5.7. For manipulating heavy loads, lifting aids like rail-mounted cranes are well established. As these objects are manipulated slowly, the control has not to be very dynamic. For weights in a range between 20 to 200 kg, lifting aids like vacuum gripper are generally used. Weights between 5 and 35 kg are quite common in industry and logistic, as a human can lift these masses on his own. On the other hand, these objects are handled fast with average pick rates of up to 600 picks per hour which adds up to 12 tons of goods per shift. As a consequence, there is a strong need to unburden the worker.

The initial idea of the body worn lifting aid is shown in Fig. 5.7. The main components are an active cable support for the arms. The cables are connected to the wrist and guided via pulleys at shoulder to the actuator. The flexible spine support takes over the cable forces and transmits the load to the body regions like pelvis and thighs. Additionally, the spine support acts like an orthotic. It supports healthy body positions to avoid incorrect postures like round back.

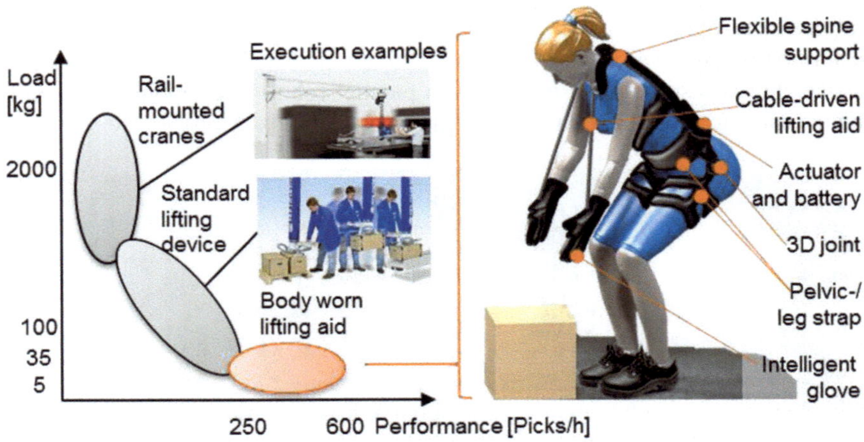

Fig. 5.7 Load and dynamic requirements (left) and concept of the cable-based lifting aid (right)

In industrial environments like production, logistics in warehouse or airports, additional requirements appear: lifting support for overhead work, which is particularly demanding a constrained operation in environments such as containers with obstacles (e.g. conveyor belts). This makes a body-fitting structure with minimal interfering geometry suitable. These requirements lead to the transition to another concept.

5.4.2 Body Worn Lifting Aid

The revised concept shown in Fig. 5.8 copies the topology of the human arm and uses rotatory joints. It provides flexible force assistance where cables can only provide forces in one direction. The basic idea is to assist in the direction of gravity and to keep the other degrees of freedom passive. This leads to two active joints per arm: in the shoulder and in the elbow joint. The alignment of the rotatory axes has to be optimized in a way, that the active joints bear the main part of the gravitational force within the possible postures of the arm.

The base of the assistive system is put on like a backpack. The hard cover back-support accommodates the transmission points at the shoulder and transmits the forces to the trunk. Textiles have been suggested to provide wearing comfort and full-area contact for force transmission. Variable connection elements to the body allow for the adaption on different anthropometries.

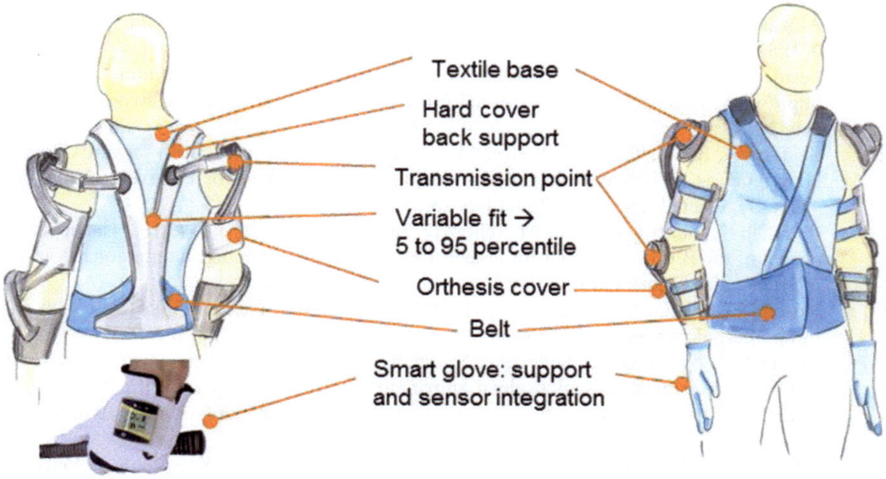

Fig. 5.8 New concept of the body worn lifting aid

Beside a lightweight design the intuitive control of the lifting aid is crucial for the user acceptance. The system will be only accepted, when it does not slow down the user during his work. The control has to distinguish between manipulation (object is gripped) and free motion of the human without any load. Without load, the lifting aid has to follow the arm movement.

During manipulation, the active joints have to support the gravitational force and move the object in the desired direction. To measure the user intention, sensors are integrated into the working glove. The next step in the current R&D-work is the finalization of the concept, building and assessment of a prototype.

Acknowledgments Research leading to the development of the Myorobotics toolkit and related results received funding from the European Union Seventh Framework Programme (FP7/2007-2013) under grant agreement n°288219 (MYOROBOTICS). Pictures of the MYO-Ganglion in Fig. 5.2 and Fig. 5.3 were provided by the Bristol Robotics Laboratory and the pictures of the monopod and quadruped robots in Fig. 5.2, as well as the right figure of Fig. 5.5 are courtesy of the Bio-Inspired Robotics Lab at ETH Zurich. Research related to the depicted results on wearable cable-driven robots has been supported by the BMBF Joint Project "Body Worn Lifting Aid" (Körpergetragene Hebehilfen) and Fraunhofer-internal project "E^3 Assistive Systems for Manufacturing" [22, 23].

5.5 *References*

[1] Dollar AM, Herr H (2008) Lower Extremity Exoskeletons and Active Orthoses: Challenges and State-of-the-Art. IEEE Transactions on Robotics, 24(1):144-158

[2] Holland O, Rob K (2006) The Anthropomimetic Principle. http://www.mindtrans.narod.ru/pdfs/cronos.pdf. Accessed 20 July 2014

[3] Alexander RM (1990) Three Uses for Springs in Legged Locomotion. Int J Robot Res 9(2):53-61

[4] Festo AG & Co. KG. Airic's_arm – Robot arm with Fluidic Muscles. http://www.festo.com/net/SupportPortal/Files/42058/Airics_arm_en.pdf. Accessed 20 July 2014

[5] Seyfarth A, Iida F, Tausch R, Stelzer M, von Stryk O, Karguth A (2009) Towards Biped-al Jogging as a Natural Result of Optimizing Walking Speed for Passively Compliant Three-Segmented Legs. Int J Robot Res 28:257-265

[6] Radkhah K, Maufroy C, Maus M, Scholz D, Seyfarth A, von Stryk O (2011) Concept and design of the BioBiped1 robot for human-like walking and running. International Journal of Humanoid Robotics 8(3):439-458

[7] Hosoda K, Sakaguchi Y, Takamaya H (2010) Pneumatic-driven jumping robot with an-thropomorphic muscular skeleton structure. Automomous Robots 28(3):307-316

[8] Mizuuchi I, Nakanishi Y, Sodeyama Y, Namiki Y, Nishino T, Muramatsu N, Inaba M (2007) An advanced musculoskeletal humanoid Kojiro. Proceedings of the 2007 IEEE-RAS International Conference on Humanoid Robots (Humanoids 2007), 294-299. 29 Nov. 2007-1. Dec. 2007, Pittsburgh, PA, USA

[9] Mizuuchi I, Yoshikai T, Sodeyama Y, Nakanishi Y, Miyadera A, Yamamoto T, Inaba M (2006) Development of musculoskeletal humanoid Kotaro. Proceedings of the 2006 IEEE International Conference on Robotics and Automation (ICRA 2006), 82-87. Orlando, FL, USA

[10] Jäntsch M, Wittmeier S, Knoll A (2010) Distributed Control for an Anthropomimetic Robot. Proceedings of the 2010 IEEE/RSJ International Conference on Intelligent Robots and Systems (IROS), 5466-5471. 18-22 Oct. 2010, Taipei, Taiwan

[11] Marques HG, Newcombe R, Holland O (2007) Controlling an Anthropomimetic Robot: A Preliminary Investigation. In: Almeida e Costa F, Rocha LM, Costa E, Harvey I, Coutinho A (ed) Advances in Artificial Life: Lecture Notes in Computer Science, Spring-er Berlin Heidelberg

[12] Grzesiak A, Becker R, Verl A (2011). The Bionic Handling Assistant: a success story of additive manufacturing. Assembly Automation 31(4):329-333

[13] Rost A, Verl A (2010) The QuadHelix-Drive - An improved rope actuator for robotic ap-plications. Proceedings of the 2010 IEEE International Conference on Robotics and Au-tomation (ICRA), 3254-3259. 3-7 May 2010, Anchorage, AK, USA

[14] Pott A, Mütherich H, Kraus W, Schmidt V, Miermeister P, Dietz T, Verl A (2013). Ca-ble-driven parallel robots for industrial applications: The IPAnema system family. Pro-ceedings of the 2013 International Symposium on Robotics (ISR). October 24-26, 2013, Seoul, Korea.

[15] Myorobotics: A framework for musculoskeletal robot development. http://www.myorobotics.eu/. Accessed 20 July 2014

[16] Marques HG, Maufroy C, Lenz A, Dalamagkidis K, Culha U, Siee M, Bremner P (2013). MYOROBOTICS: A modular toolkit for legged locomotion research using musculoskel-etal designs. Proceedings of the 6th International Symposium on Adaptive Motion of An-imals and Machines (AMAM 2013). March 11-14, Darmstadt, Germany

[17] Pratt G, Williamson M (1995) Series elastic actuators. Proceedings of the 1995 IEEE/RSJ International Conference on Intelligent Robots and Systems (IROS), 'Human Robot In-teraction and Cooperative Robots' (Volume: 1), 399-406. Pittsburgh, PA

[18] Muramatsu Y, Umehara H, Kobayashi H (2013) Improvement and Quantitative Perfor-mance Estimation of the Back Support Muscle. Proceedings of the 35th Annual Interna-tional Conference of Engineering in Medicine and Biology Society IEEE (EMBC), 2844-2849. 3-7 July 2013, Osaka, Japan

[19] Asbeck A, Dyer R, Larusson A, Walsh C (2013). Biologically-inspired Soft Exosuit. Pro-ceedings of the 2013 IEEE International Conference on Rehabilitation Robotics (ICORR), 1-8. 24-26 June 2013, Seattle

[20] Weiss TC (2014) StrongArm Technologies - ErgoSkeleton and Worker Safety. http://www.disabled-world.com/assistivedevices/ergoskeleton.php. Accessed 20 July 2014

[21] Mao Y, Agrawal SK (2012) Design of a Cable-Driven Arm Exoskeleton (CAREX) for Neural Rehabilitation. IEEE Transactions on Robotics 28(4):922-931

[22] Hebehilfe. http://www.mtidw.de/ueberblick-bekanntmachungen/mit-60-mitten-im-arbeitsleben/hebehilfe-entwicklung-und-verifikation-einer-koerpergetragenen-hebehilfe-zur-unterstuetzung-von-arbeitnehmern . Accessed 20 July 2014

[23] Fraunhofer-Leitprojekt »E³-Produktion«. http://www.fraunhofer.de/de/fraunhofer-forschungsthemen/fraunhofer-leitprojekte/e3-produktion.html. Accessed 20 July 2014

6 Capacitive Tactile Proximity Sensing: From Signal Processing to Applications in Manipulation and Safe Human-Robot Interaction

Stefan Escaida Navarro, Björn Hein, and Heinz Wörn

Karlsruhe Institute of Technology (KIT)

Abstract Recently we have shown developments on capacitive tactile proximity sensors (CTPS) and their applications. In this work we give an overview of these developments and put them into a more general perspective, emphasizing what the common grounds are for the different applications, i.e., preshaping and grasping, haptic exploration as well as collision avoidance and safe human-robot interaction. We discuss issues related to signal processing and the design of a smart skin for the robot arm and its end-effector. On a higher level we discuss the concept of proximity servoing and its use for the above mentioned applications.

6.1 Introduction

The concept of tactile proximity sensors evokes a vision of robotic systems that can be made more autonomous and safer as well as more intuitive to program for the human. Different to proximity sensing, vision and touch are two sensing modalities which are well investigated in robotics so far. Robotic applications without vision are almost unthinkable and many of them are quite robust. Tactile sensing is well established as well, especially to complement vision in grasping and interaction tasks. Still, there is a perception gap in the areas where vision is occluded and where touch is not desirable or hindering for the task. Proximity sensing closes this gap and provides continuous perception from events in the near vicinity of the robot to the touch event. Implementing tactile and proximity sensing in one sensor module allows to observe and model these events seamlessly. Because no touch can occur without a prior approach phase, be it in grasping or interaction tasks, it is promising to analyze these combined sensing in detail. In previous works, the development of a modular capacitive tactile proximity sensor was shown [3,4] enabling us to push research in this area.

For both sensing modalities, tactile and proximity, there have been many sensing principles developed and used. For tactile sensing, capacitive (e.g. [4,10]) and resistive technologies (e.g. [13]) are the most common. For proximity sensing, optical (IR, e.g. [12]) and capacitive measurement principles (e.g. [8]) are widely spread. Furthermore, multi-modal touch sensors that include proximity as well as further modalities such as temperature [15,11,9] are popular design decisions to extend the possible applications. Each technology has its advantages and disadvantages related to the physical effect exploited for measurement. The quality of signals of IR sensors are dependent on reflectance and therefore they have problems with light-absorbing and mirroring surfaces. Capacitive sensing is susceptible to shape, material and size of the objects it perceives.

Recently, many preshaping applications for robotic gripper have been shown based on proximity sensing; examples are [8,14], based on capacitive sensing and [5,6], based on IR-sensing. Common to this research is that closed-loop control methods are implemented that achieve a target relative configuration of the gripper to the object, that allows for robust grasping. Since this is a principle with a broader range of applications than just preshaping we call it *proximity servoing*, as discussed later in Section 6.3.1. Another important application is collision avoidance. Early works are [15] and the milestone work by Lumelsky and Cheung [7]. The latter demonstrates how large areas of the robot exterior can be covered by an array of IR-sensors enabling wide coverage of the robot surroundings and avoidance of obstacles. Recent work includes [12] where a mobile platform uses an array of IR-sensors to implement 360° perception for collision free navigation.

In the remainder of the chapter we want to give an overview of the areas of applications for CTPS, while also discussing aspects of the signal processing and the physical design. We want to put some results of previous works into a more general perspective as well and provide insights of how these results have to be considered for the implementation of future applications. Therefore the rest of the chapter is organized as follows: in Section 6.2 we give an overview of the signal processing. Then, in Section 6.3, we review the applications we have developed so far with our sensor system. Finally, in Section 6.4, we give concluding remarks as well as discussing in which direction our future research will go.

6.2 Signal Processing and Feature Extraction

Signal processing and feature extraction are important for us to investigate, because this topic is intertwined with the design of the sensor and has ramifications for the applications which we want to implement. Especially challenging regarding capacitive proximity sensors is the correct extraction of the desired information from the sensor signals, i.e., the relative *pose* of objects or obstacles. The main issue is the dependency of the measured signal on the object's material, size, shape and pose with respect to the sensor. As discussed in Section 6.1, alternative measuring principles like IR-based suffer from similar issues. One of the main sensor design decisions that helps in dealing with these issues is the implementation of a *spatial resolution*, i.e., arranging several sensors as arrays on the surface of the robot and within the end-effector (e.g. gripper). We call a single element of such an array *proxel*.[1] In [3] a modular sensor design was presented with which it is possible to build such an array. It should be considered *how* the sensors are arranged in order to facilitate the extraction of the desired features, while the functional aspect and mechanical design of robot arms and grippers also have to be taken into account. Fig. 6.1 shows how a planar proximity sensor array produces a *foreground image* related to the size and pose of a human hand.

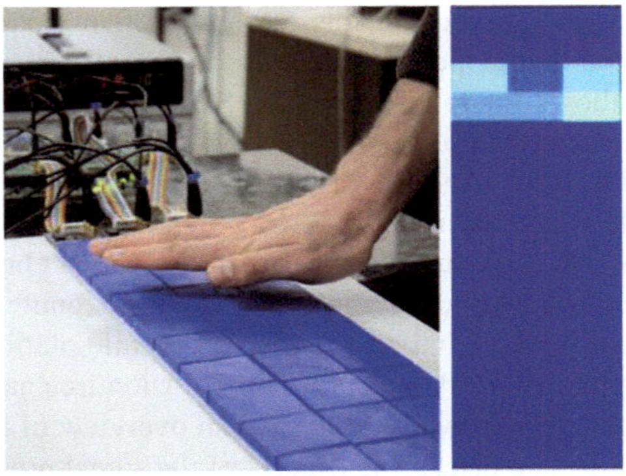

Fig. 6.1 Left: A human hand as it interacts with an arrangement of 3 × 16 tactile proximity sensors [1]. The electronics for signal processing are shown in the background. Right: Color coding of the proxels is depending on the distance of the hand to the corresponding sensor.

From these images important features such as foreground regions, center of masses, gradients (edges), region orientation, etc. can be extracted by means of traditional computer vision algorithms. Nonetheless, it must be always considered that the measurements do not strictly represent a depth-map. As discussed later in Section 6.3.1, the actual shape of objects and obstacles can be better extracted, when the relative configuration in cartesian space of the sensor array to them is known. A relevant parameter for the design of applications is the proxel size. Given by the measurement principle, bigger proxel areas will increase sensitivity and sensing range and smaller areas will allow for higher spatial resolution at the cost of range. For instance the modules shown in Fig. 6.1 have a side length of 4cm, allowing the detection at a range of up to 10cm and modeling of the hand as a whole, but not of single fingers. At last, of utmost importance is the temporal aspect of the sensing. It follows that additional temporal information such as *temporal-gradients* support the realization of applications like 3D-tracking and proximity servoing.

[1] Derived from **prox**imity **el**ements in analogy to pixels (picture elements) or taxels (tactile elements).

6.2.1 Tracking

In [1] tracking using CTPSs was investigated. Tracking delivers an object's pose over time, which is one of the central features for safe and intuitive human-robot interaction. The tracking was developed and demonstrated on an arrangement of 3×16 modules as seen in Fig. 6.2. As a tracking back-end the the *Kalman-Filter* was used. With it, the current dynamics state of the object is estimated based on state history, a noisy dynamics model for the motion of the object combined with a noisy measurement taken at each sample time. In each step the predicted state of the object, according to the dynamics model and the previous state, is combined with the measurement such that the combined uncertainty inherited from prediction and measurement is minimized.

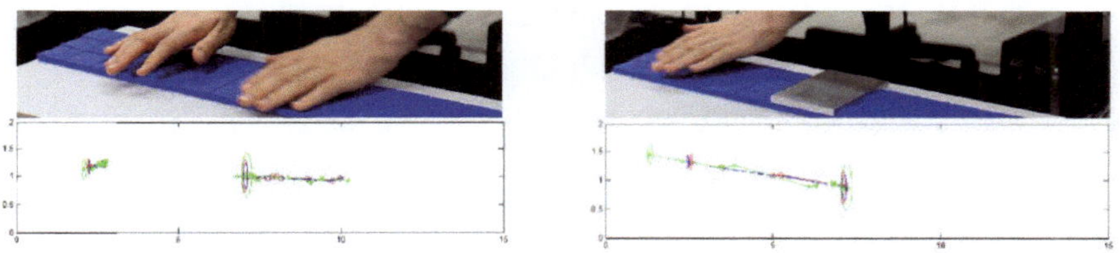

Fig. 6.2 Left: Tracking enables to distinguish two hands even if the separation in sensor data is not evident. Right: Tracking enables to handle occlusions in the proximity sensing [1]

In the same work experiments for collision prediction and detection were conducted. A robot driven movement for an object with a constant acceleration profile was used to show that the time and location of an impact on the array can be predicted. Rather than evaluating the collision prediction with an external sensing system, we evaluated it with the tactile modality of the sensor itself. With this, the adequacy for safety related applications can be assessed more thoroughly. Future developments in tracking will include a strategy that can differentiate between the types of object based on their movement profile.

6.2.2 Task and Environment Contexts for Feature Extraction

There are many scenarios in which a robot equipped with CTPS may be used purposefully. These range from static environments where the robot has to execute an autonomous task, to dynamic environments, such as human-robot cooperation, where the safety of the human operator has to be guaranteed at all time. Also, significant surrounding conditions are given by the task the robot is scheduled to execute. Most of the time a robot will move to fulfill its task, resulting in an *active safety*. On the other hand the robot may not be scheduled to perform any task at all and therefore not to move, but nevertheless should still provide *reactive safety* by avoiding collisions if possible. This two-layer classification, which is illustrated by Fig. 6.3, has important ramifications for the feature extraction. Firstly, determining in which type of environment the robot is in, allows to adjust the feature extraction according to the expectations. For instance, if the robot is cooperating with another robot, it is clear that the sensor values measured cannot correspond to a gesture performed by a human operator. Secondly, the robot's own movement is an important cue to determine robot-to-object distance and object trajectories. Generally, the confidence in the estimated dynamics of proximity events should always be higher when the robot itself is moving.

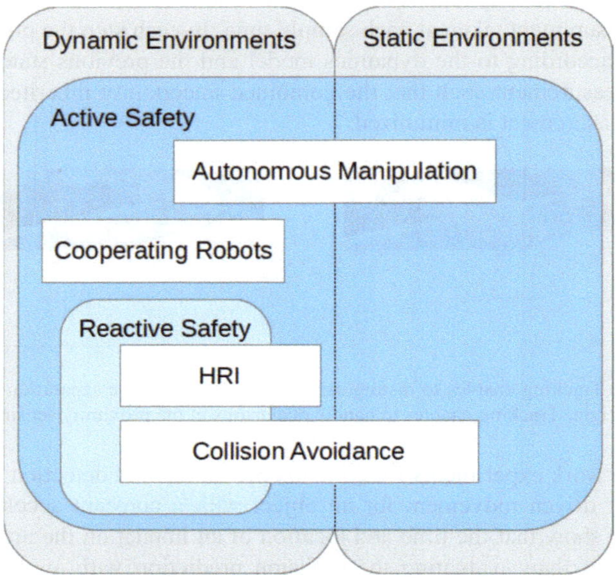

Fig. 6.3 Examples for different applications and scenarios for a robot using CTPS. The scenarios determine how features are extracted and interpreted.

For the reasons mentioned above it is important to implement situation awareness for the robot. We call this *System Behavior Strategies* which should guide the lower level signal processing. To this end a model of the internal state of the robot is needed. This model should take into account the current environment situation (presence of humans, obstacles, etc.) and the task at hand (cooperate, autonomous task execution, etc.). The goal of the strategies is then to guide the behavior, i.e., actions and reactions of the robot, with a special focus on the lower level signal processing for CTPS. Additionally, it should consider the information from other sensor modalities, such as vision.

6.3 Applications

6.3.1 Proximity Servoing

We call a closed-loop control scheme which maps proximity features to robot movements *proximity servoing*. Since proximity sensing is 3D-perception it is in principle possible to extract 6D-pose information of objects and the human from the processed sensor signals. Therefore the movement of the robot can also be influenced in all 6DOFs where the goal of the servo-controller will be to maintain a constant relative configuration to some target. This broad concept has two main occurrences: on a smart skin for a robotic manipulator and in the fingers of a gripper. On a smart skin proximity servoing provides a base for programming tasks, where proximity input can be used to drive the robot to a desired configuration. It is also the foundation on which any collision avoidance scheme will sit on top of. Within a gripper, servoing can be used to robustly preshape to an object that shall be grasped. For this latter task promising results where shown in [2].

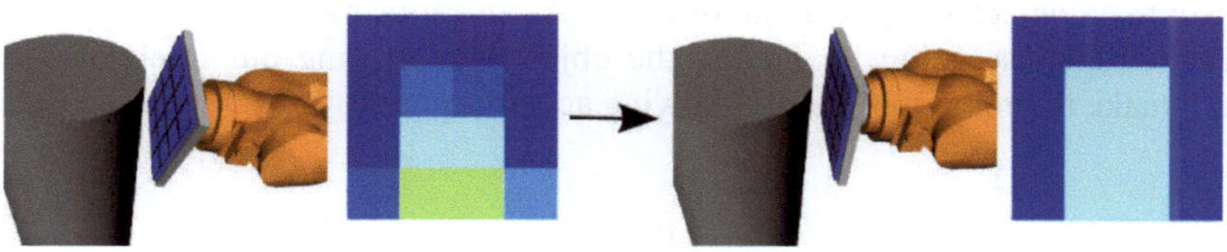

Fig. 6.4 Using an exemplary sensor array of 4 × 4 it is shown how proximity servoing aligns the sensor surface to the object by equilibrating the sensor signals along the gradient. In the aligned configuration robust information about the curvature of the object can be extracted from the signals.

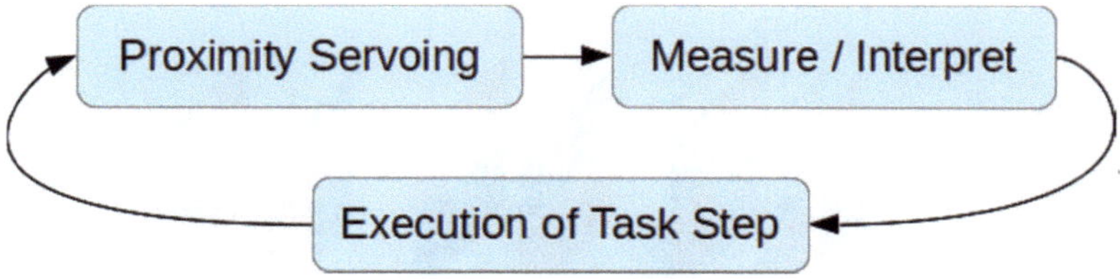

Fig. 6.5 The three steps necessary to execute a task based on CTPS when relying on proximity servoing.

In general it will not be possible to determine absolute deviations from the signal distribution on a CTPS array, which could be corrected in one step. However, a gradient can be calculated, meaning a direction of correction as well as some degree of information on the amount of deviation. It is the task of the servo-controller to continuously decrease the error along the detected gradient. When the final aligned configuration has been established, the signal distribution on the sensors will contain information about the shape of the object, as illustrated by Fig. 6.4.

From the point of view of applications based on proximity servoing it follows there is a general design principle: after servoing, shape information can be extracted, then, based on this increment in information, a next step of the task can be executed. This step will lead to a situation of possible misalignment where servoing has to be used again. This general loop is shown in Fig. 6.5.

6.3.2 Preshaping

One of the applications of proximity servoing is preshaping. We want to summarize here some general conclusions we can draw from our work in [2]. In a grasping context *preshaping* means to find a suitable configuration for the gripper and its fingers –before touch is established– that favors robustness when grasping. Fig. 6.6 shows the 2 × 2 arrangement of the sensors inside the fingers, where also groups are marked, that enable the detection of deviation of the gripper pose with respect to an object along or about one axis. As an application, preshaping basically is proximity servoing. In fact, if the object was moving our implementation would make the robot move and following accordingly.

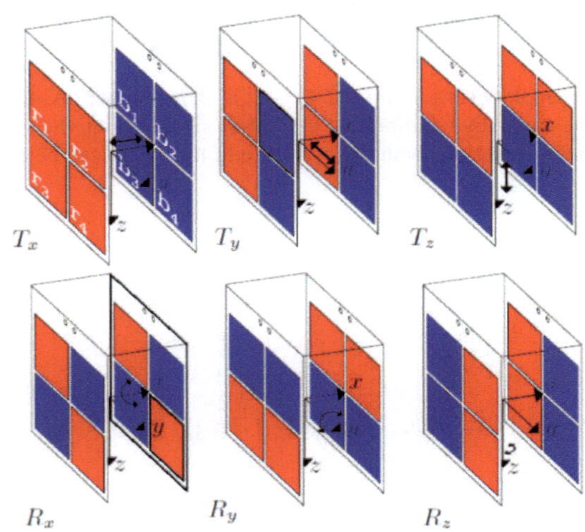

Fig. 6.6 Groups of sensors red and blue are used when comparing sensor values for finding misalignment along (T_a) or about (R_a) the cartesian axes inside a gripper [2].

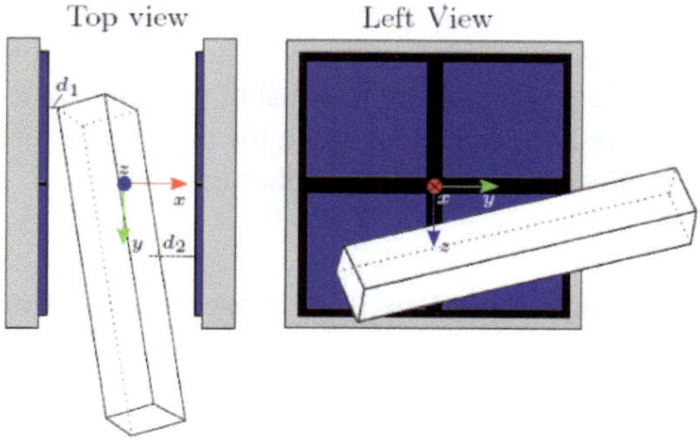

Fig. 6.7 Example for a start configuration of a misaligned object inside the gripper [2].

The basic principle of finding a misalignment is to calculate the difference between the sensor values of red and blue groups in Fig. 6.6. Any pair (r_i, b_j) for $i, j \in \{1, 2, 3, 4\}$ can in principle indicate that an adjusting is necessary. However, it is not trivial to get reliable difference values, because of the increasing noise in the measurement with increasing object-to-sensor distance and sensitivity to pose and covered area of the single sensors. As illustrated by Fig. 6.7, an object may be misaligned in several directions and the sensor closest to the corner of the object (distance d_1 in Fig. 6.7) will measure a lower signal than the sensor which has a bigger surface in front of it on the opposite side (distance d_2 in Fig. 6.7), because of the capacitive sensing principle used. It follows, that it is possible that the servo-controller moves in the incorrect direction on the x-axis in this case. On the favorable side, if the object is pre-aligned in some other axis, such as around the z-axis or along the y-axis the detection of the correction gradient becomes more robust for translation on the x-axis. Therefore, the principle of the preshape-control is to use the most reliable measurements at first, i.e., the highest sensor values, to detect the rough gradient. In a second step, considering the alignment is already somewhat stable, a majority voting scheme involving at least 2 sensors from the groups of Fig. 6.6 is used as well.

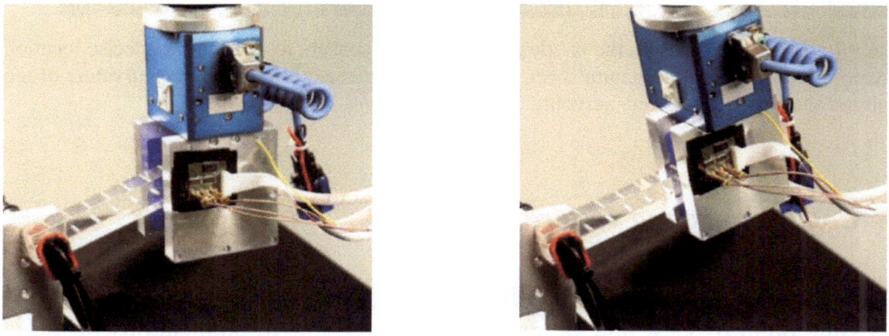

Fig. 6.8 Left: The gripper with CTPS in the start configuration before preshaping to a plexi-glass rod. Right: The gripper has finished preshaping and grasps the plexi-glass rod.

This first implementation of proximity servoing illustrates the challenges related to arranging the sensors as an array, extracting the desired information and dealing with situations specific to the capacitive sensing principle. We expect that similar solutions will be necessary when developing the algorithms for reactive path planning, collision avoidance and safe human-robot interaction.

6.3.3 Combined Haptic and Proximity-Based Exploration

Proximity servoing applied to preshaping can also be used to implement a combined haptic and proximity-based exploration. When exploring an object a strategy guides the gripper equipped with sensors to unknown areas of the workspace to elicit object features. Using touch it is possible to detect features quite precisely, but it is a challenge to approach the object prior to touching it in a way that the sensor area is aligned to the surface and the object is not displaced when being explored. The preshaping described in the previous section provides an ideal solution to these issues and leads to the combined haptic and proximity-based strategy. Even more, exploration steps based only on proximity are also possible and time efficient. In [2] we showed some results in regards to detecting edges and corners of objects as well contactless exploration of features such as curvatures. Fig. 6.9 shows the gripper executing some of the exploration skills.

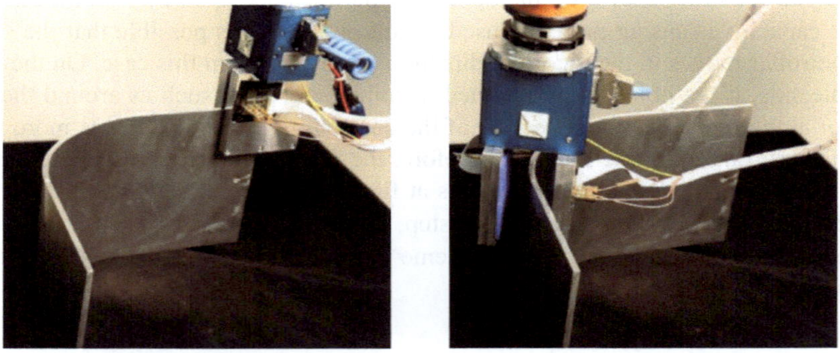

Fig. 6.9 Left: The gripper uses the tactile modality of the sensors to explore the precise location of the corner on a curved aluminum sheet. Right: The gripper explores along the curvature of the aluminum sheet using proximity servoing.

6.4 Conclusions and Future Work

In this chapter we discussed capacitive tactile proximity sensing from a signal processing point of view and we presented an overview of applications for these sensor systems, with a special focus on *proximity servoing*. In regards to the signal processing we discussed the importance of the capacitive sensing principle, the role of the sensor size and spatial and time resolution. In regards to proximity servoing we emphasized its importance by showing that it is a common base for *pre-shaping*, *haptic exploration* and *collision free path traversal*. Additionally, we discussed that touch- and proximity- sensing are complementary, especially because no touch can occur without a previous approaching event.

Based on the these results it can be stated that the proposed capacitive tactile proximity sensing concept is a very promising approach to close the existing near field perception gap in robotics and therefore plays an important role in the new field of soft-robotics. In the future we want to concentrate on developing methods for safe human-robot interaction with aim of implementing intuitive programming and human-robot collaboration scenarios. The technology of CTPSs allows for compliant robot motion and control, without having to relinquish the advantages of robot systems with high mechanical stiffness. Thus, we want to focus on the possibilities of multi-modal interaction, as illustrated by Fig. 6.10(a) and Fig. 6.10(b). A further interest is to investigate collision free path traversal (see Fig. 6.10(c)) and its relation to haptic exploration to build an environment model based on touch and proximity features.

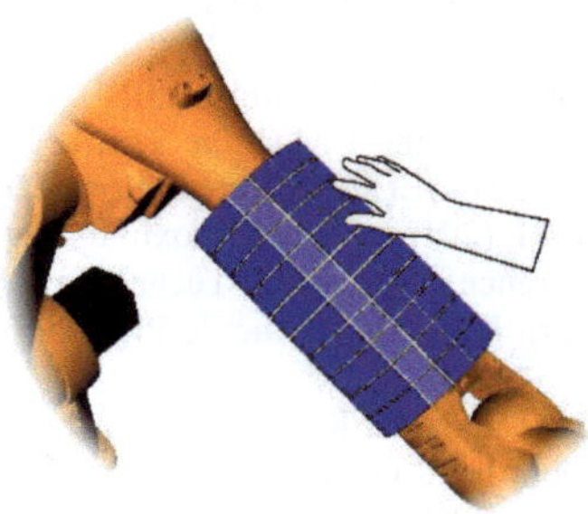

(a) Proximity-based interaction will allow for touchless, intuitive programming of robot systems.

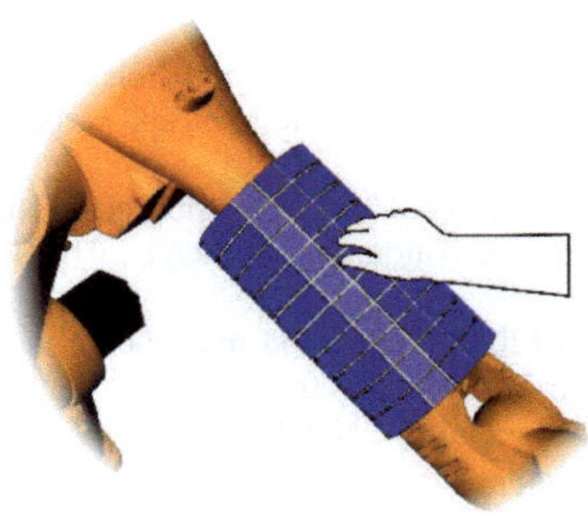

(b) Touch-based interaction with the robot skin will allow further interaction, where the simultaneous use of proximity mode is also possible.

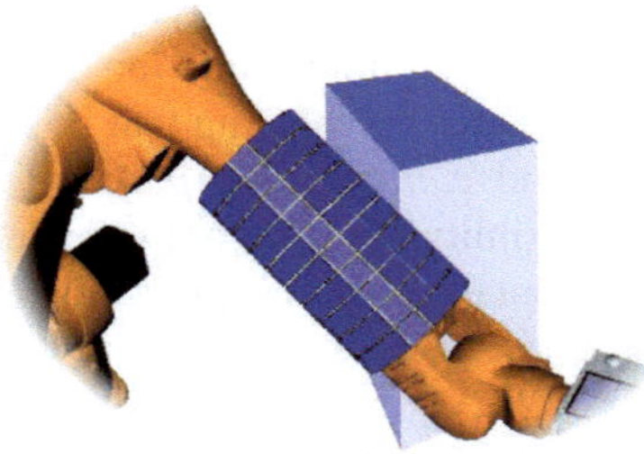

(c) An obstacle can be avoided by use of proximity sensing. The features obtained during the traversal can potentially complement an environment model for the robot.

Fig. 6.10 Possible applications for a robot endowed with a skin based on CPTS modules.

6.5 References

[1] Escaida Navarro S, Marufo M, Ding Y, Puls S, Göger D, Hein B, Wörn H (2013) Methods for Safe Human-Robot-Interaction Using Capacitive Tactile Proximity Sensors. Intelligent Robots and Systems (IROS), 2013 IEEE/RSJ International Conference on. pp. 1149–1154

[2] Escaida Navarro S, Schonert M, Hein B, Wörn H (2014): 6D Proximity Servoing for Preshaping and Haptic Exploration using Capacitive Tactile Proximity Sensors. Intelligent Robots and Systems (IROS), 2014 IEEE/RSJ International Conference on, to appear

[3] Göger D, Alagi H, Wörn H (2013) Tactile Proximity Sensors for Robotic Applications.In: International Conference on Industrial Technology (ICIT)

[4] Göger D, Blankertz M, Wörn H (2010) A Tactile Proximity Sensor. IEEE Senors 2010. pp. 589 –594

[5] Hasegawa H, Mizoguchi Y, Tadakuma K, Ming A, Ishikawa M, Shimojo M (2010) Development of intelligent robot hand using proximity, contact and slip sensing. Robotics and Automation (ICRA), 2010 IEEE International Conference on. pp. 777 –784

[6] Koyama K, Hasegawa H, Suzuki Y, Ming A, Shimojo M (2013) Pre-shaping for various objects by the robot hand equipped with resistor network structure proximity sensors. Intelligent Robots and Systems (IROS), 2013 IEEE/RSJ International Conference on. pp. 4027–4033

[7] Lumelsky VJ, Cheung E (1993) Real-time collision avoidance in teleoperated whole-sensitive robot arm manipulators. Systems, Man and Cybernetics, IEEE Transactions on 23(1), 194–203

[8] Mayton B, LeGrand L, Smith J (2010) An electric field pretouch system for grasping and co-manipulation. Robotics and Automation (ICRA), 2010 IEEE International Conference on. pp. 831–838

[9] Mittendorfer P, Cheng G (2011) Humanoid multimodal tactile-sensing modules. Robotics, IEEE Transactions on 27(3), 401–410

[10] Schmitz A, Maiolino P, Maggiali M, Natale L, Cannata G, Metta G (2011) Methods and technologies for the implementation of large-scale robot tactile sensors. Robotics, IEEE Transactions on 27(3), 389–400

[11] Stiehl W, Breazeal C (2006) A sensitive skin for robotic companions featuring temperature, force, and electric field sensors. Intelligent Robots and Systems, 2006 IEEE/RSJ International Conference on. pp. 1952–1959

[12] Terada K, Suzuki Y, Hasegawa H, Sone S, Ming A, Ishikawa M, Shimojo M (2011) Development of omni-directional and fast-responsive net-structure proximity sensor. Interna-tional Conference on Intelligent Robots and Systems

[13] Weiss K (2006) Ein ortsauflösendes taktiles Sensorsystem für Mehr-finger-Greifer. Ph.D. thesis, Universitt Karlsruhe (TH), Institut für Prozessrechentechnik Automation und Robotik

[14] Wistort R, Smith JR (2008): Electric field servoing for robotic manipulation. Intel- ligent Robots and Systems, 2008. IROS 2008. IEEE/RSJ International Conference on. pp. 494 – 499

[15] Yamada Y, Tsuchida N, Ueda M (1988) A proximity-tactile sensor to detect obstacles for a cylindrical arm. Journal of the Robotics Society of Japan 6(4), 292–300

Part III Modeling, Simulation and Control

7 Perception of Deformable Objects and Compliant Manipulation for Service Robots

Jörg Stückler and Sven Behnke

University of Bonn, Computer Science Institute VI, Autonomous Intelligent Systems

Abstract We identified softness in robot control as well as robot perception as key enabling technologies for future service robots. Compliance in motion control compensates for small errors in model acquisition and estimation and enables safe physical interaction with humans. The perception of shape similarities and deformations allows a robot to adapt its skills to the object at hand, given a description of the skill that generalizes between different objects. In this chapter, we present our approaches to compliant control and object manipulation skill transfer for service robots. We report on evaluation results and public demonstrations of our approaches.

7.1 Introduction

In today's industrial settings, robots are frequently required to execute motions fast, precisely, and reliably. The use of high-stiffness motion control can guarantee robust operation in this domain, but it also demands precise models of the dynamics of the robot mechanism and the manipulated objects. Furthermore, precautions need to be taken to prevent physical interaction with humans under any circumstances. This approach may not be applicable, e.g., in human-robot collaborative scenarios, in less structured environments, or when physical interaction with humans is unavoidable.

Generalization of robot skills is a further aspect that needs to be considered to bring robots into new applications. Often in practice, manipulation controllers need to be manually designed for each specific instance of an object class. This approach limits the range of possible applications of robotics technology by the effort that has to be taken to adapt the robot to the task, especially for service robots in our everyday environments.

We identified softness in robot control as well as robot perception as key enabling technologies for future service robots. Compliance in motion control compensates for small errors in model acquisition and estimation and enables safe physical interaction with humans. The perception of shape similarities and deformations allows a robot to adapt its skills to the object at hand, given a description of the skill that generalizes between different objects.

In this chapter, we present our approaches to compliant control and object manipulation skill transfer for service robots. We propose compliant task-space control for redundant manipulators driven by servo actuators. The actuators in our approach are back-drivable and allow for configuring the maximum torque used for position control. From differential inverse kinematics, we derive a method to limit the torque of the joints depending on how much they contribute to the achievement of the motion in task-space. Furthermore, our approach not only allows for adjusting compliance in the null-space of the motion but also in the individual dimensions in task-space. This is very useful when only specific dimensions in task-space shall be controlled in a compliant way. We utilize this compliance in several applications that require physical human-robot interaction. For instance, we demonstrate the cooperative carrying of a large object. We also use compliant control when handing objects to a human, or to guide the robot at its hand.

In many object manipulation scenarios, controllers can be described for specific object instances through grasp poses and 6-DoF trajectories relative to the functional parts of the objects. One can pose the problem of skill transfer as establishing correspondences between the object shapes, i.e., between the functional parts. Grasps and motions are then transferrable to novel object instances according to the shape deformation. We propose an efficient deformable registration method that provides a dense displacement field between object shapes observed in RGB-D images. From the displacements, local transformations can be estimated between points on the object surfaces. We apply these local transformations to transfer grasps and motion trajectories between the objects.

We develop our approaches with our service robots Cosero and Dynamaid [12,14,15]. The human-scale robots are equipped with two anthropomorphic arms each on upper bodies that can be moved on a linear actuator in the vertical direction in order to manipulate on different height levels. They move in indoor environments on omnidirectional drives with small footprints. A communication head provides the robots with human-like appearance for natural human-robot interaction. Light-weight design facilitates inherent safety of the robots.

7.2 Compliant Control for Service Robots

Task-space motion control, initially developed by Liegeois [4], is a well-established concept in robotics (see [6] for a recent survey). Common to task-space control methods is to transfer motion specified in a space relevant to a task to joint-space motion. One simple example is the control of the end-effector of a serial kinematic chain along pose trajectories in Cartesian space. For compliant motion control in task-space, acceleration- and force-based methods are frequently employed. We propose a velocity-based method. Instead of relying on redundancy resolution for compliant control, we adjust compliance for each dimension and direction in task-space as well as in the null-space of the motion when the robot kinematics is redundant for the task.

Using such compliant control, we implemented several service robot tasks that require soft and compliant interaction with objects or persons. One such task is the cooperative transport of large objects by a robot and a person. Khatib et al. [2] investigated manipulation of large objects with multiple mobile manipulators. This approach requires exact identification of the dynamics of the mobile manipulators. Yokoyama et al. [17] use an HRP2 humanoid robot to carry a large panel together with a human. The robot finds the panel by stereo vision through a model-based recognition system. The walking direction of the robot is controlled by voice commands and by force-torque sensors on the robot wrist. In our approach, the robot also recognizes the intention of the person through the motion of the table. Instead of specific force-torque sensing in the wrist, we apply compliance control to let the human move the robot's end-effectors through the table.

A further application of compliant control which we have investigated is the problem of robot guidance by a human through physical interaction. Christensen et al. [1] proposed to lead a domestic service robot around the house for initial map acquisition by taking its hand. The higher bandwidth of the arm allows for decoupling the applied forces from the robot motion. Oudeyer et al. [7] report on such following behavior emerging from compliant whole-body control of a humanoid robot. In our work, we couple compliant control of the arms with the laser-scanner based perception of the human guide.

7.2.1 Compliant Task-Space Control

We employ velocity-based task-space control and derive a control law for compliant motion of the arms. We assume that the robot actuators follow position trajectories through torque control. In our approach, we assume that the torque applied by the actuator can be limited. We derive the responsibility of each joint for the motion in task-space, and distribute a desired maximum torque onto the involved joints according to their responsibility.

Central to task-space controllers is a mapping from joint states $q \in \mathrm{R}^{\mathrm{m}}$ to states $x \in \mathrm{R}^{\mathrm{n}}$ in task-space, i.e., the forward kinematics $x = f(q)$. Inversion of the linearized relationship yields a mapping from task-space velocities to joint-space velocities $\dot{q} \approx J^{\dagger}\dot{x} + (I - J^{\dagger}J)\dot{q}^{0}$, in which secondary joint motion $\dot{q}^{0}$ can be projected into the null-space of the mapping such that the tracking behavior in task-space is not altered.

Given a desired trajectory in task-space $x_d(t)$, we derive a control scheme to follow the trajectory with a position-controlled servo actuator

$$\dot{x}(t) = K_x\left(x_d(t) - x(t)\right),\ \dot{q}(t) = K_q\left(J^{\dagger}\dot{x}(t) + \alpha(I - J^{\dagger}J)\nabla g(q(t))\right),$$

where K_x and K_q are gain matrices. The cost function $g(q(t))$ optimizes secondary criteria in the null-space of the motion, and α is a step-size parameter. Cost criteria typically include joint limit avoidance or the preference of a convenient joint state.

We set a compliance $c \in [0,1]^n$ in linear dependency of the deviation of the actual state from the target state in task-space, such that the compliance is one for small displacements, zero for large ones, and linearly interpolates in between. For each task dimension, the motion can be set compliant in the positive and the negative direction separately, allowing e.g. for being compliant in upward direction, but stiff downwards. If the task dimension is not set compliant, we wish to use high holding torques τ_i^x to position-control this dimension. If it is set compliant, the maximal holding torque interpolates between a minimal value for full compliance and a maximum torque for zero compliance.

To implement compliant control, we measure the responsibility of each joint for the task-space motion through the inverse of the Jacobian

$$R_{task}(t) := abs\left[J^{\dagger}(q(t))\begin{pmatrix} \dot{x}_1(t) & 0 & \cdots & 0 \\ 0 & \dot{x}_2(t) & \ddots & \vdots \\ \vdots & \ddots & \ddots & 0 \\ 0 & \cdots & 0 & \dot{x}_n(t) \end{pmatrix}\right],$$

where abs determines absolute values of a matrix element-wise.

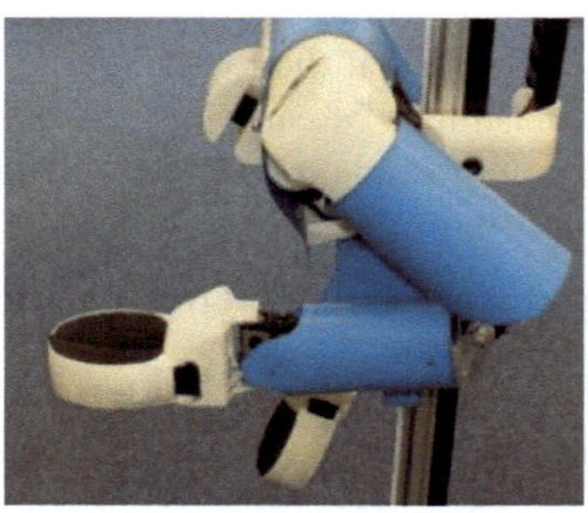

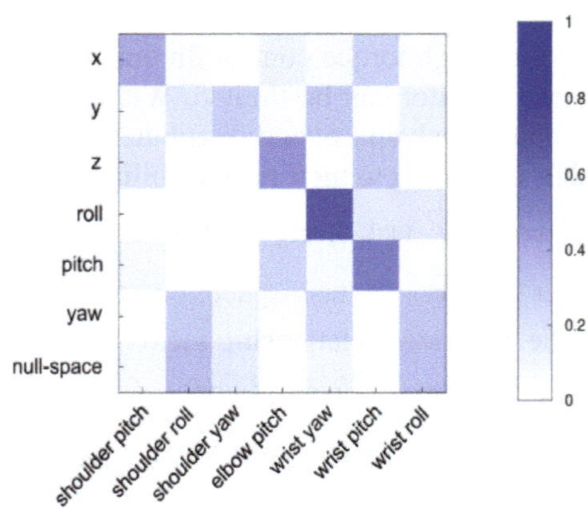

Fig. 7.1 Activation matrix in compliant control. Task-space dimensions correspond to forward/backward (x), lateral (y), vertical (z), and rotations around the axes (roll, pitch, yaw).

Each entry (i, j) of the matrix measures the contribution of the velocity of the j-th task component to the velocity of the i-th joint. In addition, we also define the responsibility of each joint for the null-space motion $R_0(t) := abs[\alpha(I - J^\dagger J)\nabla g(q(t))]$. To finally distribute our desired torque limits, we determine an activation matrix $A(t)$ by normalizing the responsibility of the joints to sum to one along each task dimension. Fig. 7.1 shows an example matrix. The task-component maximal torques are then distributed according to the activation of each joint, i.e. $\tau_q = A(t)\tau_x$.

7.2.2 Applications of Compliant Control in Everyday Environments

Object Hand-Over to a Person

Object hand-over from a robot to a person can be implemented with several strategies. For instance, object release could be triggered by speech input or by specialized sensory input such as distance or touch sensors. Through compliant control, we establish a very natural way of hand-over by simply releasing the object when the interaction partner pulls on the object (see Fig. 7.2, left). To implement this, the robot offers the object to the person and controls the motion of its end-effector compliant in forward, in upward direction, and in pitch rotation. The robot releases the object when it detects a significant displacement of its end-effector.

Guiding a Robot at its Hand

Taking the robot by its hand and guiding it is a simple and intuitive mean to communicate locomotion intents to the robot (see Fig. 7.3, left). We combine person perception with compliant control to implement such behavior: the robot extends one of its end-effectors forward and waits for the user. As soon as the user appears in front of the robot and exerts forces on the end-effector, the robot starts to follow the motion of the end-effector by driving in translational directions. The robot avoids the guide with a potential field method. It rotates its base to keep the guide at a constant angle, relative to its heading direction.

Cooperative Carrying of a Table

Cooperative transportation of large objects is a typical collaborative task in which multiple persons or robots physically interact to solve a task (Fig. 7.2 right).

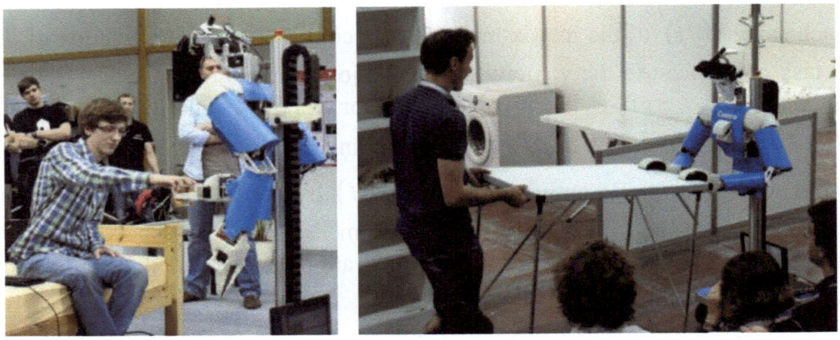

Fig. 7.2 Left: Cognitive service robot Cosero hands an object to a person. Right: cooperative carrying of a table by a person and Cosero.

We demonstrate object perception, person awareness, and compliant control in the task of cooperatively carrying a table by a person and a robot. As soon as the person appears in front of the robot, the robot approaches the table, grasps it, and waits for the person to lift it. After the robot visually perceives the lifting of the table, it also lifts the table and starts to follow the motion of the person. It sets the motion of the end-effectors compliant in the sagittal and lateral direction, and in yaw orientation. By this, the robot complies when the person pulls and pushes the table. The robot follows the motion of the person by controlling its omnidirectional base to realign the hands to the initial grasping pose with respect to the robot. The person may cease the carrying of the table at any time by lowering the table, which is also visually perceived by the robot.

Manipulation of Articulated Objects

We apply compliant control to the opening and closing of doors that can be moved without the handling of an unlocking mechanism (see Fig. 7.3, right). To open a door, our robot drives in front of it, detects the door handle with its torso laser, approaches the handle, and grasps it. The drive moves backward while the gripper moves to a position to the side of the robot in which the opening angle of the door is sufficiently large to approach the open fridge or cabinet. The gripper follows the motion of the door handle through compliance in the lateral and the yaw directions. The robot moves backward until the gripper reaches its target position. For closing a door, the robot has to approach the open door leaf, grasp the handle, and move forward while it holds the handle at its initial grasping pose relative to the robot. When the arm is pulled away from this pose by the constraining motion of the door leaf, the drive corrects for the motion to keep the handle at its initial pose relative to the robot. The closing of the door can be detected when the arm is pushed back towards the robot.

Fig. 7.3 Left: A person guides Cosero at its hand. Right: Dynamaid opens and closes a refrigerator using compliant control at RoboCup 2010 in Singapore.

7.2.3 Public Demonstrations

The tracking behavior of our compliant control method has been extensively evaluated in our prior work in [10]. In general, our approach exhibits good tracking performance in compliant mode in linear and rotational directions. If gravity needs to be compensated, a compliant motion orthogonal to the gravity direction may be slightly less accurate due to the fact that joints are involved in both counteracting gravity as well as moving into the compliant direction.

We demonstrated the applications of compliant control described in Sec. 7.2.2 with our service robot Cosero publicly at several occasions at RoboCup@Home competitions. Object hand-over occurs very frequently in the test scenarios of the competition. Our approach leads to a very natural and intuitive robot behavior that is well understood by users and has high success rates. We have demonstrated the opening and closing of a refrigerator in the final demonstration of the @Home league at RoboCup 2010 in Singapore[2]. The cooperative carrying of a table was first shown in the finals at RoboCup 2011 in Istanbul, Turkey[3]. It was also shown in combination with guiding the robot to the location of the table at RoboCup German Open in 2013. The demonstrations have been important aspects for convincing the juries of our open demonstrations and finals. We won the international RoboCup@Home competitions in 2011 [15], 2012 [14], and 2013 [12]. We also achieved 1st place in the league at RoboCup German Open competitions from 2011 to 2014.

---

[2] https://www.youtube.com/watch?v=TObO4_N0AAQ

[3] https://www.youtube.com/watch?v=nG0mJiODrYw

7.3 Object Manipulation Skill Transfer

Our approach to skill transfer can be seen as a variant of learning from demonstration. Recently, Schulman et al. [8] proposed an approach in which motion trajectories are transferred between shape variants of objects. They primarily demonstrate tying knots in rope [8] and suturing [9], while they also show examples for folding shirts, picking up plates, and opening a bottle. Their non-rigid registration method is a variant of the thin plate spline robust point matching (TPS-RPM) algorithm. We develop an efficient deformable registration method based on the coherent point drift method (CPD [5]) to align RGB-D images efficiently and accurately. We demonstrate bimanual tool-use, and propose to select tool end-effectors as reference frames for the example trajectory, where it is appropriate. In contrast to the method in [8,9], we do not assume the estimated displacement field to be valid at any pose on the motion trajectory. Instead, we design example motions relative to reference frames. These reference frames are transformed between example and new object.

7.3.1 Efficient RGB-D Deformable Registration

We propose a multi-resolution extension to the coherent point drift (CPD [5]) method to efficiently perform deformable registration between RGB-D images (see Fig. 7.4, top). Instead of processing the dense point clouds of the RGB-D images directly with CPD, we utilize multi-resolution surfel maps (MRSMaps[4] [13]) to perform deformable registration on a compressed image representation. This image representation stores the joint color and shape statistics of points within 3D voxels (coined surfels) at multiple resolutions in an octree. The maximum resolution at a point is limited proportional to its squared distance in order to capture the error properties of the RGB-D camera. In effect, the map exhibits a local multi-resolution structure which well reflects the accuracy of the measurements and compresses the image from 640×480 pixels into only a few thousand surfels.

The CPD method assumes a displacement field $v : Y \rightarrow X$ between a model point set Y and the scene points X. It aims at minimizing the squared error of the deformed points in Y with their counterparts in X,

$$\ln \mathrm{p}(\mathrm{X},\mathrm{v} \mid \sigma) = \ln \mathrm{p}(\mathrm{X} \mid \sigma, \mathrm{v}) - \lambda/2 \left\|\mathrm{v}\right\|_{\mathrm{H}}^{2}.$$

By introducing a norm on the displacement field in a reproducing kernel Hilbert space H, smoothness can be enforced. The regularized objective has a closed-form solution given the assignment of points, which requires solving a system of linear equations whose size is quadratic in the number of model points. Instead of

---

[4] Our MRSMap implementation is available open-source from http://code.google.com/p/mrsmap/ .

assuming a one-to-one mapping between the points, the CPD method explains the deformed points in Y as samples from a mixture model, in which each sample in X is a Gaussian mixture component. Since the assignment probabilities between the point sets are not known a-priori, the probabilities and the displacement field are recovered iteratively through expectation-maximization.

Color and contours in the depth image are integrated as additional point dimensions. We process RGB-D images from coarse to fine resolutions in our MRSMap representation. Since the volume covered shrinks with resolution, we constrain the borders of a fine resolution to the coarser resolution result. This objective also has a closed-form solution for the displacement field. Finally, to improve robustness and to facilitate the linear systems to be sparse, we use a compact support kernel.

Fig. 7.4 Top: deformable registration examples. Bottom: local transformation examples.

7.3.2 Skill Transfer through Shape Matching

We describe object manipulation skills as grasp poses and motion trajectories relative to an object (see Fig. 7.5). We exploit that often shape deformations induce correspondences in the functional parts between objects of the same functional object class. When the robot observes a new kind of object of a class that it knows to handle, it matches the shapes of the object at hand with the known object, and transfers grasps and motions to the new one. To this end, we apply our deformable registration method. We define grasp poses and motion trajectories in terms of local coordinate frames relative to the object's reference frame. Hence, we need a

Fig. 7.5 Skill transfer. Left: We transfer grasp and tool end-effector poses between objects through deformable registration. Right: We represent skills as motions of the tool end-effector relative to other objects, which can be transferred once the tool end-effector of a new object is known. The motion of the tool end-effector induces a motion of the grasps.

method for estimating local coordinate frame changes between the observed object and the known object.

We estimate the local coordinate frame changes from the displacement field that is recovered with our deformable registration method (see Fig. 7.4, bottom). The infinitesimal deformation at a point y is specified by the Jacobian of the displaced point, $\nabla\varphi(y) = I + \nabla v(y)$. The local rotation R is obtained through polar decomposition of the Jacobian $\nabla\varphi(y) = RU$.

7.3.3 Results

We have evaluated accuracy and run-time of our deformable registration method and compared it with plain processing of RGB-D images using CPD. On synthetically deformed RGB-D images, we achieve an average run-time of 1.29 s, plain processing requires 4.74 s. Our method also is more accurate: in average our method has a low deviation of 0.0178 m from the ground truth displacements, while plain processing yields 0.0482 m mean error. Note that for plain image processing, the original images had to be subsampled from 640×480 to 80×60 resolution. Further evaluation results can be found in [11].

We publicly demonstrated object manipulation skill transfer based on our deformable registration approach during the @Home league Open Challenge at RoboCup 2013 in Eindhoven, Netherlands[5]. The jury chose one of two new cans, while the skill was pretrained for a third instance of cans. Our robot Cosero transferred watering can manipulation skills to a novel can. Fig. 7.6 shows images taken during the demonstration. The demonstration was well received by the jury consisting of team leaders and received high scores, which was an important contribution to winning the 2013 RoboCup@Home competition.

[5] http://www.youtube.com/watch?v=I1kN1bAeeB0

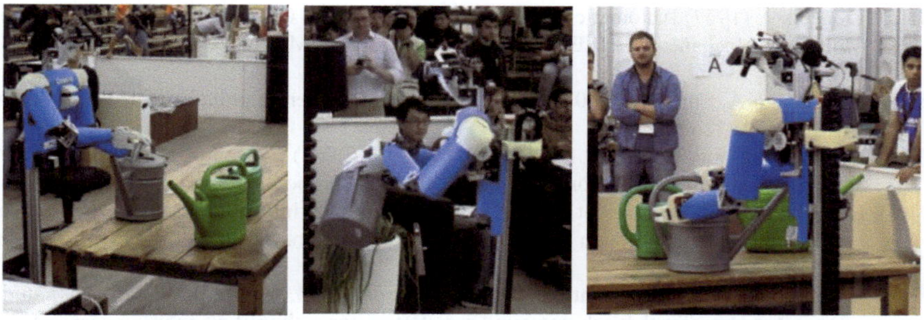

Fig. 7.6 Cosero transfers the bi-manual skill of watering a plant to a novel watering can in the final RoboCup@Home demonstration at RoboCup 2013 in Eindhoven.

7.4 Conclusions

In this chapter, we presented our approaches to compliant control and deformable registration that enable soft interaction with objects and persons, and that increase the flexibility of robot behavior in everyday environments.

We developed compliant control for the anthropomorphic arms of our service robots Cosero and Dynamaid. It has been used to demonstrate a variety of tasks in everyday environments that either require soft manipulation of objects such as opening and closing doors, or soft physical interaction with humans. For transferring object manipulation skills, we developed an efficient deformable registration method for RGB-D images. It allows for transferring grasp poses and tool end-effectors between shape variants of functional types of objects. In a public demonstration, Cosero showed bimanual handling of a novel watering can to water a plant. The reported public demonstrations have been key contributions to winning the German and international RoboCup@Home competitions since 2011.

In future work, we want to further study the modelling and manipulation of deformable objects. Through compliant control, interactive learning of deformable object models can be made possible.

7.5 References

[1] Christensen HI, Hüttenrauch H, Severinson-Eklundh K (2000) Human-Robot Interaction in Service Robotics. In: Proc of Robotik

[2] Khatib O, Yokoi K, Chang K, Ruspini D, Holmberg R, Casal A, Baader A (1996) Force strategies for cooperative tasks in multiple mobile manipulation systems. In: Proc of ISRR, 333-342

[3] Khatib O (1988) Object manipulation in a multi-effector robot system. In: Proc. of International Symposium of Robotic Research (ISRR), 137-144

[4] Liegeois A (1977) Automatic Supervisory Control of the Configuration and Behavior of Multibody Mechanisms. IEEE Transactions on Systems, Man and Cybernetics 7(12):868-871

[5] Myronenko A, Song X (2010) Point set registration: coherent point drift. IEEE Trans on PAMI 32(12):2262-2275

[6] Nakanishi J, Cory R, Mistry M, Peters J, Schaal S (2008) Operational Space Control: A Theoretical and Empirical Comparison. Int Journal of Robotics Research 27(6):737-757

[7] Oudeyer P-Y, Ly O, Rouanet P (2011) Exploring robust, intuitive and emergent physical human-robot interaction with the humanoid robot Acroban. In: Proc of the IEEE-RAS Int Conf on Humanoid Robots

[8] Schulman J, Gupta A, Venkatesan S, Tayson-Frederick M, Abbeel P (2013) A case study of trajectory transfer through non-rigid registration for a simplified suturing scenario. In: Proc of IEEE/RSJ Int Conf on Intelligent Robots and Systems (IROS)

[9] Schulman J, Ho J, Lee C, Abbeel P (2013) Learning from demonstrations through the use of non-rigid registration. In: Proc of the 16th Int Symposium on Robotics Research (ISRR)

[10] Stückler J, Behnke S (2012) Compliant Task-Space Control with Back-Drivable Servo Actuators. In: RoboCup 2011, LNCS 7416, 78-89

[11] Stückler J, Behnke S (2014) Efficient Deformable Registration of Multi-Resolution Surfel Maps for Object Manipulation Skill Transfer. In: Proc of IEEE Int Conf on Robotics and Automation (ICRA)

[12] Stückler J, Droeschel D, Gräve K, Holz D, Schreiber M, Topalidou-Kyniazopoulou A, Schwarz M, Behnke S (2014) Increasing Flexibility of Mobile Manipulation and Intuitive Human-Robot Interaction in RoboCup@Home. In: RoboCup 2013, LNCS 8371, 135-146

[13] Stückler J, Behnke S (2014) Multi-resolution surfel maps for efficient dense 3D modeling and tracking. Journal of Visual Communication and Image Representation 25(1):137-147

[14] Stückler J, Badami I, Droeschel D, Gräve K, Holz D, McElhone M, Nieuwenhuisen M, Schreiber M, Schwarz M, Behnke S (2013) NimbRo@Home: Winning Team of the RoboCup-@Home Competition 2012. In: RoboCup 2012, LNCS 7500, 94-105

[15] Stückler J, Holz D, Behnke S (2012) RoboCup@Home: Demonstrating Everyday Manipulation Skills in RoboCup@Home. IEEE Robotics & Automation Magazine 19(2):34-42

[16] Williams D, Khatib O (1993) The virtual linkage: A model for internal forces in multi-grasp manipulation. In: Proc of the IEEE Int. Conf. on Robotics and Automation (ICRA)

[17] Yokoyama K, Handa H, Isozumi T, Fukase Y, Kaneko K, Kanehiro F, Kawai Y, Tomita F, Hirukawa H (2003) Cooperative works by a human and a humanoid robot. In: Proc of IEEE Int Conf on Robotics and Automation (ICRA), 2985-2991

8 Soft Robot Control with a Behaviour-Based Architecture

Christopher Armbrust, Lisa Kiekbusch, Thorsten Ropertz, Karsten Berns

University of Kaiserslautern

Abstract In this chapter, we explain how behaviour-based approaches can be used to control soft robots. Soft robotics is a strongly growing field generating innovative concepts and novel systems. The term "soft" can refer to the basic structure, the actuators, or the sensors of these systems. The soft aspect results in a number of challenges that can only be solved with new modelling, control, and analysis methods whose novelty matches those of the hardware. We will present prior achievements in the area of behaviour-based systems and suggest their application in soft robots with the aim to increase the fault tolerance while improving the reaction to unexpected disturbances.

8.1 Introduction

Soft robotics has recently received increasing attention by researchers. The aim to create systems that can interact in a more natural way with their environment has led to the development of robots with soft bodies [19], soft actuators [1, 14], and even soft sensors [21] that hardly resemble classic, rigid robots. With these novel components, however, come new challenges concerning robot control.

Due to the complexity of the involved components, it is extremely difficult or even practically impossible to create accurate models describing the dynamics of the machines. Hence, the classic, deliberative control approaches are only applicable in a limited way. Purely reactive strategies, however, fail to grasp the full complexity of high-level control. We suggest the use of behaviour-based control systems [2, 20], which combine fast reaction times with support for complex tasks. Soft robotic systems typically consist of many interconnected parts that are difficult to model and have to react fast to (unexpected) disturbances—which are also difficult (or impossible) to model. The distributed nature of behaviour-based systems (BBS) perfectly complies with the also distributed structure of soft robots. In typical BBSes there is a high degree of redundancy, which is essential in case of hardware failures, and a significant number of reactive elements. This allows for fast reactions to disturbances.

Using the example of the behaviour-based architecture iB2C[6], we will explain the advantages of behaviour-based approaches over purely reactive or purely de-

[6] iB2C: integrated Behaviour-based Control

liberative ones in detail and will sketch our aim to apply them for controlling soft robotic systems. We have structured the remainder of this chapter as follows: In Sec. 8.2, we will introduce the main features of the iB2C and the mechanisms we have invented for supporting the development and analysis of iB2C networks. We will then explain how the iB2C can be used to realize soft control systems in Sec. 8.3. Finally, in Sec. 8.4 we will summarize the main points of this chapter and discuss our vision for future work.

8.2 The Behaviour-based Architecture iB2C

The behaviour-based architecture iB2C [22] has been implemented using the software frameworks MCA2-KL[7] and FINROC[8]. It is applied to different kinds of robots in our lab, e.g. a bipedal walking machine, a humanoid robot head, and several wheel-driven indoor and outdoor vehicles.

The basic component of the iB2C is the behaviour (see Fig. 8.1), which is defined as $B = (f_a, f_r, F)$, where f_a calculates its *activity vector* $\boldsymbol{a}$ and f_r calculates its *target rating r*. The *output vector* $\boldsymbol{u}$ is transferred from the *input vector* $\boldsymbol{e}$ together with the *activation* $\iota = s \cdot (1 - i)$ using the transfer function $F : \boldsymbol{u} = F(\boldsymbol{e}, \iota)$. The activation indicates the effective relevance of a behaviour in the network. *Stimulation s* and *inhibition* $i = \|\boldsymbol{i}\|_\infty$ are signals coming from other behaviours to gradually enable or disable the behaviour.

The activity vector $\boldsymbol{a} = (a, \underline{\boldsymbol{a}})^T$ is composed of the behaviour's activity *a* and *q* so-called *derived activities* $\underline{a}_0, \underline{a}_1, \ldots, \underline{a}_{q-1}$ with $\underline{a}_i \leq a, \ \forall i \in \{0, 1, \ldots, q-1\}$. The activity indicates the degree of influence the behaviour wants to have in the network. It is also possible to transfer only a part of the activity to other behaviours using the derived activities. A behaviour's activity is limited by its activation, i.e. $a \leq \iota$. The target rating *r* describes the satisfaction of a behaviour with the current situation. The four *behaviour signals s, i, a, r* and the internal behaviour value ι are limited to [0, 1] and describe the interface of each behaviour. By contrast, there is no limitation of ***e*** and ***u***. They can differ among behaviours.

iB2C behaviours can be connected in various ways. The most common types are stimulating and inhibiting connections, in which the activity output of one behaviour is connected either to the stimulation or the inhibition input of another behaviour. Other connection types include the combination of the outputs of a num-

---

[7] MCA2-KL: Modular Controller Architecture Version 2 - Kaiserslautern Branch

[8] FINROC: the successor of MCA2-KL. See `http://finroc.org/` for more information.

ber of competing behaviours using a fusion behaviour (see below) or the sequencing of behaviours using the special coordination behaviour CBS [4].

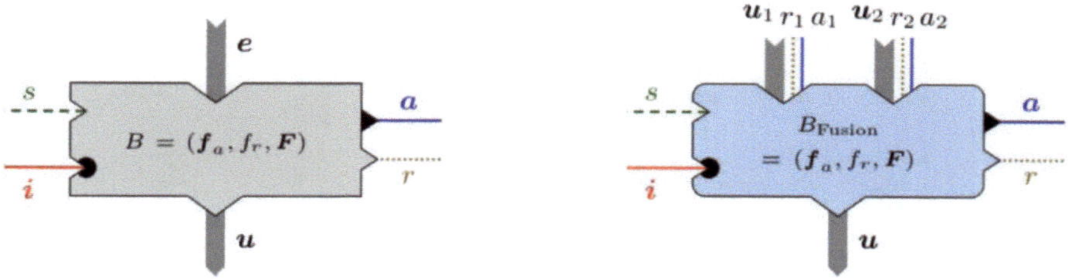

Fig. 8.1 The symbols of two basic iB2C behaviours: The general symbol of a behaviour (left), and a weighted average fusion behaviour for $n_c = 2$ competing behaviours (right).

The iB2C fusion behaviour (see Fig. 8.1) combines the outputs of several behaviours connected to it according to one of three possible fusion modes (maximum, weighted average, and weighted sum). For example, if p competing behaviours $\boldsymbol{B}_{\text{Input}_c}$ with activities a_c, target ratings r_c, and output vectors $\boldsymbol{u}_c (c \in \{0, \ldots, p-1\})$ are connected to a weighted average fusion behaviour $\boldsymbol{B}_{\text{Fusion}}$, then the outputs of $\boldsymbol{B}_{\text{Fusion}}$ are calculated as follows:

$$\boldsymbol{u}_{\text{Fusion}} = \frac{\sum_{j=0}^{p-1} a_j \cdot \boldsymbol{u}_j}{\sum_{k=0}^{p-1} a_k}, \quad a_{\text{Fusion}} = \frac{\sum_{j=0}^{p-1} a_j^2}{\sum_{k=0}^{p-1} a_k} \cdot \iota_{\text{Fusion}}, \quad r_{\text{Fusion}} = \frac{\sum_{j=0}^{p-1} a_j \cdot r_j}{\sum_{k=0}^{p-1} a_k}$$

The handling of a large number of interconnected behaviours is facilitated by behavioural groups, which encapsulate a number of behaviours or further groups and act as new behaviours in a network. To fulfil complex tasks, networks of simple behaviours are constructed and possibly combined. The challenge lies in the connection of these behaviours. As example, the control system of the biped (see Sec. 8.3) consists of over 350 behaviours, while the one of RAVON (see also Sec. 8.3) contains even more than 500 behaviours. To build up such huge networks, sound guidelines for the development and implementation are needed together with verification techniques to prove a system's correctness.

Fig. 8.2 gives an overview of the concept developed in our lab for the development and verification of behaviour networks. Soft robotic systems are typically complex and consist of many interconnected components. This makes them perfect for the application of BBSes, but this in turn results in complex software. In Sec. 8.2.1, we propose the use of a strict design concept for the development of behaviour networks that realise high-level, complex tasks. In such strongly connected systems, oscillations can easily occur. We therefore suggest a solution for oscillation detection in Sec. 8.2.2. Distributed systems like the ones found in soft robotics call for verification techniques that specifically take into account their high degree of distribution. We will present two such techniques in Sec. 8.2.3.

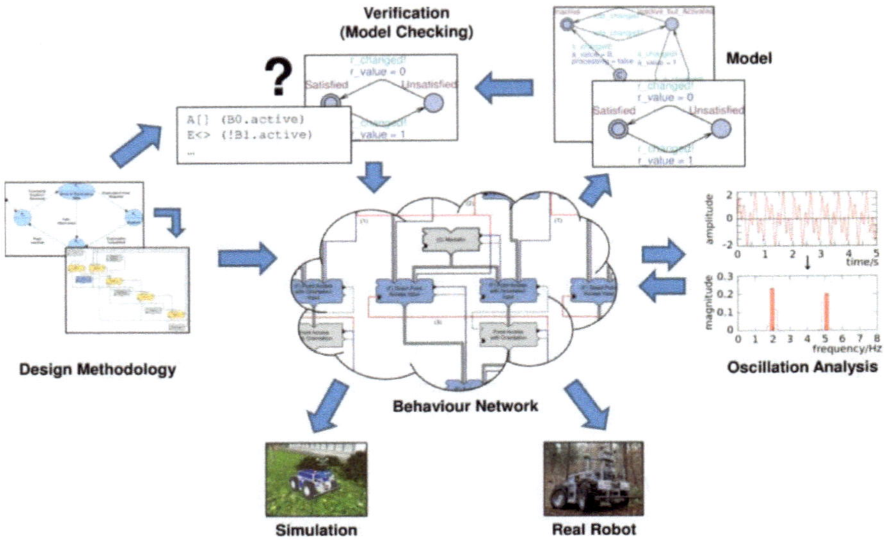

Fig. 8.2 Concept of the development and verification process for behaviour-based systems developed in our lab.

8.2.1 Design of Complex Behaviour Networks

As we have alluded in Sec. 8.1, soft robotic systems can be complex in many respects. The complexity of their hardware is reflected by the complexity of the software necessary to control a soft robot. A drawback that can often be seen in behaviour architectures is the lack of support for building large networks that are able to execute sophisticated tasks. For example, Brook's subsumption architecture [12] has been criticised for being unable to support large networks. The ability to combine iB2C behaviours into a group (see above) provides significant help in dealing with the complexity of large networks. However, research conducted at our group has shown that the quality of a behaviour network often heavily depends on the experience of the developers and their personal preferences. Guidelines [22] can mitigate this problem, but target at developers with a certain experience in designing iB2C networks. Therefore, we have invented a three-step method for supporting completely inexperienced developers during the process of designing an iB2C network for a complex task [9].

The first step is to define a complex task as a finite-state machine (FSM). Two persons are usually needed for this step: an *end-user* with detailed knowledge about the task the system shall perform and the *main developer* with detailed knowledge about existing software components and the specification of the available hardware. As soft robotics is a new field compared to industrial robotics, it is highly likely that an end-user does not have much knowledge about the capabilities of a soft robot. Hence, he shall be able to define the task in a common, nearly intuitive way while being assisted by the main developer.

The second step consists of an automatic transformation of said FSM into the skeleton of a behaviour network, i.e. a behaviour network that only contains fusion and CBS behaviours (see above), empty behaviours without functionality, and the interconnections between the behaviours. The advantage of having an automatic process here is that no person is involved. Hence, the resulting behaviour network will correspond to the FSM defining the robot's task, while a manual mapping process could easily lead to errors in the network structure.

In the third step, so-called *system specialists* manually add the core functionalities of the behaviours in the previously created skeleton network. They have detailed knowledge about specific parts of the soft robotic system and are therefore able to implement sub-components. Due to the distributed nature of BBSes, this work can easily be done in parallel by several system specialists who are supervised by the main developer. An interesting aspect here is that none of the system specialists has to be an expert for behaviour networks as the network structure has already been created automatically in the second step.

8.2.2 Oscillation Detection in Behaviour Networks

Oscillations are a typical effect in soft robotics applications. They can be enforced by the control system to perform a special movement, e.g. the walking of a biped robot or fin movements of fish robots. But they can also be highly unwanted, for example if they appear in flexible joints and complicate a correct adjustment. The detection of oscillations is therefore a very important task. In [25] we presented an approach to detect oscillations inside a behaviour-based control system during run-time, which facilitates an early reaction in case of a wrong behaviour. The oscillation detection method is based on the analysis of the frequency spectrum of an arbitrary Fourier-transformed signal. In our approach, we used the activity data of the behaviours. In short, the data is buffered, transformed, and analysed for peaks in the power spectrum indicating an oscillation. In a second step, we traced oscillations through the network to gain an overview of the path the oscillation takes through the network and to find its root cause. Future work includes the definition of desired and undesired oscillations and possible reactions based on the underlying application.

8.2.3 Verification of Behaviour Networks

We have already mentioned several advantages of using BBSes for soft robots and will go into detail about that in Sec. 8.3. A key aspect of BBSes is the distribution of the overall functionality over several components of the system. Unfortunately, this advantage comes with a downside: Determining whether a system really does what it is supposed to do can be hard as a considerable part of the intelligence of a BBS lies in its network structure, i.e. in the interaction of its behaviours. Determining whether a system operates as specified is done by methods of formal verification, e.g. deductive reasoning [15, 16] as well as model checking [13, 23]. With regard to the fact that a lot of intelligence of a BBS lies in its network, we have decided to pursue a top-down analysis approach by developing a verification technique that is especially tailored to the mostly neglected analysis of the network structure [5, 7, 6]. We have based our approach on model checking as this offers a high degree of automation and generates witnesses and counter-examples, respectively.

The idea underlying our verification concept is the modelling of iB2C networks as networks of synchronised timed automata using the model checking toolbox UPPAAL [11]. Each behaviour is represented by five automata, one for each behaviour signal and one for the activation, as shown in Fig. 8.3. If there is a connection between the signals of two behaviours, a synchronisation channel is used to connect and synchronise the corresponding automata. Using a graphical interface, we can define properties that shall hold for the behaviour signals, e.g. that one behaviour shall get active before another one does. The graphical definition of properties can be automatically transformed to a corresponding set of observer automata and queries, which can be sent to UPPAAL's verifier together with the system model. With the help of the verifier, we can then easily check whether our BBS holds the property in question and thus fulfils requirements like "the anti-collision behaviour has precedence over the driving behaviour, i.e. it can stop the robot". In recent work, we also take into account hardware failures during the verification process [17].

The use of timed automata for modelling BBSes offers several advantages like the ability of checking temporal properties (e.g. reachability) with the downside of requiring a strong abstraction in order to keep the verification process computationally feasible. By focusing on discrete system states and abstracting from state transitions, satisfiability modulo theories (SMTs) can be used to generate a more fine-grained model, which allows for checking richer properties concerning the discrete state as described in [24]. SMTs study practical methods to solve first-order logic formulae with equality in which sets of variables are replaced by predicates of the underlying theories. Examples for these background theories are the theory of real numbers, of integers, and of various data structures. The *SMT-LIB* [10] initiative has defined a standard for descriptions of background theories used in SMT systems.

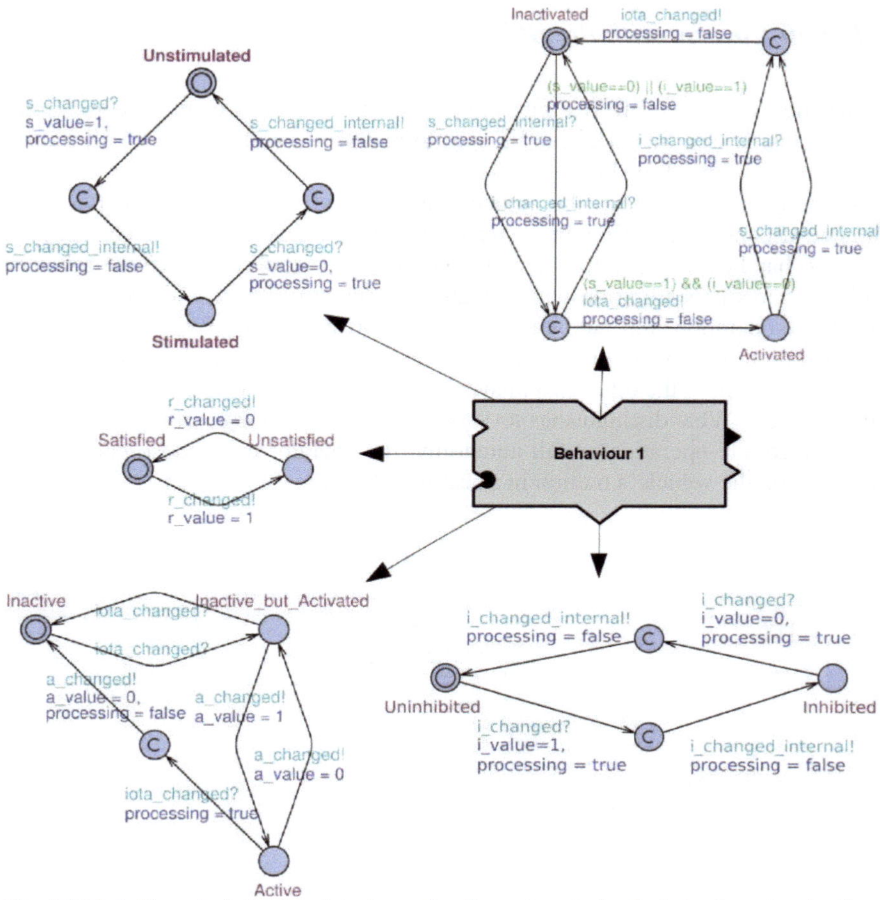

Fig. 8.3 Modelling of a behaviour interface using five automata for the behaviour signals stimulation, inhibition, activity, and target rating as well as for the activation.

Due to the support of limited non-linear real arithmetic, SMT perfectly matches the expressiveness required for modelling the behaviour interaction and coordination. The rather simple modelling process allows for automation like the modelling based on timed automata, with the advantage of having a weaker abstraction concerning the behaviour signals.

As we have explained in Sec. 8.1, soft robotic systems are very complex and it is extremely difficult to create accurate models, which makes it hard to develop suitable control systems and find appropriate control parameters. The latter verification approach is also suitable for reducing the search space for parameter identification. Therefore, properties that the overall system shall be revealing can be used to determine value ranges of parameters which guarantee them.

8.3 Soft Control with the iB2C

In the following, we will illustrate how fusion behaviours executing a weighted average fusion (see Sec. 8.2) can be used to realise a seamless, hence soft transition from one controlling behaviour to another.

In the control system of the autonomous off-road vehicle RAVON [3], a large number of connected behaviours controls the robot's movement. The control system also handles the operator's steering commands, which influence the robot in different ways depending on which degree of autonomy (pure tele-operation, assisted tele-operation, full autonomy) the operator has chosen. While during pure tele-operation the operator's commands completely bypass the robot's anti-collision system, in the other two modes they are combined with the outputs of the safety system. What distinguishes RAVON's control system from others is that during assisted tele-operation or full autonomy, the operator can choose his level of influence on the vehicle's motion in a seamless fashion.

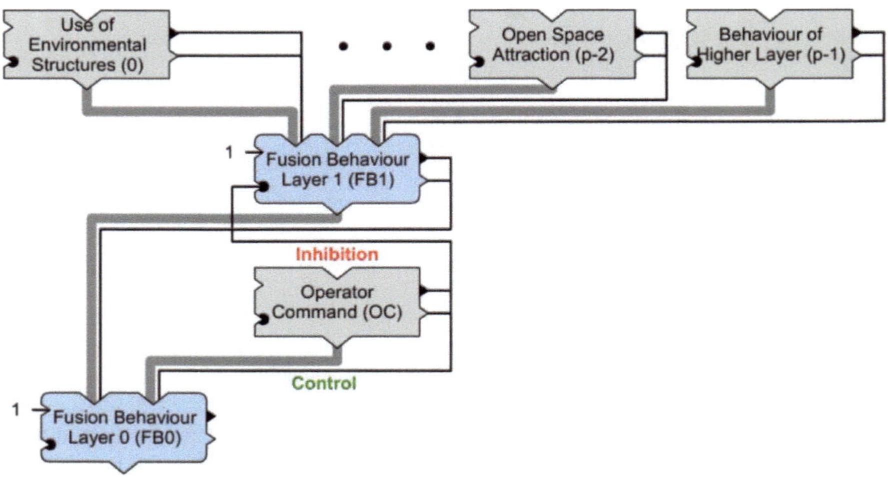

Fig. 8.4 The operator's commands inhibit commands of higher layers and are sent to lower layers.

Fig. 8.4 depicts a part of the behaviour network. The figure illustrates how an operator command is combined with outputs of p high-level navigation behaviours. The behaviour receiving the operator command (from any kind of input device) is connected via two streams with the remainder of the network: It sends the operator command along with activity and target rating down to a fusion behaviour of the lower layer and uses its activity to inhibit the fusion behaviour of the higher layer, which combines the outputs of high-level navigation behaviours. As a result, there is no binary selection of which commands are sent to the lower layers. Instead, the network can perform a soft transition from using only the high-level components over using a combination of them and the operator command to using only the latter. In [8], we provide details.

For complex soft robotic systems that shall be able to interact with their environment in a novel fashion, the ability to perform gradual, seamless transitions between two sources of control is essential. The soft integration of operator commands is only one example. Another one is a soft reaction of a robot's anti-collision system to nearby hazards: Instead of simply stopping the robot's motion, a sophisticated system will first try to move the robot around the hazard in a soft fashion. Such a functionality is also realised in RAVON's control system and allows for fast and smooth reactions to environmental disturbances.

With regard to its hardware, RAVON is a typical representative of classic (i.e. stiff) robots. The only soft hardware components are its spring-mounted bumpers. But the iB2C is also used to control a robotic system with compliant actuators, namely a simulated bipedal robot [18]. Its control system is inspired by the human locomotion system. The robot is able to perform human-like walking and can properly react to environmental disturbances (see Fig. 8.5).

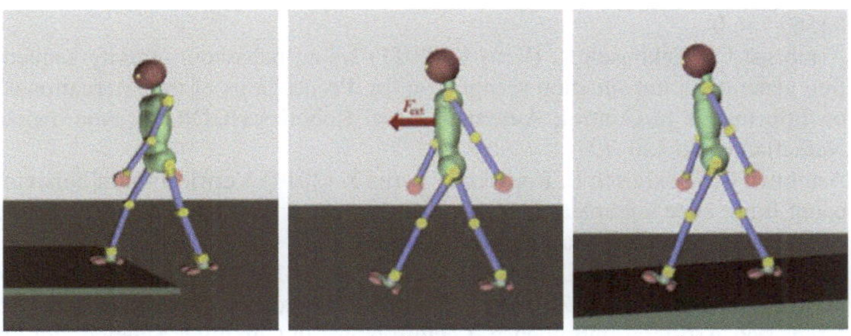

(a) (b) (c)

Fig. 8.5 The simulated biped reacting to different types of disturbances: step (a), external force acting on its torso (b), downhill slope (c) (source: [18]).

8.4 Conclusion and Future Work

In this chapter, we have motivated the application of BBSes in soft robotics and explained which techniques offer help in developing or analysing a behaviour network. As we have shown in Sec. 8.3, the behaviour architecture iB2C is perfectly suited for supporting soft interactions of different sources of control like operator commands or outputs of an anti-collision system. With this, we have demonstrated the applicability of the iB2C in novel control systems for soft robots like a bipedal robot with compliant actuators. In the context of future work, our research group is going to improve the development and analysis techniques described above, invent new ones in order to face the challenges of soft robotics, and finally apply them to soft robots.

Acknowledgments Parts of the research leading to these results has received funding from the European Union Seventh Framework Programme (FP7/2007-2013) under grant agreement number 285417.

8.5 References

[1] Albu-Schaeffer A, Eiberger O, Grebenstein M, Haddadin S, Ott C, Wimboeck T, Wolf S, Hirzinger G (2008) Soft robotics. IEEE Robotics and Automation Magazine 15(3): 20–30

[2] Arkin R (1998) Behaviour-Based Robotics. MIT Press, ISBN-10: 0-262-01165-4; ISBN-13: 978-0-262-01165-5

[3] Armbrust C, Braun T, Föhst T, Proetzsch M, Renner A, Schäfer B H, Berns K (2010) RAVON – the robust autonomous vehicle for off-road navigation. In: Baudoin Y, Habib M K (ed) Using robots in hazardous environments: Landmine detection, de-mining and other applications, Woodhead Publishing Limited, ISBN: 1 84569 786 3; ISBN-13: 978 1 84569 786 0

[4] Armbrust C, Kiekbusch L, Berns K (2011) Using behaviour activity sequences for motion generation and situation recognition. In: Proceedings of the International Conference on Informatics in Control, Automation and Robotics (ICINCO). Noordwijkerhout, The Netherlands, pp 120–127

[5] Armbrust C, Kiekbusch L, Ropertz T, Berns K (2012) Verification of behaviour networks using finite-state automata. In: Glimm B, Krüger A (ed) KI 2012: Advances in Artificial Intelligence. Springer, Saarbrücken, Germany

[6] Armbrust C, Kiekbusch L, Ropertz T, Berns K (2013) Quantitative aspects of behaviour network verification. In: Zaiane O, Zilles S (ed) Proceedings of the 26th Canadian Conference on Artificial Intelligence. Lecture Notes in Computer Science, vol 7884. Springer, Regina. Saskatchewan, Canada

[7] Armbrust C, Kiekbusch L, Ropertz T, Berns K (2013) Tool-assisted verification of behaviour networks. In: Proceedings of the 2013 IEEE International Conference on Robotics and Automation (ICRA 2013). Karlsruhe, Germany

[8] Armbrust C, Proetzsch M, Schäfer B H, Berns K (2010) A behaviour-based integration of fully autonomous, semi-autonomous and tele-operated control modes for an off-road robot. In: Proceedings of the 2nd IFAC Symposium on Telematics Applications. IFAC, Politehnica University, Timisoara, Romania

[9] Armbrust C, Schmidt D, Berns K (2012) Generating behaviour networks from finite-state machines. In: Proceedings of the German Conference on Robotics (Robotik)

[10] Barrett C, Stump A, Tinelli C (2010) The satisfiability modulo theories library (smt-lib). www.SMT-LIB.org. Accessed 16 Sep 2014

[11] Behrmann G, David A, Larsen K G (2006) A tutorial on uppaal 4.0

[12] Brooks R (1986) A robust layered control system for a mobile robot. IEEE Journal of Robotics and Automation (RA) 2(1):14–23

[13] Clarke E M, Emerson E A (1982) Design and synthesis of synchronization skeletons using branching time temporal logic. In: Kozen D (ed) Logics of Programs - Workshop. Lecture Notes in Computer Science (LNCS), vol 13. Springer, Berlin, Heidelberg, pp 52–71

[14] Deimel R, Brock O (2013) A compliant hand based on a novel pneumatic actuator. In: Proceedings of the IEEE International Conference on Robotics and Automation 2013 (ICRA 2013). pp 2039–2045

[15] Floyd R W (1967) Assigning meanings to programs. In: Schwartz J T (ed) Mathematical Aspects of Computer Science. Proceedings of Symposia in Applied Mathematics, vol 19. American Mathematical Society, Providence, Rhode Island, USA, pp 19–32

[16] Hoare C A R (1969) An axiomatic basis for computer programming. Communications of the ACM 12(10):576–583

[17] Kiekbusch L, Armbrust C, Berns K (2014) Formal verification of behaviour networks including hardware failures. In: Proceedings of the 13th International Conference on Intelligent Autonomous Systems (IAS-13). Padova, Italy

[18] Luksch T (2010) Human-like Control of Dynamically Walking Bipedal Robots. Dissertation, University of Kaiserslautern, Verlag Dr. Hut, ISBN: 978-3-86853-607-2

[19] Marchese A D, Onal C D, Rus D (2014) Autonomous soft robotic fish capable of escape maneuvers using fluidic elastomer actuators. Soft Robotics 1(1):75–87

[20] Matarić M J, Michaud F (2008) Behaviour-based systems. In: Siciliano B, Khatib O (ed) Springer Handbook of Robotics. Springer Berlin Heidelberg, pp. 891–910

[21] Park Y L, Chen B R, Wood R J (2011) Soft artificial skin with multi-modal sensing capability using embedded liquid conductors. In: Proceedings of the IEEE Sensors 2011 Conference. Limerick, Ireland, pp 81–84

[22] Proetzsch M (2010) Development Process for Complex Behavior-Based Robot Control Systems. Dissertation, University of Kaiserslautern, Verlag Dr. Hut, ISBN: 978-3-86853-626-3

[23] Queille J P, Sifakis J (1982) Specification and verification of concurrent systems in CESAR. In: Dezani-Ciancaglini M, Montanari U (ed) International Symposium on Programming - Proceedings of the 5th Colloquium. Lecture Notes in Computer Science (LNCS), vol 137. Springer-Verlag, London, UK, pp 337–351

[24] Ropertz T, Berns K (2014) Verification of behavior-based networks - using satisfiability modulo theories. In: Proceedings for the joint conference of ISR 2014 and ROBOTIK 2014. VDE VERLAG GMBH, pp 669–674

[25] Wilhelm L, Proetzsch M, Berns K (2009) Oscillation analysis in behavior-based robot architectures. In: Dillmann R, Beyerer J, Stiller C, Zöllner J, Gindele T (ed) Autonome Mobile Systeme. Informatik aktuell, Springer, pp 121–128

9 Optimal Exploitation of Soft-Robot Dynamics

Sami Haddadin

Leibniz Universität Hannover

Abstract Inspired by the elasticity contained in human muscles, elastic soft robots are designed with the aim of imitating motions as observed in humans or animals. Especially reaching peak velocities using stored energy in the springs is a task of significant interest. In this chapter, general results on maximizing a soft-robot's end-point velocity by using elastic joint energy are presented and discussed.

9.1 Introduction

Humans are capable of highly dynamic motions such as throwing or kicking. A major feature that presumably enables them to perform such tasks is their ability to store and release potential energy in the compliant elements of the musculoskeletal system in combination with inertial energy transfer between the rigid parts of the body. However, only recently, intrinsically compliant actuation and its generalization Variable Stiffness Actuation (VSA) have drawn significant attention in terms of offering novel ways to design and control [1, 3, 7, 16, 17]. The common main arguments in favor of the soft-robotics concept VSA are its robustness and expected energetic benefits, e.g. in terms of providing complex limit cycle motions for walking [18, 19]. Also, safety of compliant robots is generally argued to be better for passively compliant robots, which is however not true in general [9]. This is because of another major benefit we can expect from these novel devices: the capability to store energy in their elastic joints and release it for execution of explosive motions. These robots can potentially exceed the maximum velocity of an equivalent rigid robot with the same maximum motor velocity [15]. We explored this novel capability with the mathematical tools of optimal control. The essential problem we considered was to find the optimal motor control input for generating maximum link velocity for elastic robots. The first work on this topic was done in [10], where the optimal controls for a 1DoF elastic joint was derived. Later on, this result was extended to general VSA joints [11] and also approached by other researchers [4, 6, 13].

9.2 Problem Formulation

Generally, the Minimum Principle [12] provides necessary conditions to be satisfied by the optimal controls $\boldsymbol{u}^* \in \mathbb{U} \subset \mathbb{R}^m$ for a dynamical system expressed by first order differential equations $\dot{\boldsymbol{x}} = \boldsymbol{f}(\boldsymbol{x}, \boldsymbol{u}, t)$. The controls $\boldsymbol{u}^*$ minimizes the following scalar-valued cost functional.

$$J(\boldsymbol{u}) = \vartheta\left(\boldsymbol{x}(t_f), t_f \right) + \int_{t_0}^{t_f} L(\boldsymbol{x}, \boldsymbol{u}, t) dt. \qquad (1)$$

In (1), L is called the running cost, ϑ the terminal cost, t_f the final time, and $\boldsymbol{x} \in \mathbb{R}^n$ denotes the system state [12].

In order to gain a thorough understanding of the optimal strategies for soft robots, we start our discussion with a simple robot arm consisting of only one joint with constant joint stiffness K_J. The considered velocity controlled system dynamics ($u(t) = \dot{\theta}$) is of second order

$$M\ddot{q} = K_J(\theta - q) = \tau_J \qquad (2)$$

$$\theta = \int u \, dt, \qquad (3)$$

where q denotes the link position, θ the motor position, and M the link inertia. $\tau_J = K_J(\theta - q) =: K_J\phi$ denotes the elastic joint torque resulting from the angular deflection $\phi = (\theta - q)$. We consider the motor velocity to be bounded:

$$|u(t)| = |\dot{\theta}| \leq u_{\max} \qquad (4)$$

In order to understand the maximum performance of such an elastic joint, we investigate control strategies maximizing the link end velocity $\dot{q}(t_f)$. This basic problem can be formulated in two ways: As a final time problem, where the according maximum link velocity is to be found, or alternatively as a minimum time problem, where the maximum velocity is known in advance. More complex formulations would typically involve a running cost such as effort minimization or similar demands. The cost functional J for the first problem is simply a terminal cost

$$J = \vartheta(t_f) = -\dot{q}(t_f). \qquad (5)$$

Note that for this formulation we would need to specify a fixed final time t_f, since the velocity of the unconstrained system would otherwise tend to infinity. The investigation of this problem for various basic 1DoF systems including variable stiffness and damping can be found in [14, 20, 21, 22]. For the second problem one would typically need to know the maximum possible velocity in advance and find the minimum time trajectory to reach the desired final state. This can be derived from physical reasoning for the case of limited elastic deflection for constant and nonlinear joint elasticity, as described next.

9.3 Optimal Controls for Constrained Deflection

The most important real-world state constraint in elastic robots is the maximum angular deflection $|\phi| = |\theta - q| \leq \phi_{max}$ that uniquely determines the maximum possible link velocity $\dot{q}_{max}$. This consists of the maximum motor velocity plus an additional term $\Delta\dot{q}$, which is related to the maximum potential energy $E_{pot_{max}} = \frac{1}{2}K_J\phi_{max}^2$ that can be stored in the joint spring. For constant joint elasticity this additional velocity gain can be expressed as

$$\Delta\dot{q} = \omega\phi_{max}, \qquad (6)$$

where $\omega = \sqrt{\frac{K_J}{M}}$ is the eigenfrequency of the mass-spring system. In principle, $\dot{q}_{max}$ can be obtained with various control trajectories. Our aim is, however, to exploit the capabilities of the joint as fast as possible. Therefore, we seek for the minimum time to reach the maximum link velocity. Interestingly, some physical reasoning simplifies the problem considerably by dividing it into two separate subproblems. Going backwards in time, one has to reach a charged state at t_{ch} (fully deflected joint spring) from the final state (maximum link velocity), see Fig. 9.1. This is simply achieved by moving at $u = \dot{\theta}_{max}$ until t_{ch}. Then, one has to hit the initial state at $t = 0$ (resting position) from the charged state in minimum time.

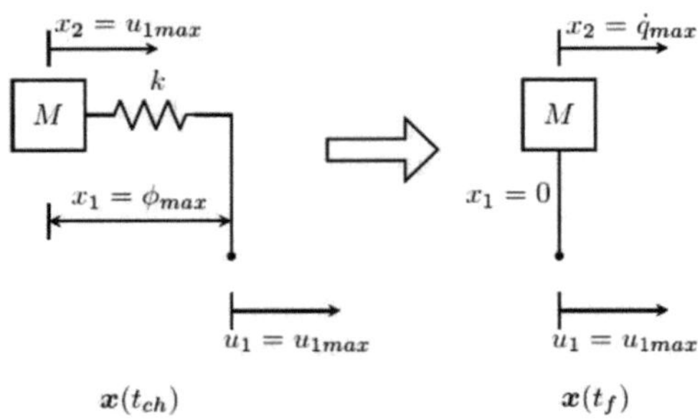

Fig. 9.1 Releasing the spring from $\boldsymbol{x}(tch)$ to reach maximum link velocity. Note that $u_1 = u = \dot{\theta}$.

The resulting optimal controls for this problem are shown in Fig. 9.2, where the system state is defined as $\boldsymbol{x} := (\phi \; \dot{q})^T$. If the maximum deflection of the spring is not sufficiently large, the state constraint can become active, before $\boldsymbol{x}(t_{ch})$ is reached (see Fig. 9.2, right). In that case we have a singular case. In the second subproblem, which solution can be shown to be of bang-bang type, a set of ellipsoidal switching curves can be constructed. Crossing these causes switching in the controls.

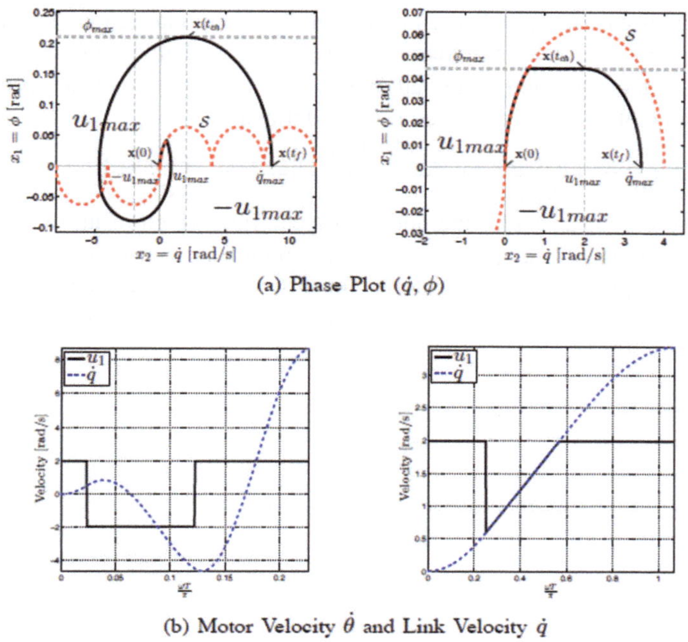

Fig. 9.2 Time optimal control strategy for maximizing link velocity.

Generally, the number of control switchings to reach $\boldsymbol{x}(0)$ depends on the spring energy and motor velocity. If the link travels at maximum motor velocity its kinetic energy is $E_{kin} = \frac{1}{2}M\dot{\theta}_{max}^2$. Let us now introduce the ratio of the maximum potential energy in the spring w. r. t. this kinetic energy

$$e_{sl} = \frac{E_{pot_{max}}}{E_{kin}} = \frac{K_J\phi_{max}^2}{M\dot{\theta}_{max}^2} = \left(\frac{\omega\,\phi_{max}}{\dot{\theta}_{max}}\right)^2. \qquad (7)$$

Eq. (7) relates directly to singularities and the number of switchings [8]. In order to express the benefit of the spring on the maximum link velocity in terms of energies, we can express the maximum link velocity as

$$\dot{q}_{max} = \dot{\theta}_{max}(1 + \sqrt{e_{sl}}). \qquad (8)$$

The term $\epsilon := \dot{q}_{max}/\dot{\theta}_{max} = 1 + \sqrt{e_{sl}}$ denotes the joint speed gain. It can be shown that the motor needs to reverse its direction of speed each time the link speed grows more than two times the motor speed, i.e.

$$n_c = \frac{\dot{q}_{max}}{2\,\dot{\theta}_{max}} = \frac{1+\sqrt{e_{sl}}}{2}, \qquad (9)$$

where n_c is the number of motor switching cycles.

The analysis can be extended easily to nonlinear joint elasticities. From an energy point of view, the maximum kinetic energy $E_{kin_{max}}$ of an elastic joint will always depend on the maximum potential energy stored in the nonlinear spring and the maximum motor velocity. e_{SL} can now be defined as

$$e_{sl} = \frac{E_{pot_{max}}}{E_{kin}} = \frac{2\int_0^{\phi_{max}} \tau_J(\phi)\, d\phi}{M\dot{\theta}_{max}^2} = \left(\frac{\dot{\phi}_{max}^2}{\dot{\theta}_{max}^2}\right)^2. \qquad (10)$$

Similar to the linear case, time-optimal trajectories, which result in the maximum link velocity $\dot{q}_{max} = \dot{\phi}_{max} + \dot{\theta}_{max}$ in minimum time can be computed. Fig. 9.3 shows phase plots of these trajectories for different e_{sl} values. The energy ratio e_{sl} was increased by decreasing $\dot{\theta}_{max}$, while keeping $E_{pot_{max}}$and thus ϕ_{max}, $\dot{\phi}_{max}$ constant.

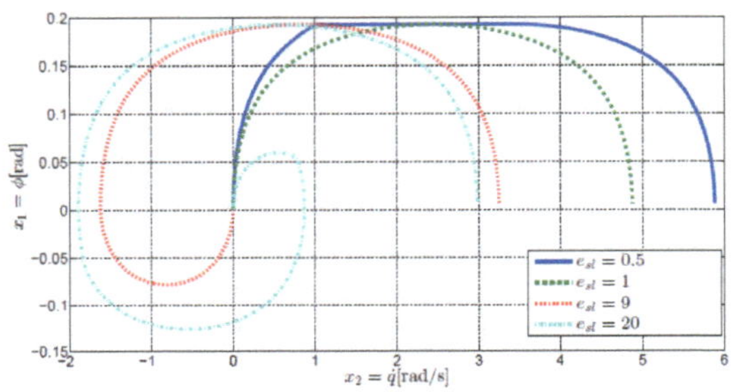

Fig. 9.3 Time optimal control strategy for nonlinear joint elasticity.

The simulated joint is chosen to have the same nonlinear exponential spring characteristics as the DLR QA-Joint [5]. The energy ratio e_{sl} can still be used as an indicator for the singularity and the number of switchings for the time-optimal trajectory. If $e_{sl} < 1$, the angular deflection constraint becomes active before $\dot{q}_{max}$ reaches $\dot{\theta}_{max}$ and a singular optimal solution is the result. Furthermore, it can be shown that for a fixed number of switches n_c, the motor velocity must always change at $\phi = 0$ to obtain the maximum values for $\dot{\phi}$ and $\dot{q}$ after the switches. Since $|\dot{\phi}(0)| = u_{1max}$ and switching u_1 changes $\dot{\phi}$ by $2u_{1max}$, the relative velocity $\dot{\phi}_{nc}$, which denotes $\dot{\phi}$ obtained with n_c motor cycles is generally bounded from above:

$$\dot{\phi}_{nc} \leq (2n_c - 1)\dot{\theta}_{max} \qquad (11)$$

Next, experimental ball throwing results with the 7DoF VSA robot *DLR Handarm System* (*HASY*) [7], taken from [21] are described. They show that our basic analysis finds its equivalence in significantly more complex multi-DoF dynamics with various state and input constraints.

9.4 Experiments

The first experiment uses one shoulder joint together with the elbow joint for fixed stiffness preset σ, i.e. it is a 2DoF throw with nonlinear joint stiffness.

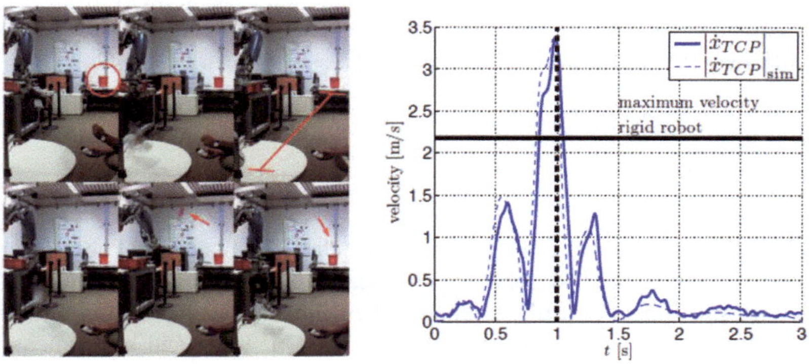

Fig. 9.4 Throwing experiment with DLR's *HASY*.

The experiment depicted in Fig. 9.4 (left) was carried out up to motor velocities of 5 rad/s for the shoulder joint and 2 rad/s for the elbow, where the robot reaches a tip velocity of $\approx 4.0\ m/s$. The throwing distance is $\approx 5.5\ m$. The ball launches in the fifth image at an angle of 45° and scores into the basket. The right plot depicts the absolute Cartesian velocity for a throw at maximum motor velocity of $\approx 2.0\ rad/s$ for both joints (which reaches almost $\approx 3.5\ m/s$ at launch).

Fig. 9.5 Throw and catch between DLR's *HASY* and *Rollin' Justin*.

The simulation and the experiment are in well accordance. If we assumed a stiff robot driving at maximum motor velocity $\dot{\theta}_{max} = 2\ rad/s$, its maximum Cartesian velocity would be $\approx 2.2\ m/s$, i.e. the elastic robot would be $\approx 50\ \%$ faster.

Fig. 9.5 shows screenshots from another experiment, where a "throw and catch" scenario involving two different soft robots is realized. The intrinsically elastic DLR *HASY* is throwing a ball using the computed OC strategy from [21] and the rigid, however actively compliant, mobile humanoid Rollin' Justin catches it using different state of the art control methods [2].

9.5 Conclusion

Making use of the energy that can be stored in the elastic joints of soft-robots is an interesting and challenging task. The related research problems open up new ways of controlling soft robots by truly exploiting their inherent dynamics and to outperform rigid robots e.g. in terms of peak velocity. Our basic and experimental analysis over a series of papers over the last years was the first attempt to systematically understand these systems by unveiling e.g. the effects of damping, constraints on the angular deflection, coupling stiffness, nonlinear dynamics, and nonlinear elastic torques with stiffness adjustment. Together with the already achieved, the continuation of this research direction might bring us significantly closer to our ultimate goal of designing soft robots with human like performance.

Acknowledgments I would like to thank my former colleagues at DLR Nico Mansfeld, Mehmet Can Özparpucu, Sebastian Wolf, Markus Grebenstein, and Alin Albu-Schäffer for their highly valued support and collaboration on this paper's topic over the last years.

9.6 References

[1] Albu-Schäffer A, Eiberger O, Grebenstein M, Haddadin S, Ott C, Wimböck T, Wolf S, Hirzinger G (2008) Soft robotics: From torque feedback controlled lightweight robots to intrinsically compliant systems. IEEE Robotics and Automation Mag: Special Issue on Adaptable Compliance/Variable Stiffness for Robotic Applications 15(3): 20 – 30

[2] Bäuml B, Schmidt F, Wimböck T (2011) Catching flying balls and preparing coffee: Humanoid rollin' justin performs dynamic and sensitive tasks. International Conference on Robotics and Automation pp. 3443–3444

[3] Bicchi A, Tonietti G (2004) Fast and soft arm tactics: Dealing with the safety-performance trade-off in robot arms design and control. IEEE Robotics and Automation Magazine 11:22–33

[4] Braun D, Howard M, Vijayakumar S (2011) Exploiting variable stiffness in explosive movement tasks. Proceedings of Robotics: Science and Systems

[5] Eiberger O, Haddadin S, Weis M, Albu-Schäffer A, Hirzinger G (2010) On joint design with intrinsic variable compliance: derivation of the DLR QA-Joint. International Conference on Robotics and Automation pp. 1687-1694

[6] Garabini M, Passaglia A, Belo F, Salaris P, Bicchi A (2011) Optimality principles in variable stiffness control: the vsa hammer. International Conference on Intelligent Robots and Systems pp. 3770 – 3775

[7] Grebenstein M, Albu-Schäffer A et al (2011) The DLR hand arm system pp. 3175–3182.

[8] Haddadin S, Krieger K, Mansfeld N, Albu-Schäffer A (2012) On impact decoupling properties of elastic robots and time optimal velocity maximization on joint level. International Conference on Intelligent Robots and Systems pp. 5089–5096

[9] Haddadin S, Albu-Schäffer A, Eiberger O, Hirzinger G (2010) New insights concerning intrinsic joint elasticity for safety. International Conference on Intelligent Robots and Systems pp. 2181 – 2187

[10] Haddadin S, Laue T, Frese U, Wolf S, Albu-Schäffer A, Hirzinger G (2009) Kick it with elasticity: Requirements for 2050. Robotics and Autonomous Systems 57:761–775

[11] Haddadin S, Weis M, Wolf S, Albu-Schäffer A (2011) Optimal control for maximizing link velocity of robotic variable stiffness joints. Proceedings of the International Federation of Automatic Control pp. 3175–3182

[12] Liberzon D (2011) Calculus of Variations and Optimal Control Theory: A Concise Introduction, Princeton University Press

[13] Mettin U, Shiriaev A (2011) Ball-pitching challenge with an underactuated two-link robot arm. International Federation of Automatic Control pp. 1–6

[14] Özparpucu M, Haddadin S (2013) Optimal control for maximizing link velocity of a visco-elastic joint. Int Conf on Intelligent Robots and Systems pp. 3035 – 3042

[15] Paluska D, Herr H (2006) The effect of series elasticity on actuator power and work output: Implications for robotic and prosthetic joint design. Robotics and Autonomous Systems 54:667–673

[16] Ham R, Sugar T, Vanderborgth B, Hollander K, Lefeber D (2009) Compliant actuator designs: Review of actuators with passive adjustable compliance/controllable stiffness for robotic applications. IEEE Robotics and Automation Mag 16(3):81–94

[17] Vanderborght B, Verrelst B, Ham Rv, Damme Mv, Lefeber D, Duran B, Beyl P (2006) Exploiting natural dynamics to reduce energy consumption by controlling the compliance of soft actuators. International Journal of Robotics Research 25(4):343–358

[18] Yamaguchi J, Inoue S, Nishino D, Takanishi A (1998) Development of a bipedal humanoid robot having antagonistic driven joints and three DOF trunk. Int. Conf. on Intelligent Robots and Systems pp. 96–101

[19] Yamaguchi J, Nishino D, Takanishi A (1998) Realization of dynamic biped walking varying joint stiffness using antagonistic driven joints. Int Conf on Robotics and Automation pp. 2022–2029

[20] Özparpucu M, Haddadin S (2014) Optimal Control of Elastic Joints with Variable Damping, European Control Conference pp. 2256-2533

[21] Haddadin S, Huber F, Albu-Schäffer A (2012) Optimal control for exploiting the natural dynamics of variable stiffness robots. Int Conf on Robotics and Automation pp. 3347-3354

[22] Haddadin S, Can Özparpucu M, Albu-Schäffer A (2012) Optimal control for maximizing potential energy in a variable stiffness joint. Conference on Decision and Control pp. 1199-1206

10 Simulation Technology for Soft Robotics Applications

Jürgen Roßmann, Michael Schluse, Malte Rast, Eric Guiffo Kaigom, Torben Cichon

RWTH Aachen University

Abstract Soft robots are implied to be inherently safe, and thus "compatible", not only with human coworkers in a production environment, but also with the "family around the house". Such soft robots today still hold numerous new challenges for their design and control, for their commanding and supervision approaches as well as for human-robot interaction concepts. The research field of eRobotics is currently underway to provide a modern basis for efficient soft robotic developments. The objective is to effectively use electronic media - hence the "e" at the beginning of the term - to achieve the best possible advance in the research field. A key feature of eRobotics is its capability to join multiple process simulation components under one "software roof" to build "Virtual Testbeds", i.e. to alleviate the dependancy on physical prototypes and to provide a comprehensive tool chain support for the analysis, development, testing, optimization, deployment and commanding of soft robots.

10.1 Introduction

Inspired by the behavior of soft-bodied animals to cope with uncertainties, the newly emerging field of robotics, soft robotics, is a new compelling and promising direction. Compared to traditional robotics, actuation and control are integrated in and distributed throughout the mechanical properties as well as the structural morphology of the robot in order to accommodate uncertainty [40]. For instance, the robot subject to external disturbance undergoes shape deformation and stiffness variation, allowing it to adjust its dynamics and morphology to the environment and to robustly complete a task.

Despite these advantages, the development of soft robots is currently facing various challenges. Among others, designing, selecting, and placing actuators [25], sensing external disturbances and controlling the structure of these highly underactuated systems in relation to very disparate environmental, operational and performance specificities, and safety requirements are some issues encountered. In view of the manifest complexity that characterizes research and application efforts in soft robotics, as well as the high economic potential and technological impact that may arise, it becomes essential to provide new perspectives for the analysis, understanding, development, testing, optimization, and employment of soft robots, regardless of the area of deployment.

Comprehensive simulation capabilites are already today the method of choice to cope with the complexity and safety issues related to advanced robotic applications [41]. Experiences show that the capability to test and optimize algorithms in simulation before the physical validation is of utmost importance [10]. Based on physically correct simulation capabilities, optimization steps can then easily and cost efficiently be applied [21]. This is crucial in hazardous environments like space, but the ideas are also beneficial when talking about cost-effectiveness and efficiency in production environments. Advanced simulation technology today makes it possible to swiftly model scenarios to be investigated, to easily estimate interaction parameters and to adjust the process very flexibly.

Considering today's robotic applications ranging from industrial robots over exploration rovers on planetary surfaces to service robots, the simulation technology appears to be available. Mechanical engineers are using FEM simulations for structural or thermal analysis, electrical engineers are using block-oriented simulations to develop sensors, actuators or controllers and both fields use kinematics or rigid body dynamics to analyze the dynamic behavior of mechanical structures. In addition, various application specific algorithms simulate special aspects like wheel-terrain interaction, optics, fluid dynamics etc. The major problem still is: The various simulation components do not yet work together well.

This chapter is meant to be a starting point to the application and combination of advanced simulation capabilities to provide a comprehensive "Virtual Testbed for Soft Robotics", i.e. a versatile virtual environment for the analysis, development, testing, optimization, deployment, commanding, and supervision of soft robots. A summary of the current state of the art in the field of modeling and simulation of "classical" and "soft robotics" in section 10.2 is followed by basic concepts of eRobotics as a systematic approach to the simulation based development of robotic applications in section 10.3. Section 10.4 addresses one of the most challenging aspects in modeling and simulation in robotics, the integration of different process simulation algorithms in order to be able to simulate complex robotic systems from system level, down to the motor windings. Section 10.5 then focuses on the simulation of actuated and controlled manipulators. Section 10.6 outlines various application scenarios for the methodology presented before. Last but not least, section 10.7 gives a short summary and an outlook into the future work currently being pursued in the field of modeling and simulation of soft robotic systems.

10.2 State of the Art

This chapter focuses on the *modeling and simulation* of soft robots, for an overview of soft/continuum robotic hardware see e.g. [28]. To get started, currently available approaches in the field of programming and simulation software tools are presented before the focus is shifted onto the transition from hard to soft robots.

10.2.1 Simulation in "Classical" Robotics

The simulation of classical robotic systems is a well-known and powerful source of information for robotic applications. A huge variety of software tools are available to plan, predict, scale, and safely test different scenarios in various disciplines in a time- and cost-efficient manner.

Methods

The methodology of classical robot simulation is usually based on kinematics and/or rigid body dynamics. Whereas kinematic approaches just focus on the path (a sequence of frames or joint angles) of a serial-link manipulator, a rigid body dynamic simulation goes one step further and adds physical interaction and realistic physical properties. Robot dynamics model the robot's reactions under the influence of applied torques and forces by integrating the equations of motion. Today, the most commonly used motion-, contact- and friction models are simplified descriptions of the physical processes [9].

Examples of the use of rigid body models for robotics are the design, analysis, and optimization of conveyor systems [38] and the virtual robot program development for assembly processes [30]. Two essential aspects of rigid body simulations are the modeling of the relevant physical effects on the one hand, and the parametrization of models on the other hand. The identification and validation of parameters is achieved in different experiment setups [35, 22, 33].

Programming and Control

The evolution of robot programming concepts reaches from early control concepts on the hardware level, via point-to-point and simple motion level languages, to motion-oriented structured robot programming languages. In general, the classical robot has a specific end effector following a preplanned path. For this, the two worlds of robotic programming - the physical world and the abstract model - have to be kept consistent by using internal and external sensors. This model is the basis for the planning and execution of automatically generated action sequences including automated generation of collision-free paths and force controlled operations [42].

Simulation Software

In the field of robotics there is a variety of efficient simulation software available. Some known, and currently used, simulation software tools are ROBCAD [36], ROBOTEXPERT [37] and PROCESS SIMULATE [26] from Siemens PLM software, DELMIA from Dassault Systèmes [7], and CIROS [32] (all used for simulation and decision making of robot-based automation), EASY-ROB [29] (robotic simulation down to joint level), V-REP [6] and GAZEBO [23] (both specialize on sensors and image-based assembly systems), ROBOTRAN (physics-based symbolic modeling of rigid body system [12], used for symbolic regression models in the identification of target-robotic-systems, like in [8]) and VEROSIM (3D simulation platform for eRobotics [34]). The application of such simulation tools led to efficient and flexible possibilities for planning and optimizing robotic factories, sensors, assembly systems, and much more.

10.2.2 Simulation in Soft Robotics

The known tools and methods of robotic simulation are not yet able to cope with various aspects in the field of soft robotics. For example, soft robotic simulation requires capabilities ranging from the simulation of single soft bodies over the self-/ and multibody-interaction to advanced robot/environment-interaction [17].

Methods

In current research there are three major aspects regarding the simulation of soft robots. The first aspect is the vast amount of degrees of freedom the simulation has to cope with when soft robots interact with their environment. In classical robotic setups the localization of the end effector is usually controlled by a well structured, nonlinear, set of equations of motion. The dynamics of soft robots can no longer be described by ordinary differential equations and needs a partial differential equations approach. A robot's configuration is then represented e.g. by a curve or surface in a specific space [40].

Secondly, a simulation of soft robots has to take nonlinear effects into account, both on the material and the geometrical side. The realization of soft robots will require compounds of multiple materials in complex shapes. To enable true heterogeneous multi-material simulation, materials with varying properties (stiffness, density, Poisson's ratio, thermal expansion coefficient, and friction coefficient) have to be dealt with as one compound object in simulation [16].

Thirdly, the parametrization of soft robots today is mainly based on estimations which are difficult to verify due to many empirically derived parameters [21].

Programming and Control

Other than the predefined motion of conventional robotic systems, soft robotic systems will have to adapt their trajectory to the changes of their body shape and structure. Continuous robots no longer have discrete joints and undergo elastic deformation. This requires novel control engineering approaches [28]. The end effector makes another difference compared to classical robots. Instead of a special end effector for a specific task, the soft robot itself can function as a universal end effector for grasping and component handling. This leads to a variety of possible grasping configurations and also to different loading outcomes due to gravity. Furthermore, new types of sensors and actuators must be simulated to be able to represent the new unconventional deformation and motion patterns. Examples for the modeling of an octopus inspired robot arm are given in [27] and path planning and shape estimation is described in [40]. Subsequently, in [39], three possible sensing methods are discussed, simulated, and validated. A non-constant curvature manipulator (canonical octopus arm) is studied in [14], where a piecewise constant curvature approximation method for forward and inverse kinematics, a continuum geometrically exact approach for forward kinematics, and a Jacobian method for inverse kinematics are proposed.

Simulation Software

In this challenging, multi-physical, field the flexibility, large deformation potential, and the non-rigidity require the simultaneous analysis of solid and fluid mechanics, electromechanics, thermodynamics, and chemical kinetics of the processes involved [40]. According to [21], there is a lack of design automation tools for soft actuation methods and control development. One approach to implement a soft dynamic simulation tool is given by Hiller and Lipson's "VoxCad" [15], where a finite element analysis (FEA) is the basis for a voxel-based mass-spring-damper model [16]. In each iteration step all internal forces are calculated and afterward all positions are updated. Nevertheless, the interaction between two soft bodies is not yet scientifically conclusive in this model. FEA simulations contain the needed information about the entire mesh and the local node properties but non-linearities, like deformations, friction, or advanced material models, lead to periodic re-meshing that in turn requires additional iterations to solve the underlying linear equations. Another tool to analyze soft body interaction is the so-called "Soft Cell Simulator", where soft multi-cellular robots can be analyzed in 2D. The underlying pressure based model, the soft behavior of single cells, scalability of multiple cells and their potential applications are discussed in [13].

10.3 The Basic Concepts of eRobotics

The state of the art section clearly illustrates the basic necessity as well as the ongoing interest in soft robot simulation technology. The research activities are in an early stage but simulation in soft robotics is a promising, emerging, field of research and addresses soft actuators, soft sensors, soft materials, soft interaction, and "soft control". Complexity and multidisciplinarity require a profound understanding of underlying mathematical models and advanced activity in implementing a comprehensive simulation environment. Alternatively, a holistic approach to simulation in soft robotics requires an encompassing concept which can act as a framework for the various developments necessary. This is where eRobotics comes into play.

The research field of eRobotics is currently an active domain of interest for scientists working in the area of e-Systems engineering. The **aim of eRobotics** is to provide a comprehensive software environment for the development of complex technical systems. Starting with user requirements analysis and system design, support for the development and selection of appropriate hardware, programming, system and process simulation, control design and implementation, and encompassing the validation of developed overall systems, eRobotics provides a continuous and systematic computer support during the entire life cycle of complex systems. In this way, the ever increasing complexity of current computer-aided technical solutions will be kept manageable, and know-how from completed work is electronically preserved and made available for further applications.

10.3.1 3D Simulation-Based Development

To provide the necessary degree of "computer support" mentioned above, eRobotics makes extensive use of 3D simulation technology, which is especially important for soft robotics to simulate continuous sensors, actuators, and joints as well as various interactions with the robots' environment. 3D simulations are used right from the beginning of the development process to test first system design studies in the concept phase. During system development, fully functioning interactive virtual prototypes allow for an efficient and goal directed development, test and verification both on the component as well as on system level at any point of time.

10.3.2 The Virtual Testbed Approach

The integration of data processing algorithms like image processing, generic controllers, or supervisors with environment models into a comprehensive simulation leads to Virtual Testbeds (VTBs). Such VTBs are the basis for simulation-based optimization, reasoning and control approaches.

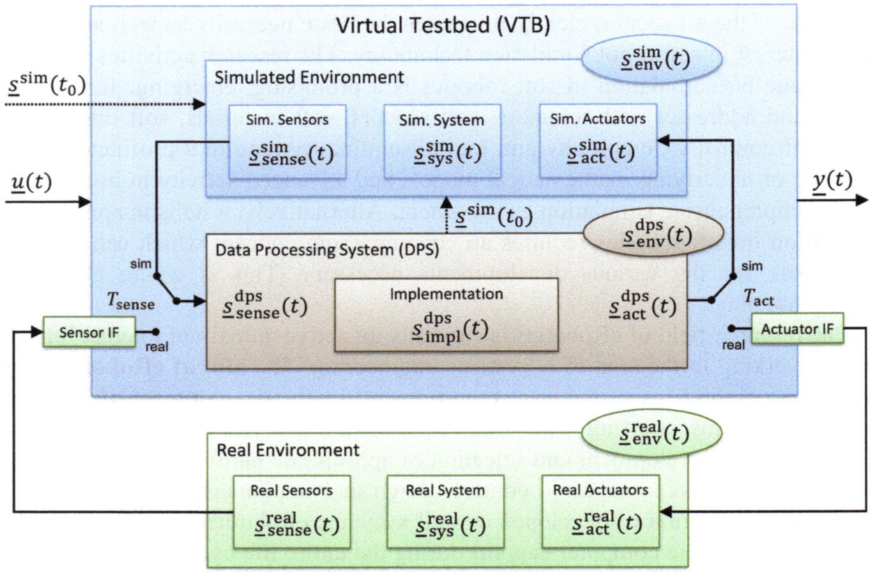

Fig. 10.1 Basic structure of a Virtual Testbed and its connection to real world systems [1]

Fig. 10.1 outlines the basic structure of a Virtual Testbed, the components involved, and their state and parameter vectors. Each vector $\underline{s}(t) = [\underline{x}(t), \underline{a}]$ consists of a dynamic part $\underline{x}(t)$ (any time-dependent state variables such as positions of moving parts, joint values, or motor currents) and a static part $\underline{a}$ (e.g. controller parameters, fixed structural dimensions, or any other constants). The core of the figure is the data processing system (DPS), which processes the sensor data or commands the robot's movements. The internal state of this DPS is called $\underline{s}^{dps}_{impl}(t)$. All sensor data input to the DPS is contained within $\underline{s}^{dps}_{sense}(t)$, whereas $\underline{s}^{dps}_{act}(t)$ contains output to actuators. Sensory input to a DPS is usually evaluated and interpreted in a certain way, which leads to an estimate about some physical quantities of the real environment, the "perceived" environment $\underline{s}^{dps}_{env}(t)$ of the DPS. For example, when a mobile robot system uses a SLAM algorithm to maintain its current location and an estimated map of its environment, then this internal map representation would be part of $\underline{s}^{dps}_{env}(t)$. In conclusion, the full state of the DPS is

$$\underline{s}^{dps}(t) = [\underline{s}^{dps}_{sense}(t), \underline{s}^{dps}_{impl}(t), \underline{s}^{dps}_{act}(t), \underline{s}^{dps}_{env}(t)] \tag{1}$$

During regular operation of the DPS in production or for evaluation, both switches T_{sense} and T_{act} are set to position "real", i.e. $\underline{s}^{dps}_{sense}(t)$ is fed from the physical sensors with $\underline{s}^{real}_{sense}(t)$ and the DPS's actuator commands $\underline{s}^{dps}_{act}(t)$ are forwarded to the physical actuators $\underline{s}^{real}_{act}(t)$.

To model the physical system (e.g. vehicle, robot, or assembly line), all its dynamic and static properties are aggregated in $\underline{s}^{real}_{sys}(t)$, which covers mechanical structure, joints, and software modules not included in the DPS - any relevant state of the real system. Finally, we have the physical environment the system is operating in. All relevant physical properties, such as e.g. geometric shapes of the ground and surroundings, lighting conditions, gravity, and so on, are represented by $\underline{s}^{real}_{env}(t)$. Hence, similar to (1), we have the full state of the physical system and its environment in

$$\underline{s}^{real}(t) = [\underline{s}^{real}_{sense}(t), \underline{s}^{real}_{sys}(t), \underline{s}^{real}_{act}(t), \underline{s}^{real}_{env}(t)]. \tag{2}$$

In order to operate the DPS in a computer simulation with both switches T_{sense} and T_{act} set to position "sim", we need to provide realistic sensor data $\underline{s}^{dps}_{sense}(t)$ and have the simulation react to the output $\underline{s}^{dps}_{act}(t)$. **This is only possible in a simulated environment which mimics all relevant aspects of the real world $\underline{s}^{real}_{\alpha}(t)$.** This is why we introduce a corresponding $\underline{s}^{sim}_{\alpha}(t)$ for each $\underline{s}^{real}_{\alpha}(t)$. Together, they make up the full state vector $\underline{s}^{sim}(t)$ (similar to (2)). Thus, our full VTB consists of the DPS's state $\underline{s}^{dps}(t)$ and the simulation's state $\underline{s}^{sim}(t)$, so

$$\underline{s}^{vtb}(t) = [\underline{s}^{sim}(t), \underline{s}^{dps}(t)] \tag{3}$$

Key idea of the VTB concept is that the DPS is a component **within** the whole VTB. This enables development, evaluation, optimization, and productive operation in the very same hard- and software environment. By turning only a single switch - i.e. T_{sense} and T_{act} combined - one can alternate between virtual and real operation. By design, the DPS is not aware whether it is being operated in a real or simulated environment.

The input signal $u(t)$ consists of commands for robot operation, controller set points, trajectories or other input. It may also be required to initialize the simulated environment and its components $\underline{s}^{sim}(t)$ with $\underline{s}^{sim}(t_0)$, which can usually be loaded from file or sometimes even be generated from $\underline{s}^{dps}(t)$, depending on the scenario.

Fig. 10.2 shows an implementation of the concept for the development of walking robots.

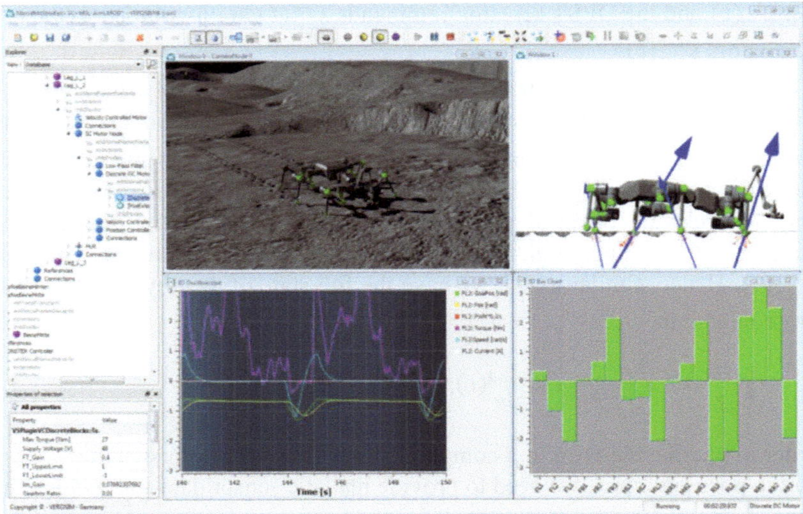

Fig. 10.2 A typical Virtual Testbed environment, here used for the development of walking robots [43] (robot: ©DFKI Bremen)

10.4 Integrating Simulation Algorithms

To simulate soft robots in Virtual Testbeds containing the robots, their environment as well as their data processing systems, it is necessary that algorithms (simulation, data processing, etc.) from multiple domains work together to mimic all relevant aspects of the real world. The goal is that the robot and its environment evolve almost identically ($\underline{s}^{real}(t) \approx \underline{s}^{sim}(t)$) with respect to the controller implementation $\underline{s}^{dps}_{impl}(t)$.

Methods for the multidisciplinary simulation of physical processes can be categorized into single-domain and multi-domain methods (see Fig. 10.3). If the system is modeled with a multi-domain language, an interpreter converts the model into a system of differential-algebraic equations (DAEs), which are then numerically solved. Another class of multi-domain simulation methods is based on the spatial discretization of the considered system. In multi-domain modeling, it is convenient for the user that he can model the overall system with a single user interface. However, the automatically generated DAE-System can be complex and unstructured and may have to be transformed to allow for a stable numerical integration.

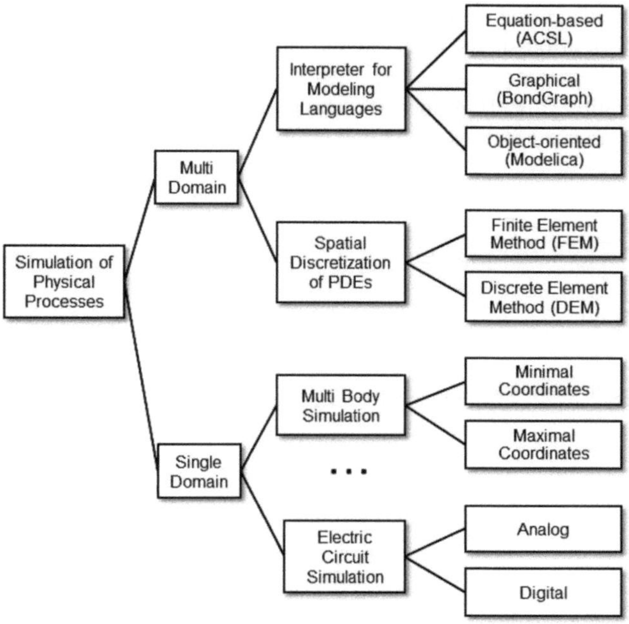

Fig. 10.3 Categorization of physical process simulation

Single-domain simulation methods that are implemented to model one specific physical process are typically more powerful in this specific domain than multi-domain tools, i.e. they cover more effects closer to reality or are computationally more efficient. On the other hand, the coupling of domain-specific simulation tools requires more effort, the user has to deal with various systems, user interfaces, programming languages, and software licenses.

10.4.1 Multi-Domain Modeling with Bond Graphs

Bond graphs are a graphical description language for physical systems [4, 3]. They can be used to model significant parts of the simulation model s_{sim} and deliver basic concepts for the integration of different algorithms (see section 10.4.2). The nodes of bond graphs are physical subsystems and the edges (called bonds) describe their interaction [4, 3]. A bond exchanges two physical quantities (called effort and flow) whose product is power (hence also called power-coupling). A bond is an ideal connection that guarantees energy conservation between the subsystems. The basic elements of a bond graph can be generalized, since the physical concepts of the various domains (electric, mechanic, hydraulic, acoustic, thermodynamic) are the same. Like physics itself, bond graphs are not causal. This has the advantage that bond graph models can be reused and hierarchically assembled to larger models. Not until the system needs to be simulated, the calculation direction in the bonds is determined in the so-called causal analysis. This corresponds to the transformation of the subsystems' equations to a DAE-system that can be numerically integrated.

For illustration, Fig. 10.4 and 10.5 show a common DC motor model in classical block diagram notation and as bond graph.

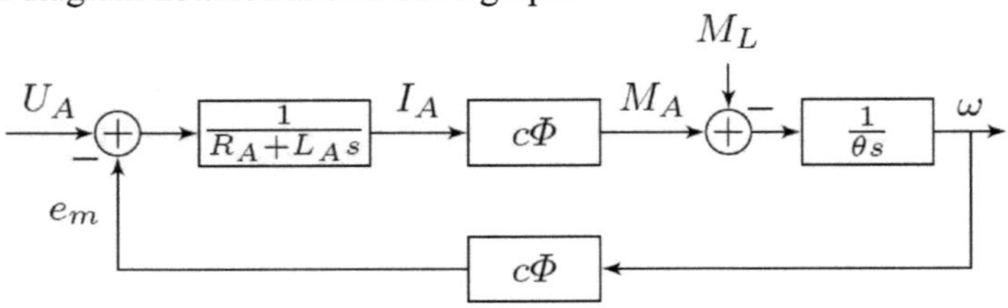

Fig. 10.4 Block diagram of a DC motor

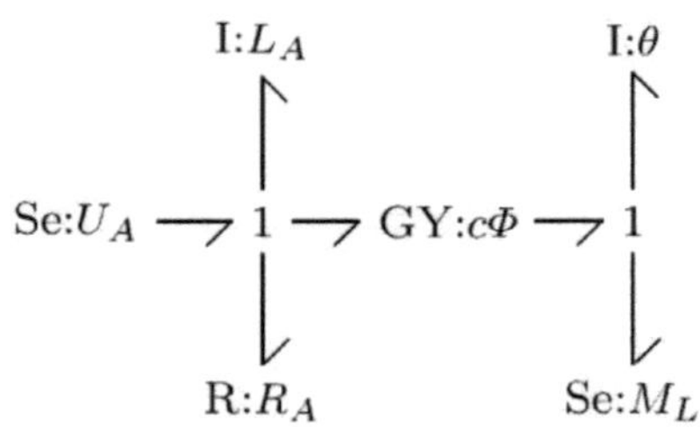

Fig. 10.5 Bond graph of the system in Fig. 10.4

10.4.2 Multi-Domain Modeling by Integrating Single-Domain Tools

In the modeling of a multi-domain system using several domain-specific simulation tools, the coupling points, the exchanged physical quantities and the numerical coupling scheme have to be chosen. Let the overall system be described with a DAE of the form:

$$\underline{\boldsymbol{A}}\,\underline{\dot{x}}(t) = \underline{f}(\underline{x}, t) \tag{4}$$

To describe one part of the system with a more efficient domain specific simulation system, it can be broken apart:

$$\begin{array}{ll} \underline{\boldsymbol{A}}_1\,\underline{\dot{x}}_1(t) = \underline{f}_1(\underline{x}_1, \underline{u}_1, t) & \underline{\boldsymbol{A}}_2\,\underline{\dot{x}}_2(t) = \underline{f}_2(\underline{x}_2, \underline{u}_2, t) \\ \underline{y}_1 = \underline{g}_1(\underline{x}_1, \underline{u}_1) & \underline{y}_2 = \underline{g}_2(\underline{x}_2, \underline{u}_2) \end{array} \tag{5}$$

To arrive at an equivalent overall system, we have to couple input and output quantities:

$$\begin{pmatrix} \underline{u}_1 \\ \underline{u}_2 \end{pmatrix} = \begin{pmatrix} \underline{\boldsymbol{L}}_{11} & \underline{\boldsymbol{L}}_{12} \\ \underline{\boldsymbol{L}}_{21} & \underline{\boldsymbol{L}}_{22} \end{pmatrix} \cdot \begin{pmatrix} \underline{y}_1 \\ \underline{y}_2 \end{pmatrix} \tag{6}$$

For the coupling of mechanical simulation models, several approaches for the exchange of physical quantities are used. This can be purely kinematic properties, like positions and velocities, or only forces or a combination of both. In the displacement/displacement coupling, i.e. the bidirectional exchange of positions or displacements, the system does not have to be separated into disjoint parts at one coupling point, but both subsystems have to contain an overlapping part [5]. The coupling has to be modeled carefully to avoid a distortion of the system properties. Furthermore, the overlapping part can become inconsistent due to integration errors which need to be synchronized. If a bidirectional exchange of velocity and force (i.e. power-coupling, see section 10.4.1) is used, a single coupling point can be chosen. This can be interpreted as a bond that is not coupling two subsystems in one DAE, but between two DAEs. This power-coupling is again energy-conserving by definition and is generic, such that it can be applied to all domains.

Besides the question of the coupling quantities, there are several approaches to the numerical integration of the coupled system. Let the system be modeled in two different simulation systems in the form of DAEs. In the strong coupling, the DAE is exported from one system to the other and the complete system is solved with one integrator. This is the most stable solution. Sometimes it can be sensible to solve the systems separately (weak coupling), for example if the system can be separated into a stiff and a non-stiff part. The non-stiff part can be solved with larger time steps and the overall simulation is more efficient. The weak coupling can be executed by embedding the first DAE with dedicated integrator into the second simulation system or with classical co-simulation. The Functional Mockup Interface [2] provides an implementation for strong and weak coupling.

There are several schemes for the coordination of the numerical integration in the weak coupling. These can be explicit (e.g. Jacobi scheme), semi-implicit (e.g. Gauss-Seidel scheme) or implicit, analogous to the integration schemes. Since the coordination scheme has no information on the DAEs, implicit coordination schemes can only work iteratively (e.g. waveform relaxation).

10.5 Simulation of Actuated and Controlled Manipulators

While the last section mainly addresses the simulation of the robot and its environment ($\underline{s}^{sim}_{sys}$), this section focuses on the development of the motion controller, an important part of $\underline{s}^{dps}_{impl}$ and the simulation $\underline{s}^{sim}_{act}$ of actuators. The developed approach aims at using active and variable compliance control to endow rigid robots with softness.

10.5.1 Simulation of Compliant Robots

From the multibody modeling point of view, a physically interacting robot is characterized by a vast amount of constraints, which range from a revolute joint up to friction, body compliance (e.g. stiffness and damping), and contacts with non-penetration requirements. In order to uniformly handle both the motion of multibody systems and related constraints, the developed approach combines the second law of Newton and the work-less enforcement of constraints [18].

10.5.2 Generation of a Compliant Trajectory

Denoting by $\underline{W_e}$ the vector of contact forces and torques and $\underline{J}$ the Jacobian matrix relating the velocities of the CoM of the links to the joint velocities, the vector $\underline{\Gamma_e} = [\tau_{1_e}, \cdots, \tau_{N_e}]$ of external joint torques follows as

$$\underline{\Gamma_e} = \underline{J}^T\,\underline{W_e}. \tag{7}$$

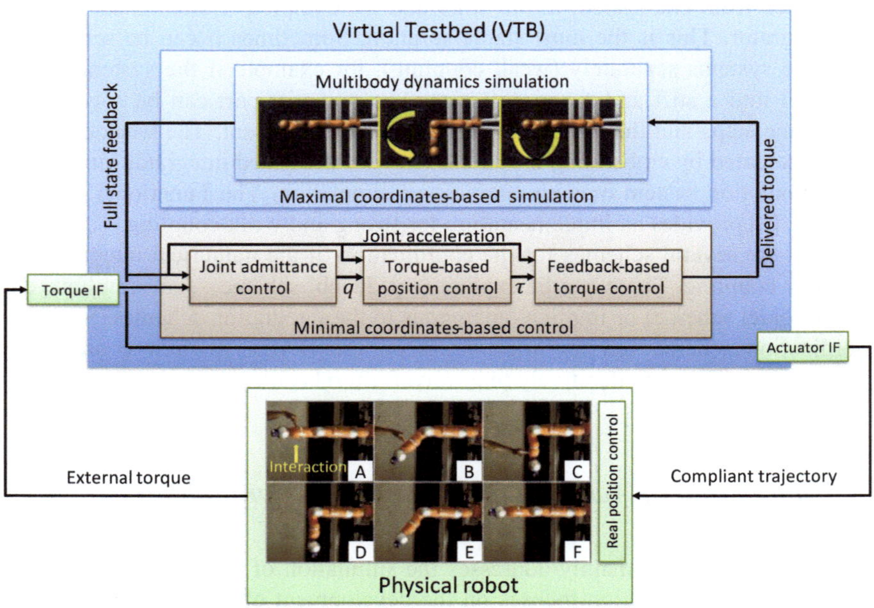

Fig. 10.6 Simplified overall hybrid simulation architecture

Admittance parameters m (mass), d (damping), and s (stiffness) are used by the joint admittance controller to generate the compliant trajectory. The latter is realized by enforcing for each joint a mass-damper-spring relationship between the sensed, external joint torque τ_e and the deviation $\Delta q = q_d - q_c(t)$ of the desired joint position q_d from the compliant, time varying joint position $q_c(t)$

$$m\Delta\ddot{q} + d\Delta\dot{q} + s\Delta q = \tau_e(t). \tag{8}$$

The time discrete solution related to (8) follows as [19]

$$\underline{\mathbf{X}}(k) = e^{\underline{\mathbf{A}}\Delta t}\,\underline{\mathbf{X}}(k-1) + \left[e^{\underline{\mathbf{A}}\Delta t} - \underline{\boldsymbol{I}}\right]\underline{\mathbf{A}}^{-1}\underline{\mathbf{F}}, \tag{9}$$

where the (2×2)-matrices $\underline{\mathbf{A}}$ and $\underline{\mathbf{F}}$ depend on m, d, s and τ_e. Further, $\underline{\mathbf{X}}(t) = [\Delta q\ \Delta\dot{q}]^T$, $\underline{\mathbf{X}}(k\Delta t) = \underline{\mathbf{X}}(k)$, $k \in \mathbb{N}$, and Δt is the time step.

In case of vanishing external joint torques, one can infer that there is no external disturbance acting on the robot. Hence, the robot has to be decelerated. For this purpose, the fixed joint admittance damping and inertia are smoothly increased and decreased as function of time with an exponential shape. In the other case of non-vanishing external joint torques, we decrease the robot resistance. To this end, joint damping and inertia smoothly decrease exponentially.

10.5.3 Torque-Based Tracking of the Compliant Trajectory

In order to track the compliant joint trajectory generated during an interaction, a feedback approach that relies on the Lagrange formalism is used to decouple and linearize the robot dynamics at acceleration level along the entire joint trajectory. Joint torques commanded for this purpose, depend on the current robot dynamics captured at each step. They are furnished by the simulated actuation, which considered the structural compliance, gear reduction, and power related to the motor voltage and current.

10.5.4 Drive Train Modeling and Simulation

An intuitive approach to model the drive train consists in apprehending its constituting electromechanical structure in a uniform fashion (see section 10.4.1). The modified nodal analysis can then systematically assemble the drive train. This results in a linear equation to be solved for the unknown currents, torques, voltages and velocities [20].

10.5.5 Torque Control

Acceleration feedback is useful to decouple the torque dynamics from the rigid body dynamics [24]. Notice that the joint state and accelerations are available on the simulator. This approach is particularly useful as it allows different torque based control techniques to be implemented on flexible joint robots. The structural flexibility of the joint is involved in the exponentially stable dynamics of the torque error.

10.6 Applications

In this section we introduce three examples which use eRobotics concepts to simulate soft robots or soft materials or to implement compliant robot behavior.

10.6.1 FESTO Bionic Handling Assistant

The FESTO Bionic Handling Assistant [11] is an innovative robotic system inspired by an elephant's trunk. With pneumatic actuators in the arm and flexible elements in the gripper, delicate objects can be manipulated very gently. The robot is inherently compliant without complex sensors or control. Despite the seemingly contradictory approaches, rigid body simulation can be used to model this soft robotic systems by partitioning the system into sufficiently small rigid parts and connecting them with appropriately parametrized spring damper constraints.

Fig. 10.7 shows an approximation of the pneumatic arm with three segments of ten links each

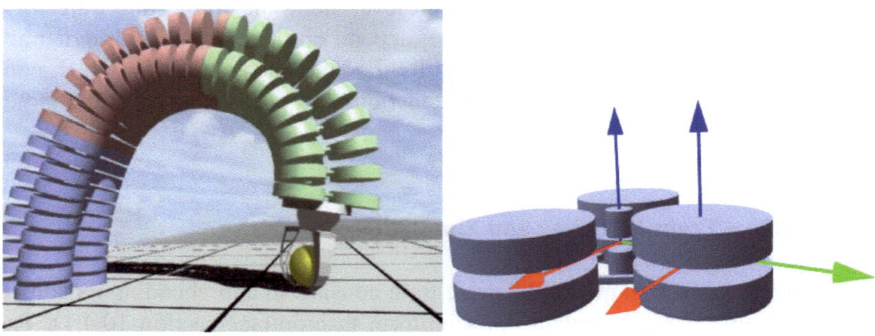

Fig. 10.7 Simulation of the FESTO Bionic Handling Assistant

10.6.2 Soft Physical Human Robot Interaction

This second example addresses the use of admittance control to decrease the felt resistance of the robot during physical interaction with the real robot. The setup is shown in Fig. 10.6. At the beginning of the experiment, the robot is fully extended straight ahead on the horizontal plane (see Fig. 10.6). Except for the fourth joint, the stiffness and damping joint values are set very high. The task performed by the operator was to first move the fourth joint from its initial position $q_4 = 0°$ to the intermediate position $q_4 = 90°$ by interacting with the sixth link. Then, the operator releases the link and the robot moves freely backwards to its initial configuration according to the chosen admittance parameters. The constant and variable admittance parameter cases were compared.

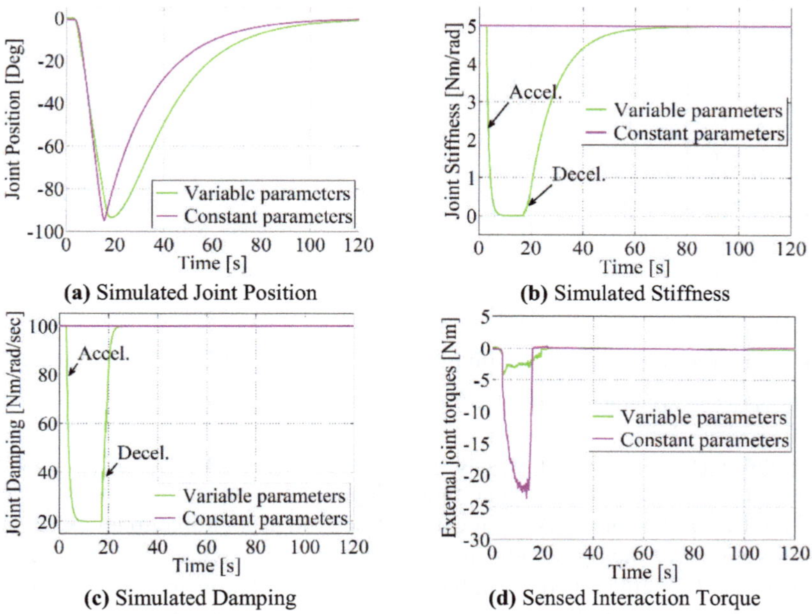

Fig. 10.8 Parameter and Dynamic Values w.r.t. the 4-th Joint

As can be seen in Fig. 10.8(a), constant admittance parameters give rise to a faster backwards joint motion when compared with the adapted parameters whose behavior is depicted in Fig. 10.8(b) and Fig. 10.8(c). The external torque amplitude difference in Fig. 10.8(d) conveys insight into the amount of robot dynamics felt during the interaction for the same angular deflection. It reveals that the operator exerts considerably less external force with adaptable parameters in order to manually move the robot. This highlights the potential of integrating simulation into compliance control to shape the resistance of a position controlled robot. It is worth to note that the approach can be applied to other robots having a positional, torque, or current interface.

10.6.3 Terramechanics

As another example for the integration of rigid and soft simulation algorithms, we present the coupling of a rigid body simulation with a soil simulation model based on cellular automata [31]. The contact between a rigid body and soft soil (e.g. regolith on the moon) can be described as a spring damper system (Fig. 10.9). The constraint of a fixed ground is depicted as a flow source with $v = 0$ (Fig. 10.9(b)). Since this does not contribute to the energy balance, the bond graph can be simplified (Fig. 10.9(c)).

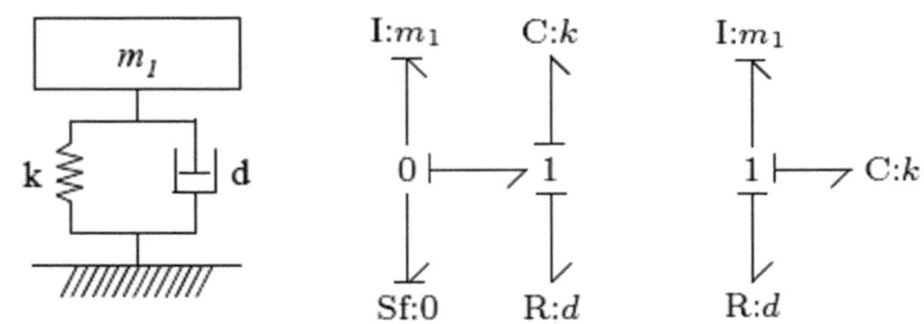

(a) Mechanical diagram **(b)** Bond graph **(c)** Equivalent bond graph

Fig. 10.9 Contact between rigid body and soft ground as spring damper system (1D)

Friction and normal forces can depend on the compression of the soil, which can be a non-linear function of the penetration depth. The tangential friction forces depend on the normal forces. All that can be described as a bidirectional exchange of flow = velocity of the rigid body and effort = restoring force of the soil. The integration of the C-Element is done in the rigid body simulation and the position of the rigid body determines the contour of the soil during contact. This is directly stored in the geometric surface model, such that visualization and simulation work on the same data. The R-Element is represented by the soil simulation model, since the dissipation energy of the contact is incorporated in the deformation of the ground. This technique is applied to realize the Virtual Testbed in Fig. 10.2 to model the contact between the legs of the robot and the lunar surface.

10.7 Conclusions and Outlook

This chapter clearly illustrates the basic necessity for powerful, versatile, and comprehensive modeling and simulation capabilities for soft robotics applications. The developments for traditional robotics can be a starting point for these developments. The approaches are able to simulate soft components to a certain extent, but to fully exploit the capabilities of soft robots in simulation, new approaches are necessary. But obviously there is a lack of modeling and simulation approaches for soft robotics. To close this gap, it is necessary to develop algorithms to simulate the various aspects of soft robots in fields like structure, materials, actuation, sensing, or interaction and to develop integration frameworks which are able to simulate soft robots on system level incorporating not only the robot itself but also its data processing part as well as its environment. eRobotics concepts seem to be able to provide the necessary engineering approaches ranging from fully functioning Virtual Testbeds up to simulation-enhanced data processing using simulation-based reasoning and control approaches. This way, simulation technology for soft robotics provide everything necessary for simulation-based development of soft robots:

- research and development of various robot structures, materials, actuators, and sensors as well as various control or interaction concepts
- consolidation of interdisciplinary developments
- test and verification of various concepts on system level
- simulations with a wide range of process variations (free choice of environment, constraints, user interactions etc.)
- hardware prototypes are no longer the bottleneck
- fast, cost effective, and goal directed evaluation of alternative designs
- transfer of the developments in practical applications
- support of technology marketing and transfer within the context of dissemination

We not only strive to deepen the above results, but also to further strengthen the symbiosis between the latest and coming progresses in robotics and advanced computer simulation to tackle current issues and open up new perspectives to future soft robotic applications, regardless of the area of deployment.

10.8 References

[1] Atorf L, Schluse M, Rossmann J (2014) Simulation-based optimization, reasoning, and control: The eRobotics approach towards intelligent robots. In: Int. Symp. on Artificial Intelligence, Robotics and Automation in Space (i-SAIRAS), to be published

[2] Blochwitz T, Otter M, Arnold M, Bausch C, Clauß C, Elmqvist H, Junghanns A, Mauss J, Monteiro M, Neidhold T, et al (2011) The functional mockup interface for tool independent exchange of simulation models. In: 8th International Modelica Conference, pp 20-22

[3] Borutzky W (2006) Bond graph modeling and simulation of mechatronic systems: An introduction into the methodology. In: 20th European Conference on Modeling and Simulation (ECMS)

[4] Broenink JF (1999) Introduction to physical systems modelling with bond graphs. URL http://www.ce.utwente.nl/bnk/papers/BondGraphsV2.pdf

[5] Busch M (2012) Zur effizienten Kopplung von Simulationsprogrammen. Kassel University Press GmbH

[6] Coppelia Robotics Software (2014) V-rep. URL http://www.coppeliarobotics.com

[7] Dassault Systems (2014) Delmia software. URL: http://www.3ds.com/de/produkte-und-services/delmia/loesungen/alle-delmia-loesungen/

[8] Dolinsky JU (2001) The development of a genetic programming method for kinematic robot calibration. Dissertation, Liverpool John Moores University

[9] Featherstone R (2008) Rigid Body Dynamics Algorithms, vol 49. Springer New York

[10] Ferraguti F, Golinelli N, Secchi C, Preda N, Bonfe M (2013) A component-based software architecture for control and simulation of robotic manipulators. In: Emerging Technologies & Factory Automation (ETFA), IEEE

[11] Festo AG & Co KG (2012) Bionic Handling Assistant. URL http://www.festo.com/net/SupportPortal/Files/42050/Brosch%5C_FC%5C_BHA%5C_3%5C_0%5C_EN%5C_lo.pdf. Accessed 2014-16-05

[12] Fisette P, Samin J (1993) Robotran: Symbolic generation of multi-body system dynamic equations. In: Advanced Multibody System Dynamics, Springer, pp 373-378

[13] Germann JM, Maesani A, Floreano D, Stöckli M (2013) Soft cell simulator: A tool to study soft multi-cellular robots. In: Int. Conference on Robotics and Biomimetics, IEEE, EPFL-CONF-196284, pp 1300-1305

[14] Giorelli M, Renda F, Calisti M, Arienti A, Ferri G, Laschi C (2012) A two dimensional inverse kinetics model of a cable driven manipulator inspired by the octopus arm. In: Int. Conf. on Robotics and Automation (ICRA), IEEE, pp 3819-3824

[15] Hiller J, Lipson H (2014) Voxcad. URL http://www.voxcad.com/

[16] Hiller J, Lipson H (2014) Dynamic simulation of soft multimaterial 3d-printed objects. Soft Robotics 1(1):88-101

[17] Iida F, Laschi C (2011) Soft robotics: Challenges and perspectives. Procedia Computer Science 7:99-102

[18] Jung TJ (2009) Methoden der Mehrkörperdynamiksimulation als Grundlage realitätsnaher virtueller Welten. Dissertation, RWTH Aachen University

[19] Kaigom EG, Rossmann J (2013) A new eRobotics approach: Simulation of adaptable joint admittance control. In: Int. Conf. on Mechatronics and Automation (ICMA), IEEE, pp 550-555

[20] Kaigom EG, Rossmann J (2013) Simulation of actuated and controlled robot manipulators. International Journal of Mechatronics and Automation (IJMA)) 3(3):191-202

[21] Lipson H (2013) Challenges and opportunities for design, simulation, and fabrication of soft robots. Soft Robotics 1(P):21-27

[22] Müller R, Esser M, Janßen C, Vette M (2010) Systemidentifikation für Montagezellen-Erhöhte Genauigkeit und bedarfsgerechte Rekonfiguration. wt Werkstattstechnik online 100(9):687-691

[23] Open Source Robotics Foundation (2014) Gazebosim homepage. URL http://gazebosim.org/

[24] Ott C, Albu-Schäffer A, Kugi A, Hirzinger G (2003) Decoupling based Cartesian impedance control of flexible joint robots. In: Int. Conf. on Robotics and Automation (ICRA), IEEE, pp 3101-3107

[25] Pfeifer R, Marques HG, Iida F (2013) Soft robotics: the next generation of intelligent machines. In: Int. Conf. on Artificial Intelligence, AAAI Press, pp 5-11

[26] Process Simulate Software (2014) Processsimulate. URL http://www.plm.automation.siemens.com/de_de/products/tecnomatix/robotics_automation/robotexpert.shtml

[27] Renda F, Cianchetti M, Giorelli M, Arienti A, Laschi C (2012) A 3d steady-state model of a tendon-driven continuum soft manipulator inspired by the octopus arm. Bioinspiration & biomimetics 7(2):025,006

[28] Robinson G, Davies J (1999) Continuum robots - a state of the art. In: Int. Conf. on Robotics and Automation (ICRA), IEEE

[29] Roos E, Behrens A, Anton S (1997) Rds-realistic dynamic simulation of robots. In: Int. Symp. on Industrial Robots, Int. Fed. of Robotics & Robotic Industries, vol 28, pp 17-27

[30] Rossdeutscher M, Zuern M, Berger U (2010) Virtual robot program development for assembly processes using rigid-body simulation. In: Int. Conf. on Computer Supported Cooperative Work in Design, IEEE, pp 417-422

[31] Rossmann J, Schluse M, Jung T, Rast M (2009) Close to reality simulation of bulk solids using a kind of 3d cellular automaton. In: Int. Design Engineering Technical Conf. & Computers and Information in Engineering Conf. (IDETC/CIE), ASME

[32] Rossmann J, Wischnewski R, Stern O (2010) A comprehensive 3-d simulation system for the virtual production. In: Int. Industrial Simulation Conference (ISC), pp 109-116

[33] Rossmann J, Steil T, Springer M (2012) Validating the camera and light simulation of a virtual space robotics testbed by means of physical mockup data. In: Int. Symp. on Artificial Intelligence, Robotics and Automation in Space (i-SAIRAS)

[34] Rossmann J, Schluse M, Schlette C, Waspe R (2013) A new approach to 3d simulation technology as enabling technology for erobotics. In: Int. Simulation Tools Conf. & EXPO (SIMEX), pp 39-46

[35] Rudolph J, Woittennek F (2008) An algebraic approach to parameter identification in linear infinite dimensional systems. In: Mediterranean Conf. on Control and Automation, IEEE, pp 332-337

[36] Siemens (2014) Robcad. URL http://www.plm.automation.siemens.com/de_de/products/tecnomatix/robotics_automation/robcad/

[37] Siemens PLM Software (2014) Robotexpert. URL http://www.plm.automation.siemens.com/de_de/products/tecnomatix/robotics_automation/robotexpert.shtml

[38] Song P, Trinkle JC, Kumar V, Pang JS (2004) Design of part feeding and assembly processes with dynamics. In: Int. Conf. in Robotics and Atomation (ICRA), IEEE, vol 1, pp 39-44

[39] Trivedi D, Rahn CD (2014) Model-based shape estimation for soft robotic manipulators: The planar case. Mechanisms and Robotics 6:021,005-1-021,005-10

[40] Trivedi D, Rahn CD, Kier WM, Walker ID (2008) Soft robotics: Biological inspiration, state of the art, and future research. Applied Bionics and Biomechanics 5(3):99-117

[41] Vukobratovic M (2009) Dynamics and robust control of robot-environment interaction, vol 2. World Scientific

[42] Wahl FM, Thomas U (2002) Robot programming-from simple moves to complex robot tasks. Institute for Robotics and Process Control, Technical University of Braunschweig

[43] Yoo YH, Jung T, Römmermann M, Rast M, Kirchner F, Rossmann J, & Center RI (2010). Developing a virtual environment for extraterrestrial legged robot with focus on lunar crater exploration. In Proceeding of 10th International Symposium on Artificial Intelligent, Robotics and Automation in Space, vol. 29, No. 01.9.

11 Concepts of Softness for Legged Locomotion and Their Assessment

Andre Seyfarth, Katayon Radkhah, Oskar von Stryk

Technische Universität Darmstadt

Abstract In human and animal locomotion, compliant structures play an essential role in the body and actuator design. Recently, researchers have started to exploit these compliant mechanisms in robotic systems with the goal to achieve the yet superior motions and performances of the biological counterpart. For instance, compliant actuators such as series elastic actuators (SEA) can help to improve the energy efficiency and the required peak power in powered prostheses and exoskeletons. However, muscle function is also associated with damping-like characteristics complementing the elastic function of the tendons operating in series to the muscle fibers. Carefully designed conceptual as well as detailed motion dynamics models are key to understanding the purposes of softness, i.e. elasticity and damping, in human and animal locomotion and to transfer these insights to the design and control of novel legged robots. Results for the design of compliant legged systems based on a series of conceptual biomechanical models are summarized. We discuss how these models compare to experimental observations of human locomotion and how these models could be used to guide the design of legged robots and also how to systematically evaluate and compare natural and robotic legged motions.

11.1 Biomechanics of Legged Locomotion

Computer simulation models can be very powerful tools for analyzing and describing human and animal locomotion. In the last years sophisticated human motion simulation environments have become widely accessible to the research community both for forward dynamics (e.g. OpenSim [5]) as well as for inverse dynamic calculations (e.g. AnyBody [4]). These software tools can be used to describe human (or animal) motion dynamics as the result of the interaction between body mechanics and actuator (muscle) dynamics. However, because of the very large number of parameters and submodels involved their validation and calibration is highly challenging and not yet reasonably solved. In contrast to these detailed models there is an increasing number of conceptual models which are designed to understand the basic mechanisms of legged locomotion. These models are mostly based on the inverted pendulum model (IP) [1] or on the spring-loaded inverted pendulum model (SLIP, originally introduced by Blickhan [2] and

McMahon and Cheng [22]). Both models can be considered as template models [11] as they describe the key mechanics of the center of mass (CoM) during legged locomotion at a most reduced level of detail.

In the IP (inverted pendulum) model, the distance between the CoM and the contact point at the ground during stance phase is assumed to be constant. Hence, the CoM is travelling on a circular arc over this support point at the ground. In contrast, in the SLIP (spring-loaded inverted pendulum) model — a two-dimensional spring-mass model — the leg operates similar to a linear prismatic spring. Here, the distance of the CoM to the support point is not constant. For instance, in running the leg first compresses and finally decompresses until the end of contact. The characteristic loading-unloading cycle of the conceptual leg spring is associated with a typical sinusoidal pattern of the ground reaction force (GRF). This basic leg mechanics can also be found in other bouncy gaits such as human hopping [10] and jumping [29] or quadrupedal trotting [15].

An analysis of experimental data of human running reveals that the leg indeed compresses with leg force being about proportional to the amount of leg shortening [20]. This experimental observation has led to the concept of a leg spring. Here the leg spring constant, called *leg stiffness,* describes the ratio between the maximum leg force and maximum leg compression (the amount of shortening of leg length during contact). It is important to note that this ratio only describes an overall spring-like leg *behavior* represented in the force-length relationship and does not necessarily reflect the action of a single mechanical (passive) or controlled (active) spring. For example, also a mass which is moved up- and downward by being attached to circulating disc will generate a spring-like behavior, meaning that the force generated in a harmonic manner will lead to a 180° out-of-phase relation between force and position of the mass. Hence, the concept of the leg spring is less representing a real spring but rather the mechanics of a harmonic oscillation, which may be the result of quite different complex mechanisms (e.g. kinematic program, compliant structures, muscle-tendon-skeleton dynamics).

In case of the human locomotor function, leg forces are generated through muscle activation. The muscle forces are transferred to the joints (as joint torques) through tendons, which operate in series to the muscle fibers. Hence, compliant leg function is determined by the highly elastic properties of the tendons of which many span multiple leg joints. For instance, in the human Gastrocnemius muscle – the large calf muscle spanning knee and ankle – the muscle fiber length is only about 10% of the total length of the muscle-tendon complex [18]. Thus, at higher forces and resulting loading dynamics, the muscle fibers can only partially contribute to overall muscle-tendon work. As a result, the force-length profile generated by the muscle-tendon system is largely determined by the function of the elastic tendon [30]. This elastically shaped muscle function in bouncing gaits (e.g. running, jumping, hopping) translates into the joint function and finally into a leg function, which force-length relationship is very similar to that of a linear spring. The resulting slope of this relationship is often compared to (and sometimes even interpreted as) the "leg stiffness". This comparison (and phrasing) has to be made very carefully, as the observed “leg stiffness” is a combination of both elastic and nonelastic mechanisms, which need to be separated from each other.

In Table 11.1, selected body and leg spring parameters in human and animal locomotion are summarized providing key reference data for legged robots to achieve comparable motion performance. Please note that the stiffness values provided by Herr et al. [16] are derived based on a computer simulation model to match experimental gait patterns. The leg stiffness values estimated by Lipfert et al. [20] take shifts in the center of pressure (CoP) position during contact into account. Farley et al. [8] found that across different animals, leg stiffness scales to body mass according to $k_{LEG} = 0.715m^{0.67}$. In most of the analyzed animals in this study, leg stiffness was found to be largely independent of speed.

11.2 Legged Locomotion in Robotics

Since the beginning of robotics more than 50 years ago, robots for practical applications are predominantly designed by the principle of kinematic chains of rigid joints and links [28]. Mechanical elasticity has been considered harmful. Limited positioning accuracy caused by link deflection due to compliance, which cannot be avoided by robot design, are usually handled by augmenting position control algorithms with models of link deflection in order to compensate for them. The rigid kinematic chain paradigm facilitates a corresponding modular robot design from largely independent building blocks (i.e. joint level actuation, sensing and control). Robot tasks are usually formulated in operational space (e.g., as trajectories of the end effector in world coordinates). In practice, operational space control is commonly approximated by coordinated decentralized, single-input-single-output joint space controllers [3]. Such an approach is feasible for conventional robot designs with relatively high stiffness and low compliance. Elastic and compliant behavior of a rigid robot in a contact task can be achieved by advanced control methods, e.g., impedance control [34].

These concepts for the design and control of powerful robotic arms provided the models for the currently most common four-legged and bipedal robot designs [17]. The Zero-Moment-Point (ZMP) is the best known and commonly used scheme to implement stable bipedal walking for such robots. Virtual compliance, e.g. based on impedance control, and improved actuators provides some of the locomotion performance of humans like jogging type motion with small flight phases of both feet. This approach cannot recover much energy between steps and is therefore highly energy inefficient. It also requires full feedback and sufficiently low latency of the control loop.

Table 11.1 Selected spring-leg parameters for four-legged animals and humans.

species	mass (kg)	leg length (m)	stiffness fore-limb (kN/m)	stiffness hind-limb (kN/m)	speed (m/s)	gait	Reference
human	73.4	0.97	—	12	2.5	run	Farley and Gonzalez, 1996
human	70.9	0.95	—	23.5	1.55	walk	Lipfert et al., 2012
human	70.9	0.98	—	16.5	2.59	run	Lipfert et al., 2012
dog 1	5.1	0.20	1.9	1.2	1.9	trot	Herr et al. 2002
dog 2	23.9	0.50	2.9	1.9	2.9	trot	Herr et al. 2002
goat	25.2	0.48	4.9	2.7	2.8	trot	Herr et al. 2002
horse 1	134	0.75	18	9.1	2.7	trot	Herr et al. 2002
horse 2	676	1.5	37	22	2.9	trot	Herr et al. 2002

The locomotor system of humans and animals is following another design approach. The biological system would not be capable to realize these state-of-the-art control concepts used in engineering. Its motor system is highly redundant and compliant with many actuators (muscles) spanning one or multiple leg joints and many (individually controlled) motor units with different actuator properties sharing the work within one muscle. At the same time biological signal processing and actuator dynamics are slow. The resulting latencies in the control loop only enable feedforward control of fast motions. Compliant structures (e.g. tendons, ligament, titin) are largely shaping the forces acting on the body.

With increasing motion speed the contribution of sensory feedback to motor control reduces and the system and actuator dynamics are becoming key players for motion generation. Sophisticated control approaches such as ZMP or hybrid zero dynamics (HZD) [33] could not operate on the biological system.

Recently the development of novel variable impedance actuators has gained strong momentum in robotics [37]. These provide promising abilities for compliant robot design including capabilities to store energy and to passively support push off for the next step and to instantaneously compensate for shocks from collisions of the feet with the ground. However, new design and control concepts need to be investigated to fully utilize the potential of these new compliant actuators for legged robots [38] including systematic assessment of actuation and control with muscle-tendon units spanning multiple joints versus compliant single joint actuation only.

With the help of biomechanical template models [11], key parameters of biological motion patterns can be identified and matching control approaches can be derived. For the transfer into technical systems, the key challenge remains to define to what extent the biomechanical template can and must to be represented by the design of the mechanical system and the actuators and what is the contribution of control to shape the body dynamics and motion performance during selected motion tasks.

11.3 Biomechanical Concepts for Legged Locomotion

The design process of technical legged systems like legged robots or leg prostheses which mimic the design and function of the biological counterpart can be strongly supported by biomechanical concepts. However, often these concepts are focusing only on specific features of legged locomotion. In order to be useful for the design of technical legged systems, several of these conceptual models consequently need to be extended and combined properly to reach a sufficient level of detail and complexity. In the following we will address how specific design features of the biological system can be combined in a more comprehensive conceptual approach for the design of legged robotic systems.

As pointed out in the introduction, human leg function is largely shaped by elastic properties of muscles and other soft tissues of the human body. Though, the tendon properties in the human leg are well defined, the resulting leg function is largely dependent on the leg geometry (e.g. whether the joints are extended or flexed) and on the muscle-tendon dynamics and their interactions during a specific movement. For example, in human standing calf muscles can operate (lengthen and shorten) out-of-phase to the Achilles tendon [36]. For instance, with increasing muscle force, the tendon lengthens while the muscle fibers shorten. As a result, the overall muscle-tendon system acts stiffer compared to the tendon stiffness. In order to separate the effects of muscle-tendon dynamics and leg geometry (leg segmentation and joint angles) in adjusting the overall leg function stiffness (e.g. leg stiffness), models with different levels of complexity have been developed.

In a study of Geyer et al. [12] the architecture of the human leg was reduced to a two-segment model with an extensor muscle describing the repulsive leg function during bouncing tasks (e.g. hopping). In this model it was shown, that based on the characteristic properties of muscle fibers (Hill-type muscle model), a corresponding muscle activation pattern is required to generate cyclic vertical jumps. These patterns can be provided as a feedforward command [13] or as a combination of a constant activation and a modulating feedback signal based on proprioceptive sensory signals. As an outcome of this simulation study, a positive feedback of muscle force was predicted to be best suited in order to achieve stable hopping cycles. Hopping frequency and hopping height could be adjusted based on the feedback parameters (gains, delays). Even though tendon elasticity largely improves hopping height, stable hopping was also possible without any tendons in series to the muscle fibers.

In recent studies of Häufle et al. [13, 14], this model was further reduced to a one-dimensional muscle model operating as a virtual "leg muscle" supporting the body mass. With this model the effects of muscle dynamics were considered independent of leg segment dynamics. The results are in line with the findings of Geyer et al. [12] that muscle function for stable hopping can be achieved by both feedforward and feedback control of the muscle. Interestingly, the combination of both actuation schemes provided the best results regarding energy stability in cyclic hopping.

These two models illustrate that the key mechanisms for achieving repulsive leg function as observed in human locomotion do not rely on elastic elements. The muscle dynamics can be exploited by feedforward and feedback activation schemes to result in the observed spring-like leg operation. Hence, leg compliance is an option (which provides many benefits, e.g. for energetics, stability, and shock resistance) but not required to achieve stable hopping and gaits.

11.4 Radial and Tangential Leg Function

The sagittal-plane leg function in human locomotion can be divided into two directions: the radial leg function (e.g., the function represented by the leg spring) and the tangential leg function. The latter includes force contributions, which are directed outside the leg axis. For instance, tangential leg forces can be used to redirect the leg force from the CoM in order to stabilize body posture. In contrast to the radial leg function which is the focus of the IP and SLIP model, tangential leg function has received more attention only recently.

Radial leg function is required to direct the forces outside the leg axis. This is important when you want to kick the ball with your foot. In legged locomotion radial leg function is required in order to adjust the leg angle during swing phase [6] in order to prepare for the next ground contact (e.g. foot placement). At the same time, radial leg function is needed to achieve postural stability (balance) during standing, walking, and other gaits. In human walking, leg forces during stance phase point – in contrast to the bipedal SLIP model – to a point above the CoM. The intersection point is also called virtual pivot point (VPP) [21], as it mimics the function of a virtual support point of a physical pendulum. The extension of the SLIP model by a rigid trunk shows that postural stability in walking and running can be achieved when the leg forces are deviated to a fixed VPP point at the supported body. By shifting the horizontal location of the VPP relative to the body axis (line through CoM and hip), the resulting hip torques will accelerate or decelerate the gait. Interestingly, the hip torques predicted by the VPP model are very similar to those observed in human walking.

The radial leg function is not only a key to redirect leg forces during ground contact (as described by the VPP concept) but may also align the leg angle during swing. A simple model to describe the swing-leg dynamics in locomotion could be a spring-loaded pendulum, which is supported by the upper body (pelvis). Recently Song et al. [32] found that this model is well able to represent the experimental hip joint forces caused by the swing leg during human walking, if swing leg stiffness is adjusted to walking speed. The pendulum length may, however, not only result from the distance of the swing leg CoM to the hip joint but also well tuned by two-joint thigh muscles which can tune the rotational stiffness of the swing leg. Both the rotational swing-mass system and the pendulum are sharing similar dynamics for moderate angular displacements. Hence, the rotational elastic oscillator can mimic and thus tune the virtual pendulum length of the swing leg. With that both balance and swing-leg function could be represented by a similar template model, a virtual pendulum. In contrast to the VPP concept, this pendulum shares spring-like leg properties (as in the SLIP model) also in the radial leg function. Thus swing-leg function can be considered as a superposition of repulsive leg function (virtual leg spring) and balance (virtual pendulum), leading to a spring loaded pendulum model (SLP). These three fundamental subfunctions for legged locomotion (balance, repulsive leg function, and swing-leg function) and the corresponding template models (VPP, SLIP, SLP) are summarized in Fig. 11.1.

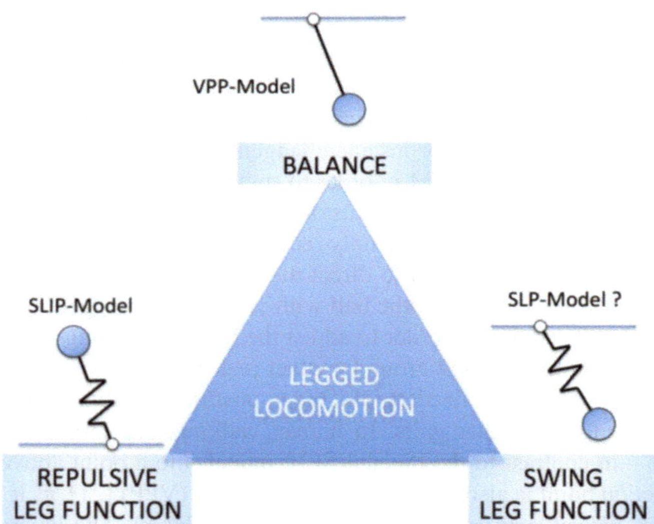

Fig. 11.1 Legged locomotion can be considered as a combination of three elementary subfunctions: balance, repulsive leg function, and swing leg function. For each of these subfunctions, mechanical template models can be identified which describe the dynamics of the center of mass (SLIP model), of the body orientation (VPP model), and of the swing-leg center of mass. The template model for the swing leg function is a topic of current research and could be a spring-loaded pendulum (SLP), as suggested by recent findings of Song et al. 2014 [32].

11.5 Leg Segmentation and Multi-Joint Structures

Conceptual models based on the SLIP template show that both the radial and the tangential leg function are complementing each other in order to generate stable gaits and to maintain postural stability during locomotion. However, in the segmented leg, individual joint torques at hip, knee, and ankle will influence both leg functions. How could the biological body take advantage of the matching properties of the both underlying leg functions?

In order to resolve this issue, it would be extremely helpful if the tangential leg function could be accessed independently from the radial leg function. One possible solution is taking advantage of the two-joint (biarticular) structure of some of the leg muscles. These muscles are able to provide specific combinations of joint torques and to exchange energy between leg joints. It turns out that with a proper lever arm design (e.g. hip to knee lever arm ratio of 1:2) these two-joint muscles can generate force contributions, which are perpendicular to the leg axis [27]. Hence, by activating these muscles, leg force contributions will be created which merely contribute to the tangential leg function but not to the radial leg function. With an arrangement of a pair of antagonistic two-joint thigh muscles (like rectus femoris and hamstrings), it is possible to implement the VPP concept independent of the function of the single-joint muscles (which could contribute to the radial leg function). Hence, the architecture (geometrical arrangement) of leg muscles could be a key in providing differential access to the tangential leg function as a complementation of the radial leg function.

Currently, these ideas and concepts are substantiated by taking compliant properties of two-joint muscles into account. This research may lead to insights how radial and tangential leg functions might be adapted to each other and to what extent the control of body posture can be solved in a generalized way that includes a free selection of a large variety of motion tasks ranging from standing, walking, and running gaits.

11.6 From Biomechanical Concepts to Robots

In the BioBiped project (www.biobiped.de) the focus is on exploring the roles of the musculoskeletal actuator arrangement in a humanoid robot with segmented legs. The two built prototypes BioBiped1 and BioBiped2 feature three-segmented legs with nine active and passive, human-like muscle-tendon units per leg spanning one or two joints [39]. Based on the human lower limb system, the hip, knee, and ankle joints are spanned by a pair of monoarticular antagonist-agonist, series elastically actuated (SEA) or passive tendons, as shown in Fig. 11.2(c). The biarticular muscles found in humans are realized as passive tendons with built-in extension springs connecting two segments.

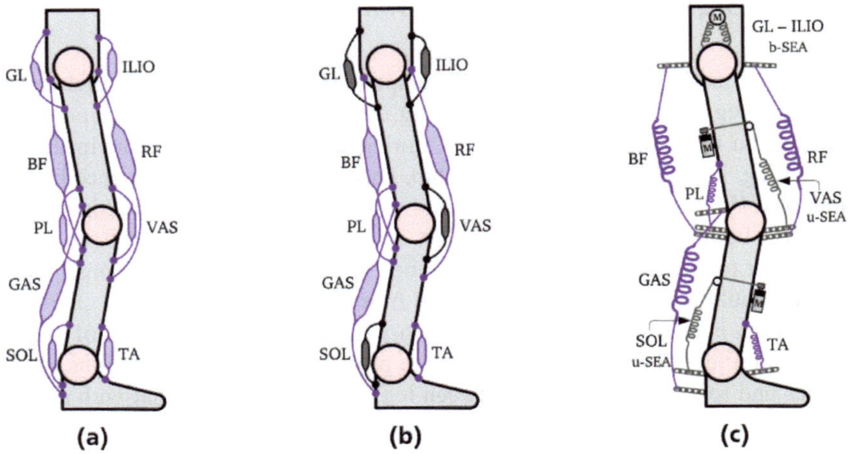

Fig. 11.2 Evolution of BioBiped1's actuation concept: (a) human musculoskeletal leg system; (b) selection of actuators and passive elements: the hip joint requires both active flexion and extension, while in the knee and ankle joints only actuated extensor muscles are required (highlighted); c) implementation of passive and active elements as motors and springs [23].

With this novel, specifically selected musculoskeletal design a number of research questions can be addressed. This hardware-based research approach is complemented by a sufficiently realistic modeling and simulation methodology [19,23,24]. Passive rebound studies in simulation investigating different actuation designs demonstrated that dynamic and energy-efficient locomotion cannot be achieved through stiff actuation without causing critical damage to the motor gearboxes [26]. More importantly, it was shown that the energy restitution ratio increases with joint compliance. However, exceeding a specific leg compliance will negatively affect the energy restitution and also the dynamic performance, e.g., for hopping the hopping height and duty factor. Thus, applying a kind of "cascaded optimization" to optimize, first, the actuation with respect to energy restitution and other selected performance criteria such as hopping height and ground clearance and, subsequently, to optimize the controller gains to keep the torques of the motors as low as possible is an essential requirement for an optimal use of the leg actuation design [23]. Open-loop controlled motions revealed that omitting a careful selection of all regulating parameters of the design space, i.e., rest lengths, attachment points, spring stiffness, may lead to timing issues of the tendons interfering with each other. The simulation results also showed that actuated biarticular tendons can further reduce the complexity of the leg actuation design and enhance energy savings, while preserving the desired dynamic locomotion behavior.

Demonstrating the importance, an earlier insight from biomechanics, known as the Lombard paradox, was rediscovered and explored using detailed multibody system dynamics simulations [25]. According to the Lombard paradox, biarticular muscles have even more sophisticated functionalities than usually assumed. They are responsible for additional actions during dynamic locomotion. For example, the Gastrocnemius muscle is not only responsible for flexing the knee joint. During the last part of ground contact phase during sprinting, it also acts as synergist extending the knee joint at angles above a specific flexion degree [35]. Such muscle action, labeled by Lombard in 1903 the "paradoxical" function of biarticular muscles, was also observed to be true for the hamstrings. Applying this paradox, a novel bipedal locomotion model could be established that is capable of dynamic hopping motions without the need of a knee motor, leading to energy savings of more than 60 %. In an earlier work it was suggested that an active knee is not required for level-ground walking [7].

In summary, these findings are encouraging for advancing musculoskeletal robot designs with enhanced locomotion capabilities. By subsequently refining the robot's design and control, biomechanical concepts can be demonstrated, validated technical legged systems and new insights (e.g. hidden paradoxical findings) can be revealed.

11.7 Assessment of Locomotor Function in Biomechanics and Robotics

The development of proper conceptual models for human locomotion is key in separating underlying task-specific subfunctions required to achieve stable gaits. For legged locomotion, three functional requirements need to be fulfilled [31], as shown in Table 11.2.

Table 11.2 Functional requirements for legged locomotion.

	Repulsive leg function	Body balance	Leg swing
Purpose	Counteract gravity by rebounding body vertically during stance	Counteract gravity by aligning body axis vertically	Position swing leg for next touch-down
Underlying template	Virtual inverted elastic pendulum (SLIP)	Virtual pendulum (P)	Virtual elastic pendulum (SLP)
Key properties	Leg stiffness, leg length, leg angle	Virtual center of rotation (VPP position)	Pendulum length, leg stiffness
Parameters for assessment	External work on CoM, leg length, energy, power, elastic capacity, leg lengthening	Internal work on body pitch, body angular momentum, pitch excursions	Internal work on leg swing, leg swing amplitude, swing leg shortening

These subfunctions required for legged locomotion can be used as a basis for a more functional description of locomotor function in humans, animals, and robotic systems. This is extremely helpful as the design and the ways of actuation as well as the materials of these systems may be quite different. With this matrix of functional requirements the identification of deficits in current legged robotic systems in comparison with their biological counterparts could be largely facilitated. It remains for further research to identify which of these elements are most important for comparison of different locomotor systems. Also, the list of the task-specific subfunctions might be incomplete which would require an extension of the conceptual models.

Aside from this biomechanical approach, it is crucial to develop and apply measures to rate key criteria of locomotion performance across different models, i.e., simulation models of motion dynamics of humans or robots. This goal can be achieved within two steps and represents a milestone for developing generally applicable benchmarks to foster the progress in the robotics community.

The first task is to specify essential aspects of human locomotion to be transferred to robot systems by mathematical models. These aspects are expected to be partially complementary and competitive criteria to each other (such as locomotion speed and energy consumption). A comprehensive catalog of locomotion performance criteria for various gaits should include preferably dimensionless criteria from relevant categories: (1) energy-efficiency (e.g., mechanical and electrical energy consumption, energy restitution ratio [23], kinetic and potential energy fluctuations), (2) dynamic mobility [23] (e.g., altitude difference of the center of mass (CoM), duty factor, speed), (3) control efforts (i.e., proportion of feedback versus feedforward control influence, sensory information resolution and processing speed), (4) postural stability and (5) robustness against disturbances.

It is hypothesized that by validating the three above suggested biomechanically motivated functional requirements for legged locomotion (Table 11.2) as well as the application of an aggregation of mathematical models of the aforementioned categories will help to better understand biomechanics of locomotion and to use these insights to advance the design and motion dynamics of technical legged systems toward human-like locomotion in appearance and performance.

11.8 Outlook

Although compliant leg function is an obvious key feature of human and animal locomotion such as walking, running or jumping it is still not sufficiently understood to directly transfer it to legged robots or leg prostheses with similar motion performance. Originally attributed mainly to the axial leg function, it became clear that also non-axial force contributions are shaped by elastic components. The origin of elastic leg function can be found in the design of the muscle-tendon units, in compliant structures (e.g. ligaments) in the human body. It requires an appropriate neural control to result in the required muscle activation patterns for elastic, repulsive operation of the leg. The leg function cannot be simply implemented by compliant structures as they cannot respond to unexpected changes in the environment (e.g. uneven ground, slopes, pushes) or in the body adequately (e.g. changed body mass when carrying weights).

In order to be able to achieve versatile compliant leg function in a variety of tasks and conditions, a careful design of body, actuator and control properties arranged in a segmented body is required. Here, the underlying mechanisms are still largely unclear and require further research, e.g. regarding following aspects:

What kind of muscle properties needs to be implemented in a technical system (e.g. serial/parallel compliance, damping, activation dynamics) in order to realize comparably efficient and versatile tasks?

How can complex scenarios of motion and interaction with multiple contacts be realized (e.g. with hands and feet contacts)?

What is different between leg function and arm function regarding their motor function capabilities?

How do biomechanical templates relate to biological control concepts? Are there also neuromuscular control templates matching the biomechanical templates?

Do we need new kind of soft materials and actuator properties to mimic human-like locomotion in technical systems? What are proper models for human- or animal-like motion performance?

Currently, a new technology of 3D-printed elastic materials is evolving, e.g. for designing custom-made shoe insoles (www.rsprint.be), orthoses (Ekso Bionics) or prostheses. These technologies need be further developed to achieve adjustable compliance also during operation (e.g. in response to changed environments or subject conditions).

Acknowledgments This research has been supported by the German Research Foundation (DFG) under grants SE1042/6 and STR 533/7.

11.9 References

[1] Alexander RM (1976) Mechanics of bipedal locomotion. Perspectives in experimental biology, Oxford, UK Pergamon Press, pp 493-504

[2] Blickhan R (1989) The spring-mass model for running and hopping. Journal of Biomechanics 22:1217-1227

[3] Chung W, Fu LC, Hsu SH (2008) Motion Control. Chapter 6 of Springer Handbook of Robotics. Ed. by B. Siciliano, O. Khatib, pp 133-159

[4] Damsgaard M, Rasmussen J, Christensen ST, Surma E, de Zee M (2006) Analysis of musculoskeletal systems in the AnyBody modeling system. Simulation Modelling Practice and Theory 14:1100-1111

[5] Delp SL, Anderson FC, Arnold AS, Loan P, Habib A, John CT, Guendelman E, Thelen DG (2007) OpenSim: Open-Source software to create and analyze dynamic simulations of movement. IEEE Transactions on Biomedical Engineering 54(11):1940-1950

[6] Desai R, Geyer H (2012) Robust swing leg placement under large disturbances. IEEE Intl. Conf. on Robotics and Biomimetics (ROBIO) pp 265-270

[7] Endo K, Herr H (2009) A model of muscle-tendon function in human walking. IEEE International Conference on Robotics and Automation (ICRA) pp 1909-1915

[8] Farley CT, Glasheen J, McMahon TA (1993) Running springs: speed and animal size. Journal of Experimental Biology 185(1):71-86

[9] Farley CT, Gonzalez O (1996) Leg stiffness and stride frequency in human running. Journal of biomechanics, 29(2), 181-186

[10] Farley CT, Morgenroth DC (1999) Leg stiffness primarily depends on ankle stiffness during human hopping. Journal of biomechanics, 32(3):267-273

[11] Full R, Koditschek D (1999) Templates and anchors: neuromechanical hypotheses of legged locomotion on land. Journal of Experimental Biology 202:3325-3332

[12] Geyer H, Seyfarth A, Blickhan R (2003) Positive force feedback in bouncing gaits?. Proceedings of the Royal Society of London Series B: Biological Sciences 270(1529):2173-2183

[13] Haeufle DFB, Grimmer S, Seyfarth A (2010) The role of intrinsic muscle properties for stable hopping—stability is achieved by the force-velocity relation. Bioinspiration and Biomimetics 5(1):016004

[14] Haeufle DFB, Grimmer S, Kalveram KT, Seyfarth A (2012) Integration of intrinsic muscle properties, feed-forward and feedback signals for generating and stabilizing hopping. Journal of The Royal Society Interface 9(72):1458-1469

[15] Herr HM, McMahon TA (2000) A trotting horse model. The International Journal of Robotics Research 19(6):566-581

[16] Herr HM, Huang GT, McMahon TA (2002) A model of scale effects in mammalian quadrupedal running. Journal of Experimental Biology 205(7):959-967

[17] Kajita, S., Espiau, B. (2008) Legged Robots. Chapter 16 of Springer Handbook of Robotics. Ed. by B. Siciliano, O. Khatib, pp 361-389

[18] Kubo K, Kanehisa H, Takeshita D, Kawakami Y, Fukashiro S, Fukunaga T (2000) In vivo dynamics of human medial gastrocnemius muscle-tendon complex during stretch-shortening cycle exercise. Acta Physiologica Scandinavica 170(2):127-135

[19] Lens T, Radkhah K, von Stryk O (2011) Simulation of dynamics and realistic contact forces for manipulators and legged robots with high joint elasticity. International Conference on Advanced Robotics (ICAR) pp 34–41

[20] Lipfert SW, Günther M, Renjewski D, Grimmer S, Seyfarth A (2012) A model-experiment comparison of system dynamics for human walking and running. Journal of Theoretical Biology 292:11-17

[21] Maus H M, Lipfert SW, Gross M, Rummel J, Seyfarth A (2010) Upright human gait did not provide a major mechanical challenge for our ancestors. Nature Communications 1:70

[22] McMahon TA, Cheng GC (1990) The mechanics of running: how does stiffness couple with speed?. Journal of Biomechanics 23:65-78

[23] Radkhah K (2013) Advancing Musculoskeletal Robot Design for Dynamic and Energy-Efficient Bipedal Locomotion. PhD Thesis, TU Darmstadt, CS Dept.

[24] Radkhah K, Lens T, von Stryk O (2012) Detailed dynamics modeling of BioBiped's monoarticular and biarticular tendon-driven actuation system. IEEE/RSJ Intl. Conf. on Intelligent Robots and Systems (IROS) pp 4243-4250

[25] Radkhah K, von Stryk O (2013) Exploring the Lombard paradox in a bipedal musculoskeletal robot. International Conference on Climbing and Walking Robots and the Support Technologies for Mobile Machines (CLAWAR) pp 537-546

[26] Radkhah K, von Stryk O (2014) A study of the passive rebound behavior of bipedal robots with stiff and different types of elastic actuation. IEEE International Conference on Robotics and Automation (ICRA) pp 5095-5102

[27] Rode C, Seyfarth A (2013) Balance control is simplified by musculoskeletal leg design. Dynamic Walking conference

[28] Scheinman V, McCarthy JM (2008) Mechanisms and Actuation. Chapter 3 of Springer Handbook of Robotics. Ed. by B. Siciliano, O. Khatib, pp 67-86

[29] Seyfarth A, Friedrichs A, Wank V, Blickhan R (1999) Dynamics of the long jump. Journal of Biomechanics 32(12):1259-1267

[30] Seyfarth A, Blickhan R, Van Leeuwen JL (2000) Optimum take-off techniques and muscle design for long jump. Journal of Experimental Biology 203(4):741-750

[31] Seyfarth A, Grimmer S, Häufle D, Kalveram KT (2012) Can robots help to understand human locomotion? at - Automatisierungstechnik 60(11):653-660

[32] Song H, Park H, Park S (2014) Swing leg kinetics can be described by springy-pendulum in human walking, Dynamic Walking conference

[33] Sreenath K, Park HW, Poulakakis I, Grizzle JW (2011) A compliant hybrid zero dynamics controller for stable, efficient and fast bipedal walking on MABEL. The International Journal of Robotics Research 30(9):1170-1193

[34] Villani L, De Schutter J (2008) Force Control. Chapter 7 of Springer Handbook of Robotics. Ed. by B. Siciliano, O. Khatib, pp 161-185

[35] Wiemann K, Tidow G (1995) Relative activity of hip and knee extensors in sprinting—implications for training. New studies in Athletics 1 10(29–49)

[36] Loram, I.D., Maganaris, C.N., Lakie, M. (2004) Paradoxical muscle movement in human standing. The Journal of Physiology, vol. 556, pp 683-689

[37] Vanderborght, B. et al. (2013) Variable impedance actuators: A review. Robotics and Autonomous Systems, vol. 61, pp 1601–1614

[38] Moro, F.L., Tsagarakis, N.G., Caldwell, D.G. (2014) Walking in the resonance with the COMAN robot with trajectories based on human kinematic motion primitives (kMPs). Autonomous Robots, vol. 36, no. 4, pp 331-347

[39] Radkhah, K., Maufroy, C., Maus, M., Scholz, D., Seyfarth, A., von Stryk, O. (2011) Concept and design of the BioBiped1 robot for human-like walking and running. International Journal of Humanoid Robotics, Vol. 8, No. 3, pp. 439-458

12 Mechanics and Thermodynamics of Biological Muscle - A Simple Model Approach

Syn Schmitt and Daniel Haeufle

University of Stuttgart

Abstract Macroscopic muscle models allow for a detailed analysis of the mechanic and thermodynamic function of biological muscles. Here we summarize results from various simulation studies which emphasize the extraordinary design features of biological muscles. Discussed are the benefits resulting from (1) wobbling masses and the muscles soft-tissue inertia effects, (2) biological damping, (3) internal mass distribution, (4) stabilising properties of active muscles in upright stance and periodic hopping, (5) reduced control effort due to these stabilising effects. We present approaches to systematically transfer these results to technical actuators and exploit these properties in the next generation of functional artificial muscles.

12.1 The Biological Muscle Drives the Animal Motion

Biological movement in the animal kingdom is driven by skeletal muscle, a biological soft actuator. A glance at the complexity and variety of the generated movements shows that muscle is a versatile, powerful, and flexible actuator [1]. This is achieved because muscle can operate in different modes depending on the contraction dynamics and the structural implementation [2, 3].

Research in muscle biomechanics is a vital and broad field for far more than a century now (see e.g [4]). The work of A.V. Hill in the first quarter of the 20th century is still the basis for many muscle models, as he described the macroscopic function and design of real biological muscles [5]. The key to success is that, at least, two typical muscle properties, the force-velocity and the force-length relationship, can be found in a broad range of species (rat, cat, frog, human) [6]. One step further, by using established muscle models at hand, it was shown that the mechanical efficiency and the thermodynamic enthalpy rate add to the typical properties of biological muscle, as well. Nevertheless, there still remains much to be understood when it comes to structure and function of the biological muscle.

A completely different approach to animal movement was published in 1990, when the concept of passive dynamic walking was introduced [7]. Human-like walking turned out to be producible by a purely passive approach. Solely a small decline in the walking terrain accounted for a constant energy supply of the walking machine, the so-called 'gravity walker'. In the same reductionistic approach, but in theory, the spring-mass model for hopping and running [8] and for walking [9] were introduced. Both spring-mass models, again, use a virtual constant energy supply in the form of constant horizontal velocity, to maintain continuous motion, as does the gravity walker.

In real life, however, the biological locomotion system obtains its movement energy through the skeletal muscles. In that sense, the muscle can be considered to be at the heart of biological motion. Thus, understanding the biological muscle's function and design is a prerequisite to understanding biological motion.

12.2 The Biological Muscle's Various Design Features

Detailed analysis of mechanical muscle properties, contraction dynamics, structural arrangement, and movement generation reveal, that the muscle is a soft, highly non-linear, tunable and variable actuator with many obscure, and at first sight, cumbersome properties in comparison to technical actuators. In this section we highlight some of these properties and demonstrate that these are in fact design features.

12.2.1 The Biological Muscle's Passive Mechanic Characteristics

In terrestrial locomotion, high accelerations on the body occur, for example, induced by legs impacting with the ground. Soft-tissue masses of the body undergo damped oscillations and shock waves travel through the body after ground impact.

Wobbling mass models: Appropriate biomechanical models to account for highly dynamical movements describe gross soft-tissue dynamics by "wobbling masses". Fig. 12.1 shows the calculated mechanical energy balances of shank and thigh wobbling masses of the stance leg for the first 90 ms after touch-down in human heel-toe running [11].

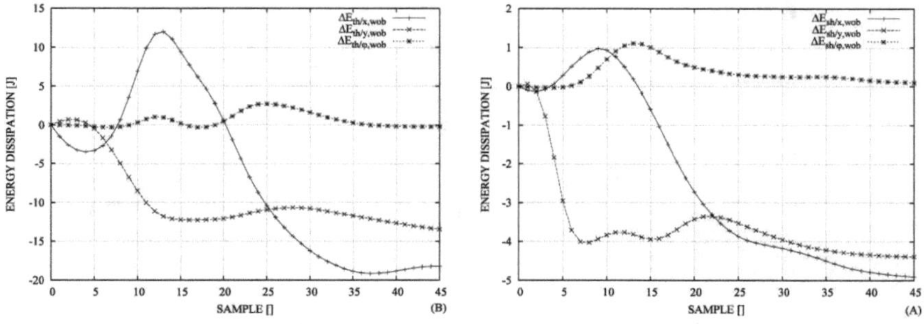

Fig. 12.1 The calculated mechanical energy balances of thigh (left side) and shank (on the righthand side) wobbling masses of the stance leg for the first 90 ms after touch-down in human heel-toe running. Figures courtesy of [11].

The wobbling mass kinematics were non-invasively obtained by acquiring the motion of grids of lines painted on the skin of the corresponding muscle masses with high-speed cameras [10]. Despite the overall energy loss per step cycle, wobbling masses are very special design features enabling fast movement. Also, they potentially allow for a faster signal traveling of the instant of impact to the head – faster than the nervous system would allow a sensor signal to travel up to the brain.

Biological damping: High-frequency vibrations can provoke damage within tendons eventually even leading to rupture. Low but significant damping of the passive material in series to the contractile machinery may well suffice to damp these hazardous vibrations. Fig 12.2 shows a modified Hill-type model and the comparison of three cases: no additional damping, parallel damping, and serial damping. The study revealed that serial damping at a physiological magnitude suffices to explain damping of high-frequency vibrations of low amplitudes [12]. Delocalised damping spread all over the whole muscle seems to be a design feature to suppress harmful oscillations.

Mass distribution: It is state of the art that muscle contraction dynamics is adequately described by a hyperbolic relation between muscle force and contraction velocity (Hill relation), thereby neglecting muscle internal mass and inertia (first-order dynamics). Such first-order contraction dynamics, however, interacts with muscle internal mass distribution and the resulting inertia effects (Fig. 12.3). It may be indispensable in the future to introduce second-order contributions into muscle models to understand high-frequency muscle responses, particularly in bigger muscles. By that, muscular contraction could be better understood, locally within the muscle and of the muscle as a whole, in response to a variety of realistic acceleration scenarios such as impacts on the ground. It is to be expected, that the mass distribution within a muscle is a design feature.

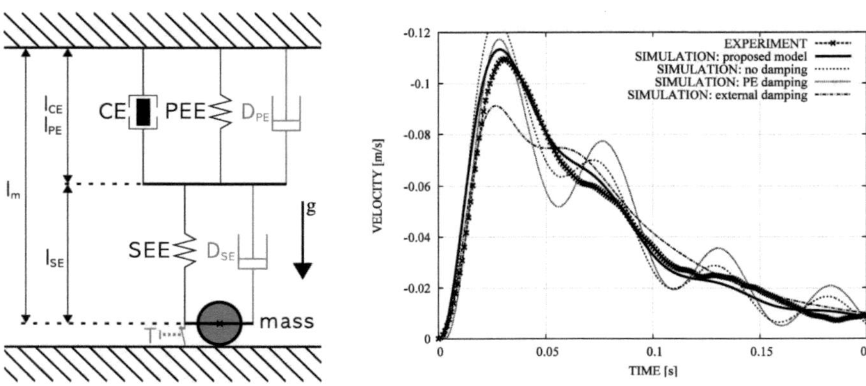

Fig. 12.2 A modified Hill-type model and the comparison of three cases: no additional damping, parallel damping, and serial damping. The study revealed that serial damping (DSE) at a physiological magnitude suffices to explain damping of high-frequency vibrations of low amplitudes. Figure courtesy of [12].

12.2.2 Active Muscle and Stability

The distinct mechanical and thermodynamical properties of the biological muscle simplify motor control. This effect was demonstrated in several simulation studies [14, 15, 16, 17]. It appears that muscle properties are able to compensate for perturbations and facilitate the convergence of dynamic explosive or repetitive cyclic movements. The authors concluded that the intrinsic muscle properties, represented by the force-length-velocity relation, act as a zero time delay peripheral feedback system, termed "preflex" [18]. These muscle properties are a design feature of biological muscles.

Stability of upright stance: In a simulation study of a two-segment inverted pendulum model (Fig. 12.4, AB), it could be demonstrated, that a muscle-driven system outperforms a system driven by a direct torque generator with linear characteristics [19]. Fig 12.4 on the right depicts the results with α: Model reactions to perturbations in foot orientation and β=0°: Control target (upright posture). The left column shows the reaction to a ramp perturbation, the middle column to a 1 Hz, and the right column to a 0.1 Hz sinusoidal perturbation. The top row shows the results without feedback, middle row with a simple P controller (direct torque controller gain: P 500; muscle controller gain: P 1), and bottom row with a PID controller (direct torque controller gains: P 500, I 50, D 500; muscle controller gains: P 1, I 0.3, D 0.3). In all, the muscle-driven model did not even topple without feedback (Fig. 12.4, first row on the right). Also, when using the simple P controller (middle row) the muscle-driven model performed better during all perturbations and was able to cope better with a feedback delay of Δt=0.1 s.

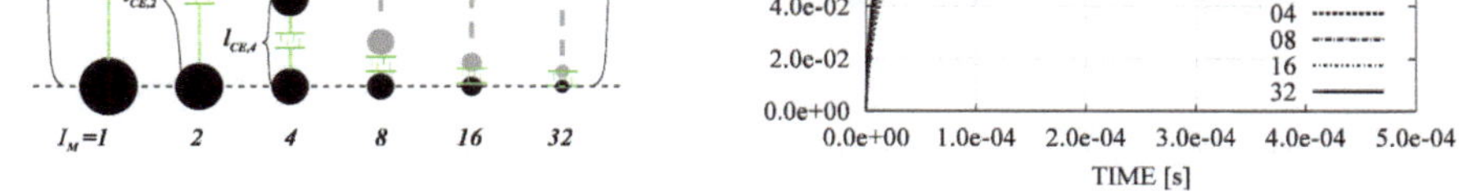

Fig. 12.3 Simulated accelerated contractions of alternating sequences of Hill-type contractile elements and point masses (left). It was found that in a typical small muscle the force levels off after about 0.2 ms, contraction velocity after about 0.5 ms (right). Figures courtesy of [13].

Stability in hopping: Further investigations of the muscle contribution to the stabilization of periodic hopping movements with respect to perturbations in landing height were accomplished [20, 21]. For this purpose, periodic hopping movements were simulated and the complexity of the force-length and force-velocity curves was stepwise increased from constant (=1, thus has no effect) to linear to non-linear phenomenological fit of the biological data. The stability of the hopping patterns was analyzed using return maps (Poincaré map) of the apex. With this approach it was possible to show that the force-velocity relation is responsible for the stabilization of periodic hopping patterns. A faster convergence after perturbations with increasing complexity of the force-velocity relation was found. The characteristics of the force-length relation only marginally influenced hopping stability.

Quantifying control effort of biological movement: If the muscle properties allow simple control and intrinsically increase stability, one would expect a reduced neural load on the system, i.e., less control effort. Previous definitions of control effort in engineering were based on system specific values e.g., output signal voltage of a controller [22], motor armature voltage [23], motor torque [24], actuation voltage in polymer actuators [25], or pressure in pneumatic actuators [26]. In some studies, control effort was also associated with muscle activation [27, 28] or muscle electromyography (EMG) signal [29]. If control effort is measured in voltage, current, pressure, and muscle activation, different actuation principles can hardly be quantitatively compared.

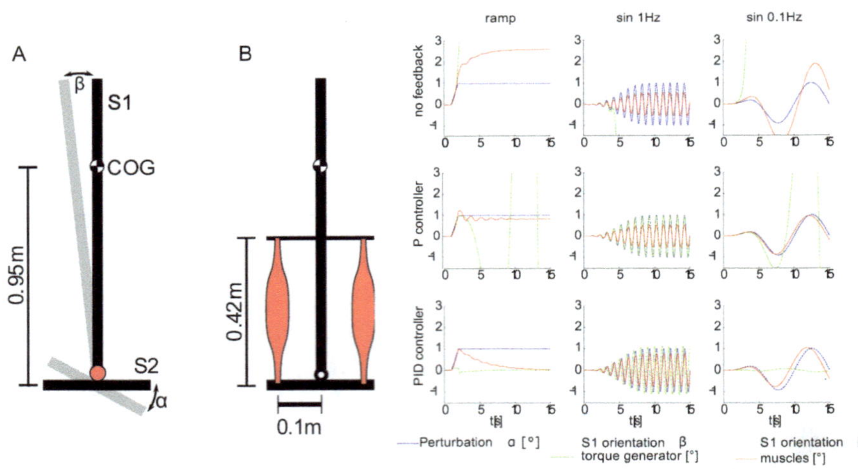

Fig. 12.4 Left: Model of the inverted pendulum. S1 represents the leg-trunk segment, S2 represents the foot. COG indicates the center of gravity location of S1. α is the angle of the foot (perturbation) and β the deviation from the upright position of S2. A: the joint is actuated by a direct torque generator with linear characteristics. B: the joint is actuated by two antagonistic muscles. Right: Model reactions to perturbations in foot orientation α. Figures courtesy of [19].

To allow such a comparison, we developed a new measure for control effort, applicable to and comparable across completely different actuators and control approaches [17]. In a nutshell, we propose to quantify the minimal sensor information required to control a certain movement. From Shannon's information entropy [30], we derived the information provided by a sensor. It depends on the duration of the movement T, the time resolution Δt, and the sensor signal limits $u_{max}-u_{min}$ and resolution Δu:

$$I = \frac{T}{\Delta t} log_2 \left(1 + \frac{u_{max} - u_{min}}{\Delta u}\right) \tag{1}$$

This simple measure can be applied to almost any type of sensor and requires only a discretized sensor output and discrete repeated measurements. By varying the resolutions, one can find the minimally required information, which we identify with control effort.

The novelty of our approach is to measure the information entropy of the movement control process as a function of physical structure and control method:

$$I_{movement} = I_{movement}(structure, control) \tag{2}$$

By varying the (bio-)mechanical structure of the actuator in the hopping simulations mentioned above, we could show that a muscle driven by a simple reflex control scheme requires an order of magnitude less control effort compared to a DC-motor and a proportional differential (PD) controller (I=32bits vs. I=660bits). These first examples indicate the enormous benefit a biomechanical system can gain from the specific properties of its actuators. This work has to be extended to more complex systems, e.g. walking simulations, in the future.

12.2.3 Mechanical Efficiency and Thermodynamic Enthalpy Rate

As the first and often only criterion for verifying biology-like properties of muscle, its isotonic force-velocity relation is investigated. This design feature is common to all muscle models in biomechanics, defined by the so-called Hill "constants" A and B, the two asymptotes of the concentric Hill relation. Albeit, an obvious second criterion: the muscle's mechanical efficiency, and a third criterion: the muscle's enthalpy rate, have been shown to discriminate between different muscle model approaches, more accurate. In terms of biological validity, the second and the third criterion verify the muscle model's biology-like properties far better, than an evaluation using only the force-velocity relation would do [31, 32]. Fig 12.5 depicts a comparison between different model predictions and experimental data of the mechanical efficiency and the enthalpy rate during muscle contraction.

12.3 Designing a Technical Actuator from the Biological Prototype

In prosthetics and bio-inspired robotics it would be desirable to have an artificial actuator with the design features of the biological muscle [36, 6]. On the technical side, actuators for machines are very well known since a long time, for example, the first electro-magnetic actuators are known since 1834 [37]. These classical actuators, however, fail to reproduce the passive characteristics of biological muscles [38]. Technical actuators yet including bio-inspired features, for example, elasticity, fail to incorporate one or more of typical biological muscle characteristics: energy density or power to weight ratio, passive characteristics, scalability, contraction paths, controllability [39, 40, 36]. Nevertheless, technical, bio-inspired actuators also known as artificial muscles are constantly developing [41, 42, 36, 43] and the construction of artificial muscles, nowadays, is one of the most challenging developments in biomedical science [44, 40].

Currently, the development of artificial muscles is driven by technological leaps. To actually reproduce the key design features of biological muscle, we propose to develop artificial muscles based on bio-mechanical muscle models. Microscopic muscle models are able to explain some characteristics of biological muscle on a molecular level [45, 46, 47, 48]. However, such molecular-based actuators are not likely to be constructible by technology and scalable to animal- or human-like size in the near future.

On the other hand, macroscopic muscle models represent the entire actuator. The key problem here, so far, was that their contraction dynamics were based on phenomenology, i.e. force dependencies such as the force-length or the force-velocity relation are adopted as best-fit functions to experimental data. In a new model approach [31, 49], the force-velocity relation – a key design feature of biological muscle – was traced back to a simple mechanical structure (Fig. 12.6). It was shown that this simple structure exhibits operating points with a hyperbolic force-velocity dependency (Fig. 12.7). Based on this concept, we developed a numerical model of an actuator and a technical proof of concept, which demonstrated its real world functionality (Fig. 12.6 right) [49, 50].

In this approach, the force-velocity relation is no longer a phenomenological outcome of a black box (i.e. the CE) but rather a physical outcome of the interaction of three elements AE, PDE, and SE. Therefore, this model can be interpreted as a basic engineering design template for the CE of a Hill-type artificial actuator.

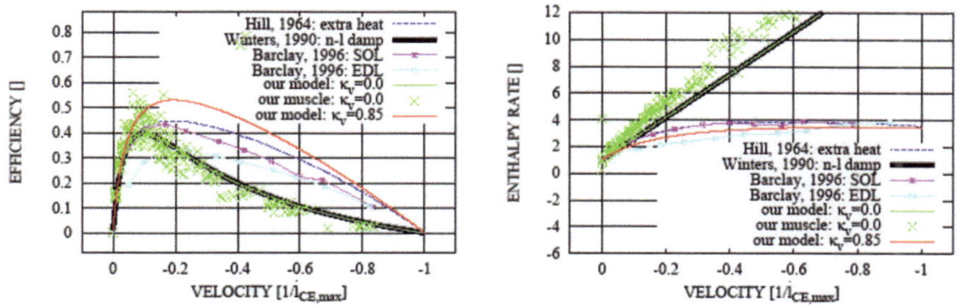

Fig. 12.5 Comparison of mechanical efficiencies and enthalpy rates. Experimental data (Barclay, 1996: SOL: M. soleus and EDL: M. extensor digitorum longus) [33] are compared to those predicted by our model, choosing κ_V=0.85. Additional data plotted are from isotonic experiments of an artificial muscle ("our muscle", see section 12.3), from our model case κ_V=0.0 , and from Hill's refined fit to his measurements [35] (compare also [31]). Figures courtesy of [32].

12.4 Next Generation of Bio-inspired and Bio-like Actuators

In industry robotics, high accuracy and repeatability are required for precise and reliable assembly lines. This is achieved through stiff systems with high performance and fast control architecture. For this purpose, actuators are required to be stiff and linear (in the sense of input-output). Drawbacks are the resulting danger if they would interact with humans and unpredictable environments and the opposite design principles compared to the elastic biological system. To mimic the biological elasticity in the drive, fast torque controlled robots were equipped with implicit elastic behavior through control [52]. In theory, the whole muscle function could be mimicked just by controlling the actuator force according to a muscle model [36, 6, 53]. Problems arise, however, from the limitations of sensor and actuator precision, control loop time delays, and sampling time [52].

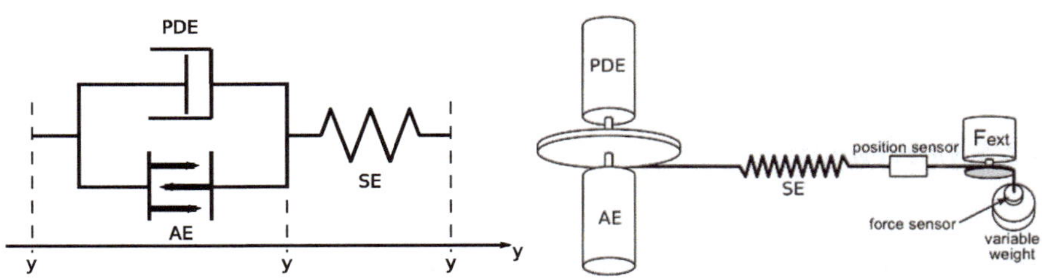

Fig. 12.6 Left: the dynamic properties of the contractile fibers in a muscle can be reproduced by this arrangement of mechanical structures. Right: the real world functionality was proven in this technical setup, where current controlled CD motors generated the force of the AE, PDE, and optionally a constant external force. For more details see [51, 50, 32].

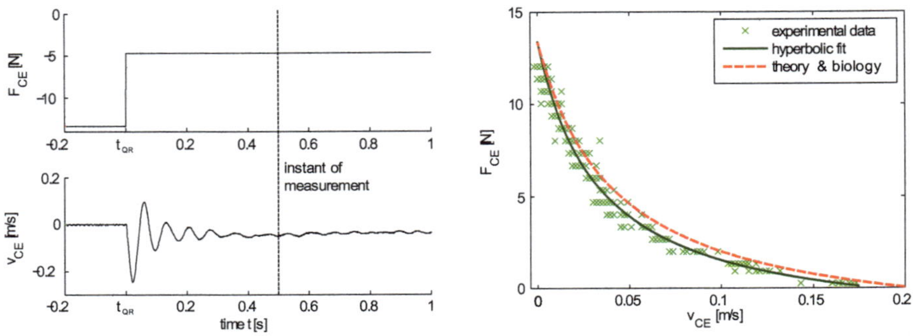

Fig. 12.7 Predicted and experimentally measured force-velocity relation. Left: Time trace of quick release experiments against a constant external force (isotonic). Force and velocity were measured 0.5s after t_{QR} and plotted in the force-velocity graph (right). Figure courtesy of [50, 32].

From a biological and control effort point of view, it is clear, that the muscle properties have to be explicitly built into the actuator. This has been studied and applied extensively for the series elasticity [38, 52, 54, 55] and results in shock tolerance, lower reflected inertia, more accurate and stable force control, less damage to the environment, and energy storage capabilities, very similar to the biological system [38].

The biological damping characteristics of the passive distributed viscosity of the actuator and the active damping dynamics of the force-velocity relation, however, have so far rarely been implemented. Of course, they are similar to the controlled damping in standard PD-controllers, but again, have to be implemented as structural properties. First approaches in this direction seem to be very promising [56].

As good as this progress is, the knowledge transfer from biology to technology has to be continued. From our point of view, the next generation of bio-inspired and bio-like actuators has to incorporate explicit damping properties. We propose for this purpose a muscle model which accounts both for the mechanical as well as for the thermodynamical characteristics of biological muscle [50, 32].

12.5 References

[1] T.A. McMahon, Muscles, reflexes, and locomotion (Princeton University Press, 1984)
[2] C.N. Maganaris, Acta Physiol Scand 172(4), 279 (2001)
[3] A.N. Ahn, R.J. Full, Journal of Experimental Biology 205(3), 379 (2002)
[4] R. Heidenhain, Mechanische Leistung, Wärmeentwicklung und Stoffumsatz bei der Muskelthätigkeit (Breitkopf und Härtel, Leipzig, 1864)
[5] A.V. Hill, Proceedings of the Royal Society of London. Series B 126(843), 136 (1938)
[6] G.K. Klute, J.M. Czerniecki, B. Hannaford, The International Journal of Robotics Research 21(4), 295 (2002)
[7] T. McGeer, The International Journal of Robotics Research 9(2), 62 (1990)

[8] R. Blickhan, Journal of Biomechanics 22(11/12), 1217 (1989)
[9] H. Geyer, A. Seyfarth, R. Blickhan, Proceedings of the Royal Society B: Biological Sciences 273(1603), 2861 (2006)
[10] M. Günther, V. Sholukha, D. Keßler, V. Wank, R. Blickhan, Journal of Mechanics in Medicine and Biology 3(3/4), 309 (2003)
[11] S. Schmitt, M. Günther, Archive of Applied Mechanics 81(7), 887 (2011)
[12] M. Günther, S. Schmitt, V. Wank, Biological Cybernetics 97(1), 63 (2007)
[13] M. Günther, O. Röhrle, D.F.B. Haeufle, S. Schmitt, Comput Math Methods Med 2012, 848630 (2012)
[14] A.J. van Soest, M.F. Bobbert, Biological Cybernetics 69(3), 195 (1993)
[15] M.M. van der Krogt, W.W. de Graaf, C.T. Farley, C.T. Moritz, L.J.R. Casius, M.F. Bobbert, Journal of Applied Physiology 107(3), 801 (2009)
[16] H. Geyer, A. Seyfarth, R. Blickhan, Proceedings of the Royal Society of London. Series B, 270(1529), 2173 (2003)
[17] D.F.B. Haeufle, M. Günther, G. Wunner, S. Schmitt, Phys Rev E 89(1), 012716 (2014)
[18] I.E. Brown, S.H. Scott, G.E. Loeb, Society of Neuroscience, Abstracts 21, 562 (1995)
[19] S. Schmitt, M. Günther, T. Rupp, A. Bayer, D. Häufle, Comput Math Methods Med 2013, 570878 (2013)
[20] D.F.B. Häufle, S. Grimmer, A. Seyfarth, Bioinspiration & Biomimetics 5(1), 016004 (2010)
[21] D.F.B. Haeufle, S. Grimmer, K.T. Kalveram, A. Seyfarth, Journal of the Royal Society, Interface 9(72), 1458 (2012)
[22] K. Goher, M. Tokhi, in Proceedings of the 22nd European Conference on Modelling and Simulation, vol. 5 (2005), vol. 5, pp. 3–6
[23] R.J. Wai, IEEE Transactions on Industrial Electronics 53(4), 1328 (2006)
[24] J.E. Bobrow, B. Martin, G. Sohl, E.C. Wang, F.C. Park, J. Kim, Journal of Robotic Systems 18(12), 785 (2001)
[25] Y. Fang, X. Tan, G. Alici, IEEE Transactions on Control Systems Technology 16(4), 600 (2008)
[26] J. Lilly, P. Quesada, IEEE Transactions on Neural Systems and Rehabilitation Engineering 12(3), 349 (2004)
[27] A.D. Kuo, IEEE Transactions on Biomedical Engineering 42(1), 87 (1995)
[28] A.C. Schouten, E. de Vlugt, F.C.T. van der Helm, G.G. Brouwn, Biological Cybernetics 84(2), 143 (2001)
[29] D.B. Lockhart, L.H. Ting, Nature Neuroscience 10(10), 1329 (2007)
[30] C.E. Shannon, Bell System Technical Journal, reprint with corrections 27(7,10), 379 (1948)
[31] M. Günther, S. Schmitt, Journal of Theoretical Biology 263(4), 407 (2010)
[32] S. Schmitt, D.F.B. Haeufle, R. Blickhan, M. Günther, Bioinspir Biomim 7(3), 036022 (2012)
[33] C. Barclay, The Journal of Physiology 497(Pt 3), 781 (1996)
[34] J.M. Winters, Multiple muscle systems: biomechanics and movement organization (Springer-Verlag Berlin and Heidelberg GmbH & Co. Kg., 1990), chap. Hill-based muscle models: a systems engineering perspective, pp. 69–93
[35] A. Hill, Proceedings of the Royal Society of London B 159, 1297 (1964)
[36] B. Hannaford, K. Jaax, G. Klute, Autonomous Robots 11(3), 267 (2001)
[37] M. von Jacobi, Mémoire sur l'application de l'électromagnetisme au mouvement des machines. Tech. rep., Potsdam (1835)
[38] G.A. Pratt, M.M. Williamson, Proceedings 1995 IEEE/RSJ International Conference on Intelligent Robots and Systems. Human Robot Interaction and Cooperative Robots pp. 399–406 (1995)
[39] J.D. Madden, Science 318(5853), 1094 (2007)
[40] R. Baughman, Science (New York, NY) 308(5718), 63 (2005)

[41] H.F. Schulte, in The application of external power in prosthetics and orthotics (Publication 874 of the National Academy of Sciences, 1961), pp. 94–115
[42] D. Caldwell, G. Medrano-Cerda, M. Goodwin, Control Systems, IEEE 15(1), 40 (1995)
[43] Y. Bar-Cohen, in Proceedings of the SPIE's 6th Annual International Symposium on Smart Structures and Materials, vol. 3669, ed. by Y. Bar-Cohen (1999), vol. 3669, pp. 1–414
[44] G. Klute, J. Czerniecki, B. Hannaford, Advanced Intelligent Mechatronics, 1999. Proceedings. 1999 IEEE/ASME International Conference on pp. 221–226 (1999)
[45] V. Lombardi, G. Piazzesi, M. Ferenczi, H. Thirlwell, I. Dobbie, M. Irving, Nature 374(6522), 553 (1995)
[46] M. Reconditi, M. Linari, L. Lucii, A. Stewart, Y. Sun, P. Boesecke, T. Narayanan, R. Fischetti, T. Irving, G. Piazzesi, M. Irving, V. Lombardi, Nature 428(6982), 578 (2004)
[47] I. Telley, J. Denoth, K. Ranatunga, Advances in Experimental Medicine and Biology 538, 481 (2003)
[48] H. Huxley, European Journal of Biochemistry 271(8), 1405 (2004)
[49] D.F.B. Häufle, M. Günther, R. Blickhan, S. Schmitt, Rehabilitation Robotics (ICORR), 2011 IEEE International Conference on pp. 1–6 (June 29 2011-July 1 2011)
[50] D.F.B. Häufle, M. Günther, R. Blickhan, S. Schmitt, Applied Bionics and Biomechanics 9(3), 276 (2012)
[51] D.F.B. Häufle, M. Günther, R. Blickhan, S. Schmitt, Journal of Bionic Engineering 9(2), 211 (2012)
[52] A. Albu-Schäffer, O. Eiberger, M. Grebenstein, S. Haddadin, C. Ott, T. Wimbock, S. Wolf, G. Hirzinger, IEEE Robotics & Automation Magazine 15(3), 20 (2008)
[53] A. Seyfarth, K.T. Kalveram, H. Geyer, in Proceedings of Fachgespräche Autonome Mobile Systeme (Springer, 2007), p. 294300
[54] J. Hurst, A. Rizzi, IEEE Robotics & Automation Magazine 15(3), 42 (2008)
[55] R. Ham, T. Sugar, B. Vanderborght, K. Hollander, D. Lefeber, IEEE Robotics & Automation Magazine 16(3), 81 (2009)
[56] E. Garcia, J. Arevalo, G. Munoz, P.G. de Santos, Robotics and Autonomous Systems 59(10), 827 (2011)

Part IV Materials, Design and Manufacturing

13 Nanostructured Materials for Soft Robotics – Sensors and Actuators

Raphael Addinall[1], Thomas Ackermann[1, 2] and Ivica Kolaric[1]

[1] Fraunhofer Institute for Manufacturing Engineering and Automation IPA, Stuttgart

[2] Graduate School of Excellence advanced Manufacturing Engineering (GSaME), University of Stuttgart

Abstract The advances in nanotechnology during the past two decades have led to several breakthroughs in material sciences. Ongoing and future tasks are related to the transfer of the unique properties of nanostructured materials to the macroscopic behaviour of composite structures and the system integration of novel materials for improved mechanical, electronic and optical devices. Nanostructured carbons, especially carbon nanotubes, are promising candidates as novel material for future applications in several fields. One of the big aims is the utilisation of the unique intrinsic mechanical and electronic properties of carbon nanotubes for sensing and actuation devices. The combination of excellent electrical conductivity and mechanical deformation makes carbon nanotubes ideal for applications in sensors and actuators and opens new possibilities in construction design of next generation robotic systems, which can be built with soft, bendable and stretchable materials. This chapter gives a brief overview on the properties of carbon nanotubes and their potential for actuators and sensors in soft robotics.

13.1 Introduction

Nanostructured carbon materials – first and foremost CNTs (Carbon Nanotubes) and graphene – are the fastest growing research domain in nanotechnology. Since their discovery in 1991 [1], CNTs have sparked huge interest as they exhibit outstanding mechanical and electronic properties [2, 3]. Formally, CNTs can be regarded as a rolled-up graphene sheet (Fig. 13.1). The roll-up vector C_h results from the vector addition $C_h = na_1 + ma_2$ of the graphene unit vectors a_1 and a_2. The descriptors n and m are integers and define the nomenclature of the CNT (n,m). As a result of the geometry and a model from solid state physics – the tight binding model – CNTs can be either metallic or semiconducting. Here we limit the description to the rule that CNTs are metallic if (n-m)/3 is an integer, e.g. (7,4), (5,5) or (9,0). Other CNTs are semiconducting, e.g. (6,5), (7,3) or (8.0). For further reading we refer to [2, 3] and the references therein. Besides single-wall carbon nanotubes (SWCNTs) there also exist multi-wall carbon nanotubes (MWCNTs), which consist of several rolled layers in a concentric alignment [2].

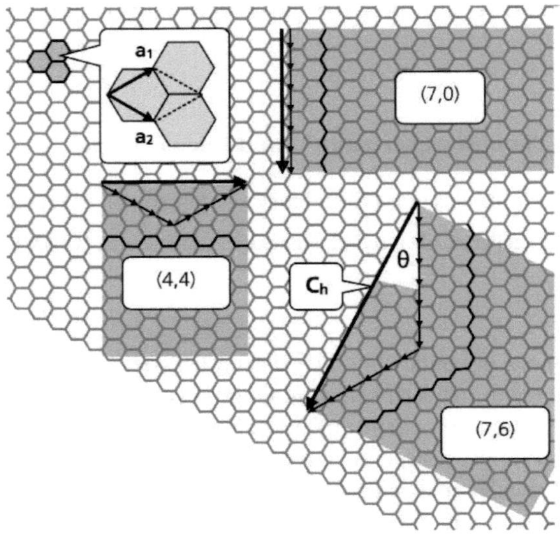

Fig. 13.1 Formal classification of three different carbon nanotubes based on the roll-up vector C_h

Hence, with regard to their electrical and optoelectrical properties, semiconducting CNTs and metallic CNTs offer different potential applications. Semiconducting CNTs are frequently discussed as single-electron transistors for future microelectronic components. In 1998, the IBM research division presented a single-electron transistor based on individual semiconducting CNTs [4]. By contrast, metallic CNTs are rather used as an ensemble in conductive films. In theory, CNTs can carry an electrical current density up to roughly $4 \cdot 10^9$ Acm^{-2}, which is about a thousand times higher than for copper before breakdown due to electromigration [5]. The extraordinary high intrinsic conductivity occurs due to the strong chemical bonding of in the CNT walls.

The aforementioned potential benefits of carbon nanotubes have had a huge impact on the amount of work, which has been carried out and published in several fields for the past two decades. However, when it comes to the utilisation of these beneficial properties for macroscopic components, either as an ensemble consisting only of a CNT network or as a composite material such as CNT/polymer composites, things become challenging.

Basically there are two major problems. At first, today's production techniques for CNTs deliver soot that contains CNTs of various diameters and consequently a mixture of metallic and semiconducting CNTs [6]. For high purity CNT samples – either metallic- or semiconductive-enriched – further purification methods such as ultracentrifugation [7, 8] or chromatography [9] have to be applied. These methods are the bottleneck in the development of components as they are only scalable to some extent. The second challenge concerns the integration of the CNTs into a polymer matrix or onto a substrate. As an example, despite the aforementioned high electrical conductivity of metallic CNTs, a transparent network of metallic carbon nanotubes still does not reach the low sheet resistance of indium tin oxide or silver nanowires within the same transparency range [10]. This is due to the high resistances of the junction points in the CNT network. However, during the past years the performance of these films could be improved and the sheet resistances of films containing carbon nanotubes have reached values, which come closer to the industrial demands in display industry [11, 12]. Furthermore the films are bendable and stretchable. This property is a significant advance over conventional materials such as brittle indium tin oxide when it comes to the usage in mechanically flexible components for soft robotic devices. Here we will report on the potential of CNTs for actuators followed by a section on CNT-based sensors for applications in soft robotics.

13.2 Actuators

Actuators are mechanical devices, which can be used for the purpose of inducing strain into a system in order to generate a movement or a change of shape. Conventional actuators function based on pneumatic, electric or hydraulic principles. With regard to applications in soft robotics, these actuator types are not always suitable as weight, size, restrictive shapes and stiff materials limit the freedom of component design. Stretchable and bendable polymers can overcome these problems. Electroactive polymers (EAPs) are a relatively new class of actuator materials. They can change their shape or size as a response to electric stimuli [13]. Besides the mechanical flexibility, EAPs offer several other major benefits such as light weight, structural versatility, easy material processing and usually low costs.

Various classes of EAPs can be used as actuators to be integrated into robotic systems. More importantly the soft nature of the polymer based actuator is intrinsically suited for the next generation of robotics: soft robotics. EAPs are commonly classified in two major classes (Fig. 13.2). Ionic EAPs are activated by an electrically-induced transport of ions and/or molecules. The intercalation of ions into a host material such as a CNT network induces a change of the electric structure of the chemical bonding [13].

The changing of the electronic structure generates a deformation of the bonds and consequently expansion or shrinking. In dielectric EAPs the actuation is induced by electrostatic interactions between two electrodes, which encircle the polymer. Electronic EAPs can generate large strain at reasonable rates while demonstrating large displacements (strain %). However the use of high voltages makes them not particularly well suited for mobile applications such as soft robotics and is thus not discussed in much detail within this chapter.

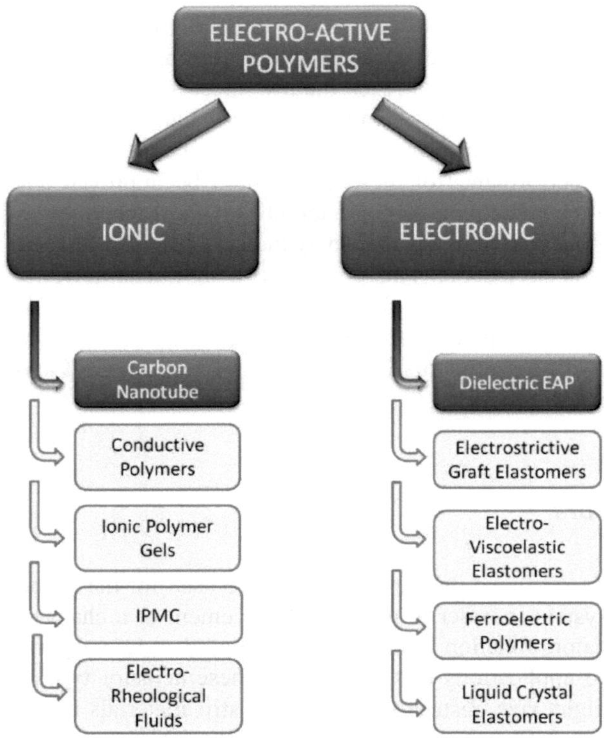

Fig. 13.2 Classification of materials for actuators based on electroactive polymers

When contemplating the use of EAP materials for soft robotics, one major consideration arises, which is the relation between the applied electrical stimuli and the resulting displacement, force and reaction rate. This is one of the fundamental question in which the answer will govern the possible application. From this, both ionic EAPs and electronic EAPs have their advantages and disadvantages. In the case of ionic EAPs, their low voltage operation could make them ideal candidate for soft robotic applications if it was not for the fact that – in most cases [14] – they need to be hydrated at all times. Ionic/polymer/metal-composites, being one example of ionic EAPs, show promising properties with respect to biomimetic uses. Ionic polymer/metal composites consist of a thin ionomeric membrane with noble metal electrodes plated on its surface. It also has cations to balance the charge of the anions fixed to the polymer backbone [15]. However, the force generated with these types of actuators is relatively small and typically in the mN or single N unit range.

CNTs have shown promising characteristics when applied to actuator technology. They can either be used alone as an actuating devise (nano tweezers, gate systems) or as a filler material within polymers and ionic liquid mixtures, enabling the creation of a layered actuating structure exhibiting displacements in the cm range, forces in the Newtonian range and reaction rates in the s to ms range. The principle of actuation is illustrated in Fig 13.3. Upon injection of electrodes from an external source, the ions within the polymer outer layers and supplied by the ionic liquid within, separate due to the repulsion between ion and electron. The separation leads to a concentration of ions on one side of the actuator layer, effectively swelling the material through ion intercalation. This swelling results in the laminate structure to bend. The direction is governed by the polarity of the external voltage supplied.

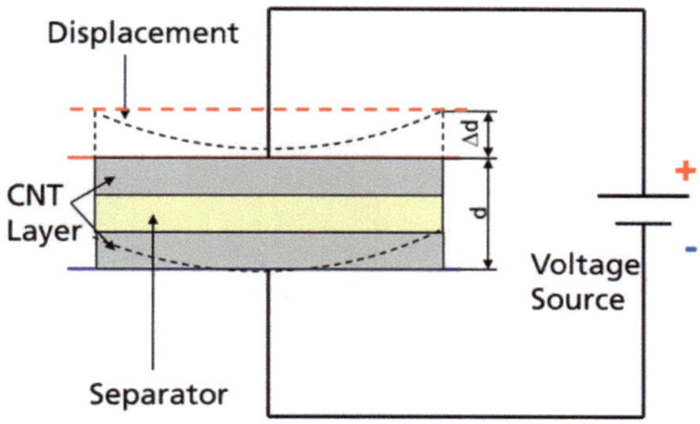

Fig. 13.3 Schematic principle of an actuator based on ionic electroactive polymers

Tri layer CNT based ionic actuators require low voltages (2V) and the strain produced can be in the range of 5.8% (triple layer solid state bending type actuator) [16, 17]. These actuators (triple layer solid state bending actuators) differ from the double layer bending actuators [18, 19] as they do not need to be immersed in an electrolyte solution since the electrolyte is embedded within the polymer matrix. However at the moment these demonstrate poor reproducibility from the point of view of actuation displacement. Current work undertaken at Fraunhofer IPA is focused on such CNT polymer hybrid actuator systems. The actuator class called A3D (Actuating Three Dimensional) actuators, which uses volumetric change generated by ion intercalation within the polymer chains and CNTs as well as the quantum-chemical based expansion due to electrochemical double-layer charging. From the three types of CNT based ionic polymer actuators described in this chapter (bilayer, trilayer and A3D) the only variation is the geometry and electrode material set up. For example long thin actuators generate a larger displacement, circular provide more force and ultrathin actuators move at a much higher frequency. Furthermore, the incorporation of many actuators into stacks with an aim of multiplying any given characteristic (force, displacement, speed) has also been studied. Lastly the interaction between contact electrodes and multiple actuator systems has been a governing parameter often neglected. A systematic approach (Fig. 13.4) must be taken in order to test, characterise and optimise each type of variable with an aim of producing, from an engineering point of view, a working functional model mimicking real life applications. Thus by incorporating scientific findings with engineering principles the possibility to adopt CNT-polymer actuators for use in future soft robotics applications will be one step closer.

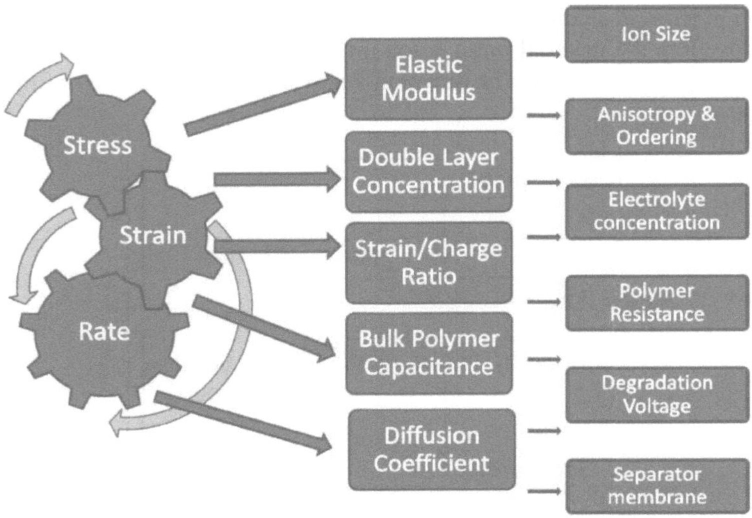

Fig. 13.4 Systematic approach for the testing, characterisation and optimisation of ionic electroactive polymers

As can be concluded the use of electronic and ionic EAPs materials bring many challenges when used alone. It is therefore important to realise that the combination of EAPs with CNTs can lead to new and improved properties which could be specifically designed or tailored for use in soft robotics. The addition of highly electrically conductive and extremely strong CNT within a polymer based matrix would inevitably increase electrical conductivity and the Young's modulus of the resultant actuator material (conductivity from 500 S/cm to 700S/cm) and mechanical strength (from 170MPa to 255 MPa), properties which are highly desirable when selecting suitable materials for soft actuator systems. In addition, the increase in stiffness will decrease the creep tendency of polymers, enabling for more reproducible and accurate actuation [20]. Although further research is needed within the domain of CNT based ionic actuators, the potential has already been shown mainly through functional models: bending actuators (Fig. 13.5a), breaking systems (Fig. 13.5b), suction and dispensing capability (Fig. 13.5c) all validate their functionality and applicability within the soft robotic domain.

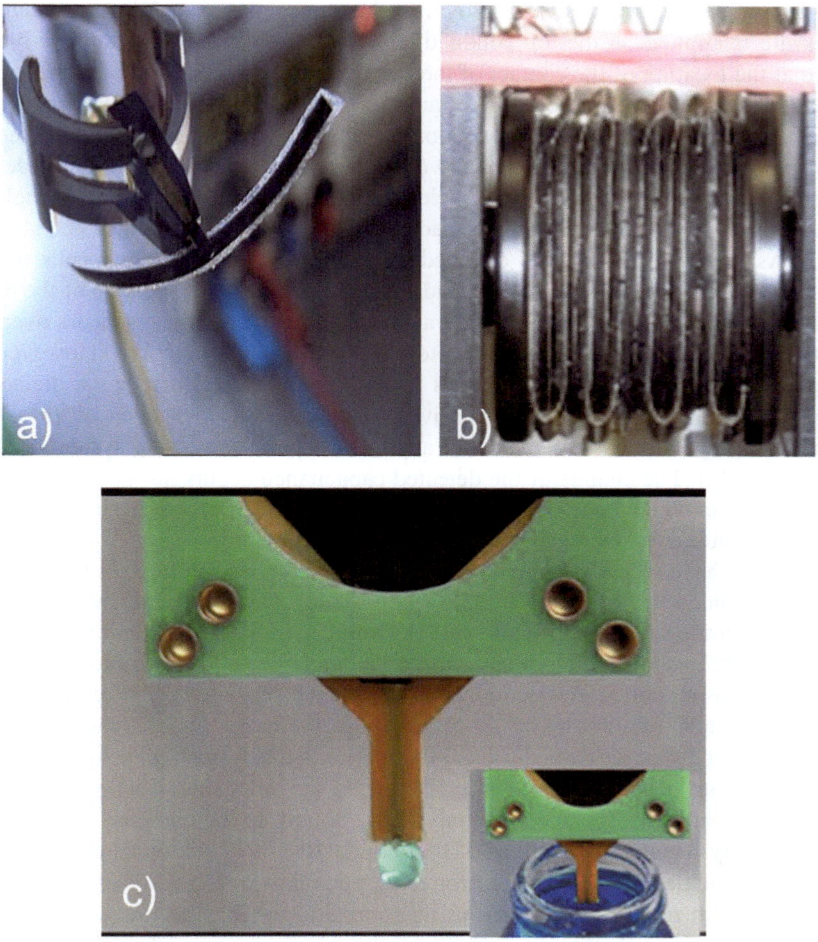

Fig. 13.5 a) Trilayer CNT based actuators mimicking gripping functionality b) Stack of round trilayer CNT-ionic actuators integrated to function as a disk brake c) CNT based actuator system for liquid handling devices; developed at Fraunhofer IPA and AIST Kansai

13.3 Touch Sensors

In a certain way a touch sensor can be described with the inverse function principle of an electric actuator. Whereas an electric actuator moves or changes its shape due to electrical stimuli, a touch sensor converts an externally generated mechanical pressure on its surface into and electronic signal. However this statement is only partially correct as it describes a resistive touch sensor where two electrically conductive films are separated by a thin insulating layer. Voltage is applied to one of the conductive films and sensed by the other conductive film. When mechanical pressure is applied on the top conductive layer, the electrical resistance is measured at the location of the contact of the conductive layers. For applications in displays the conductive layer has to be transparent. Since a few years another type of touch sensor is used in many applications such as smartphones and tablet computers: capacitive touch sensors. Capacitive touch sensors are stimulated by the detection of a close object, which has a dielectric different from air. In touch panels in display devices the human body capacitance of the finger generates the stimuli [21]. As the electrical field is changed by the iron in the red blood cells of the nearby finger, a capacitance meter inside the device detects the change of the electrical field lines. One of the major advantages of capacitive touch sensors over resistive touch sensors is the possibility to allow multi-touch functions. However, in applications for soft robotics, there is not always an object that has a dielectric other than air. There also are capacitive touch sensors, which react on pressure. The key component is the same as for any other capacitive sensing system: a plate capacitor [21]. The change of the detected capacitance occurs not due to an external field of an object but due to the change of the distance between the two capacitor plates caused by the mechanical pressure of the object.

Lu et al. have recently given a comprehensive overview on flexible and stretchable electronics for the usage in soft robotics [22]. One of the potential applications is a tactile sensing artificial skin (electronic skin, E-skin). For soft robotics it is obligatory that such components base on a bendable and stretchable carrier material such as silicones. As mentioned above, touch sensors consist of conductive layers, which should not be brittle or heavy for the usage in soft robotic components. Indium tin oxide does not fulfil these criteria due to its high brittleness. Metallic wires are bendable but not stretchable. On the other side, CNT networks exhibit excellent performance not only with regard to bendability but also to expansibility. Hence, the coating of thin conductive CNT networks onto flexible substrates such as silicones offer promising potential for the realisation of haptic sensing in soft robotics. However, as mentioned in the introduction section, there are still enormous challenges concerning the purifications of the CNTs and the coating process onto the substrate. Due to the resistances at the junctions in a CNT network, the overall sheet resistance is much higher as one expects from the high intrinsic conductivities of metallic CNTs. Hybrid structures could overcome the drawbacks of the sole materials. Recently we have developed a transparent flexible film based on silver nanowires (AgNWs) and CNTs [23]. It is reported that the mechanical flexibility as well as the electrical performance of silver nanowire networks can be improved by a top layer of CNTs [24]. Fig. 13.6 shows an atomic force microscope image of a conductive AgNW/CNT hybrid film. The CNTs enhance the overall conductivity by creating electrical bridges between the AgNWs.

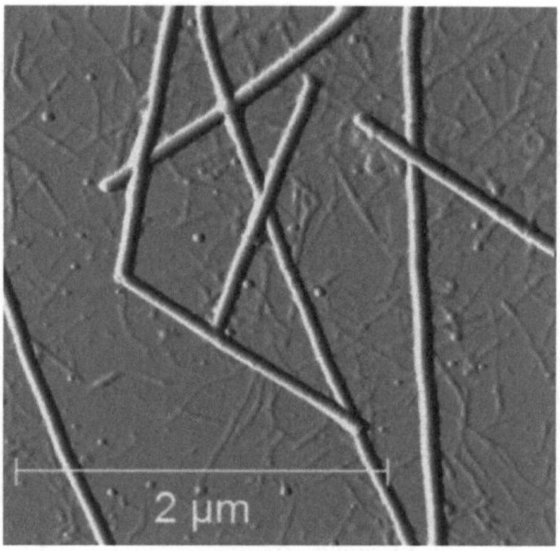

Fig. 13.6 Atomic force microscopy image of a silver nanowire/carbon nanotube hybrid network

13.4 Conclusions and Perspectives

Components for soft robotic devices demand novel materials with improved mechanical and electrical properties. With conventional stiff and brittle materials the goal of creating soft robots cannot be achieved. Therefore novel materials have to be investigated in an interdisciplinary research environment. These materials will be one of the essential driving forces for the development of future robotic components. Due to their extraordinary mechanical robustness and outstanding electronic properties, nanostructured carbon materials offer high potential for the production of mechanically flexible components in soft robotics. The biggest challenge for the realisation of these devices is the integration of the materials into or onto the carrier material. However, recent developments have shown that both new actuators and touch sensors can be developed and improved with the usage of carbon nanotubes.

Acknowledgments The authors thank the German Research Foundation (DFG), the Federal Ministry of Education and Research (BMBF) and the Fraunhofer Society for the financial support of their work.

13.5 References

[1] Iijima S (1991) Helical microtubules of graphitic carbon. Nature 245:56-58
[2] Saito R, Dresselhaus G, Dresselhaus M (1998) Physical properties of carbon nanotubes. Imperial college press

[3] Louie SG (2001) Electronic properties, junctions and defects of carbon nanotubes. In: Carbon nanotubes – synthesis, structure, properties and applications, Springer

[4] Martel L, Schmidt T, Shea HR et al (1998) Single- and multi-wall carbon nanotube field-effect transistors. Appl. Phys. Lett. 73(17):2447-2449

[5] Hong S, Myung S (2007) A flexible approach to mobility. Nature Nanotechnology 2:207-208

[6] Szabo A, Perri C, Csato A (2010) Synthesis methods of carbon nanotubes and related materials. Materials 3:3092-3140

[7] Arnold MS, Stupp SI, Hersam MC (2005) Enrichment of carbon nanotubes by diam ter in density gradients. Nano Lett. 5:713-718

[8] Arnold MS, Green AA, Hulvat JF et al (2006) Sorting carbon nanotubes by electronic structure via density differentiation. Nature Nanotechnology 1:60-65

[9] Flavel SF, Moore KE, Pfohl M et al (2014) Separation of single-walled carbon nanotubes with a gel permeation chromatography system. ACS Nano 8(2):1817-1826

[10] De S, King PJ, Lyons PE et al (2010) Size effects and the problem with percolation in nanostructured transparent conductors. ACS Nano 4(12):7064-7072

[11] Dan B, Irvin GC, Pasquali M (2009) Continuous and scalable fabrication of transparent conductive carbon nanotube films. ACS Nano 3(4):835-843

[12] Mirri F, Ma AWK, Hsu TT (2012) High-performance carbon nanotube transparent conductive films by scalable dip coating. ACS Nano 6(11):9737-9744

[13] Kosidlo U, Omastova M, Micisik M et al (2013) Nanocarbon based ionic actuators – a review. Smart Mater Struct 22: 104022

[14] Nemat-Nasser, S, Thomas C. (2001) Electroactive polymer (EAP) actuators as artificial muscles. Reality, potential and challenges. SPIE Press Monograph 139–191.

[15] Qu L., Peng Q, Dai L et al (2008) Carbon nanotube electroactive polymer materials: opportunities and challenges. MRS Bulletin 33:215–224.

[16] Fukushima T, Asaka K, Kosaka A et al (2005) Fully Plastic Actuator through Layer-by-Layer Casting with Ionic-Liquid-Based Bucky Gel. Angew. Chem. Int. Ed. 44(16):2410-2413

[17] Vohrer U, Kolaric I, Haque MH et al (2004) Carbon nanotube sheets for the use as artificial muscles. Carbon 42(5-6):1159-1164

[18] Gao M, Dai L, Baughman RH et al (2000) Electrochemical properties of aligned nanotube arrays: basis of new electromechanical actuators. In: Proc. SPIE – Int. Soc. Opt. Eng. 3987:18-24

[19] Fraysse J, Minett AI, Jaschinski O. et al (2002) Carbon nanotubes acting like actuators. Carbon 40, 1735-1739

[20] Spinks GM, Mottaghitalab V, Bahrami-Samani M et al (2006) Carbon-Nanotube-Reinforced Polyaniline Fibers for High-Strength Artificial Muscles. Adv. Mater 18(5):637-649

[21] Dahiya RS, Valle M (2013) Robotic tactile sensing. Springer

[22] Lu N, Kim D-H (2014) Flexible and stretchable electronics paving the way for soft robotics. Soft Robotics 1(1): 53-62

[23] Ackermann T, Sahakalkan S, Zhang Y et al (2014) Improved performance of transparent silver nanowire electrodes by adding CNTs. In: 8th IEEE NEMS, in print

[24] Tokuno T, Nogi M, Suganuma K (2012) Hybrid transparent electrodes of silver nanowires and carbon nanotubes: a low temperature solution process. Nanoscale Research Letters 7:281

14 Fibrous Materials and Textiles for Soft Robotics

M. Milwich[1, 2], S.K. Selvarayan[1], G.T. Gresser[1]

[1]Institute of Textile Technology & Process Engineering Denkendorf, [2]Hochschule Reutlingen

Abstract Soft, mechanically compliant robots are developed to safely interact with a "human environment". The use of textiles and fibrous (composite-) materials for the fabrication of robots opens up new possibilities for "Softness/Compliance" and safety in human-robot interaction. Besides external motion monitoring systems, textiles allow on-board monitoring and early prediction, or detection, of robot-human contact. The use of soft fibers and textiles for robot skins can increase the acceptance of robots in human surroundings. Novel topology optimization tools, materials, processing technologies and biomimetic engineering allow developing ultra-light-weight, multifunctional, and adaptive structures.

14.1 Introduction

To-date, industrial robots are mainly designed for the handling of materials, parts, tools, etc. and to perform heavy, hazardous, and highly repetitive tasks. For those jobs, robots are made of massive and rigid materials to perform powerful and precise operations. Therefore, there remains a certain level of threat to the humans who operate them or work close to them.

For applications such as human assistance in industrial surroundings, health care, handling of soft materials and field explorations, mechanically compliant "soft" robots are developed to safely interact with those critical environments. Soft robots can elastically deform, operate in highly constrained environments without causing damage to themselves and the things they interact with, and can even prevent their human co-workers from harm. Hence, the soft robots differ from their "hard" counterparts in terms of design, capabilities, and properties by using adaptive, compliant, and lightweight, topology-optimized fibrous materials, and sophisticated fiber lay-up. Textiles can provide smooth, soft, and pleasant surfaces. Textile surfaces can display information or entertainment and give certain compliance and reaction/stopping time before the inner robot skeleton hits sensitive matter or soft human skin. Further "softness" can be reached in using multifunctional sensing and active materials to mimic multifunctional biological systems. Fibrous or textile materials are especially suitable to fulfill biomimetic requirements for the construction of soft robots.

14.2 Fibrous Materials: Properties and Architecture

The material requirements for soft robots are sometimes conflicting: extreme light-weight design will provide a good mass/working load ratio, but the material should allow for robustness, high damping, high speed and high repeat accuracy. This can be achieved by using fibrous and textile composite materials with enhanced passive vibration damping in combination with new active fibrous actuation technologies.

Fiber reinforced composites consist of polymeric, ceramic, or metallic fibers which are embedded in polymeric, ceramic or metallic matrices. Different textile processes e.g. weaving, braiding, knitting, or embroidery result in different textile architectures and different mechanical properties.

Composites, and ultra-light-weight design, are not an invention of mankind. In nature, nearly all tissues of living organisms are composed of fibers embedded in matrix systems. There are only four natural fibers and few matrix systems. The astonishingly wide range of properties of natural organisms is mainly the result of a sophisticated fiber lay-up following the load paths in the part - in often very confined spaces. Advanced textile processes like 3D-braiding, 3D-fabric weaving and Embroidery (Fig. 14.1) additionally have the potential to mimic those optimized natural fiber lay-ups and provide the possibility to manufacture near-net shaped structures.

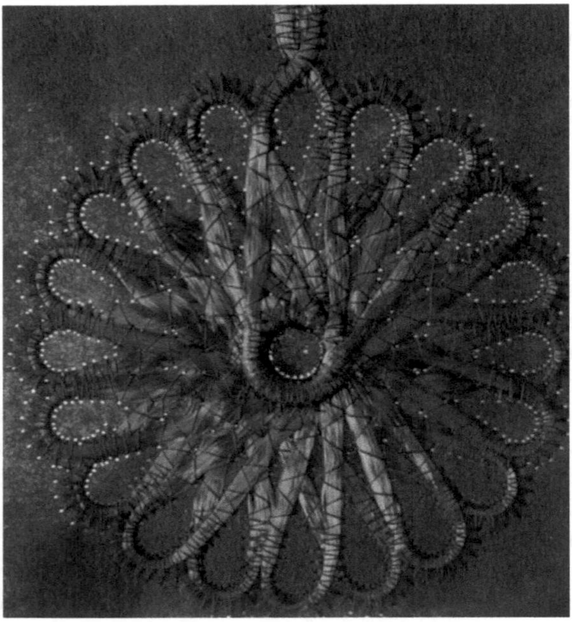

Fig. 14.1 Embroidery of Carbon Fibers to place the fibers along the load paths in the part. (Courtesy ACC Technologies, FS Software&Konstruktionen GmbH)

These advanced textile processes individually, or in combination with each other, allow the properties and functionalities of the produced part to be tailored, since the properties of textile materials and composite parts depend on the direction in which the fibers/filaments are aligned.

Figure 14.2 illustrates the effect of the fiber directions on the elastic moduli of a fiber reinforced composite. Both samples can carry nearly the same load. However, the 0°/90°-arrangement of the fibers result in a high stiffness in z-direction, whereas the composites with ±45° arrangement of fibers are relatively flexible in horizontal direction. This flexibility will e.g. lead to fewer injuries when the robot arm contacts human tissue.

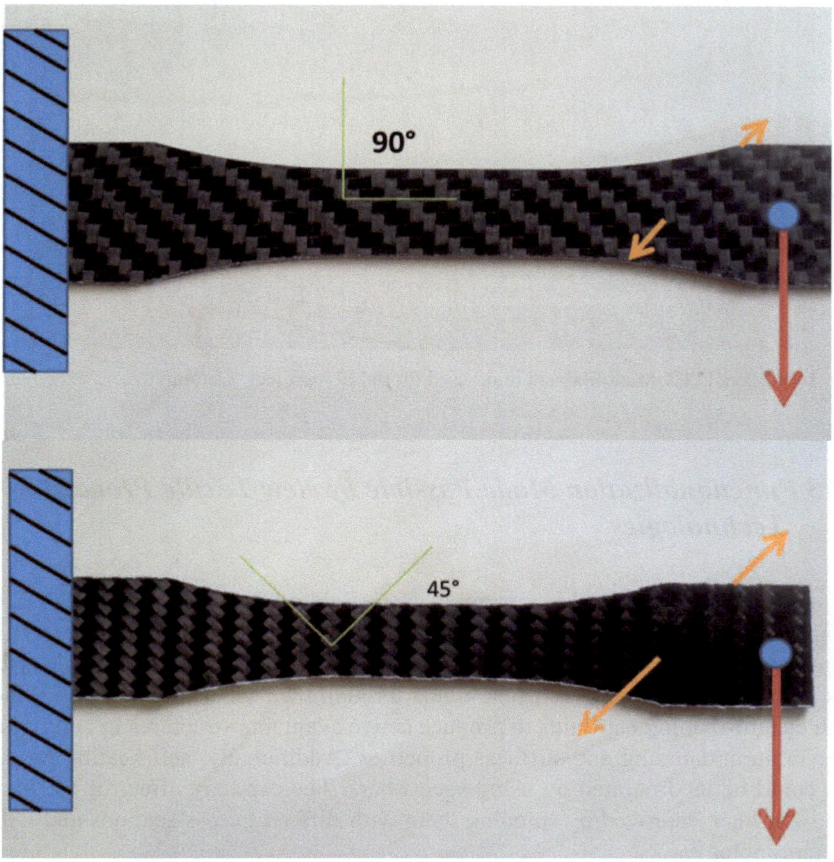

Fig. 14.2 Composite stiffness variation based on fiber angle

Figure 14.3, again, shows one of the many possibilities of textile composite materials. A biaxial PET Polyethylene terephthalate monofilament braid, made by ITV Denkendorf with no change of fiber angle along its tapered length, is embedded in an elastomeric matrix creating a highly flexible robot arm. This special braid was used as scaffold for the robot arm developed in the European project "Octopus".

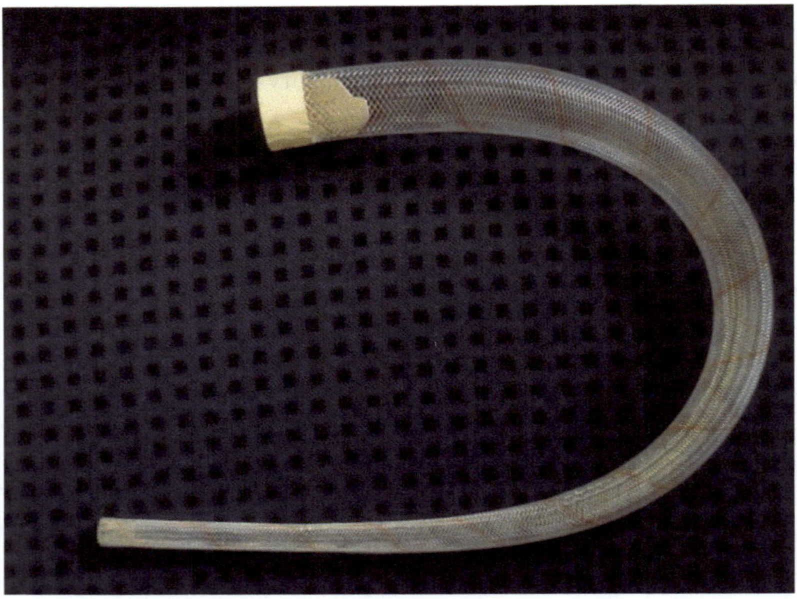

Fig. 14.3 Biaxial PET Monofilament braid used for the EU- project "Octopus"

14.3 Functionalization Made Possible by New Textile Processing Technologies

By new developments, and innovations, in the field of processing technologies, fibrous materials can be targeted to have multi-functional properties. For example, spinning of bi- or multi-component filaments can lead to new functionalities. Hollow fibers can be used to transport fluids through their channels, or can be filled with electro-rheological fluids to produce active adaptable structures in controlling the vibration damping and stiffness properties. Additionally, self-healing properties could be incorporated by using such fibers. The capillary effect of the fibers can be further improved by spinning them with different cross-sections and internal morphology.

Coating/sizing with nano-particles like CNTs, nano-functionalized silica or piezo-materials together with surface modification processes like laser, plasma, or gamma ray irradiation are used to generate functionalized surfaces, e.g. self-illuminating, conducting electricity, sensing, enhanced damping, or improved fracture behavior.

New textile processes such as Dornier Open Reed Weaving (Fig. 14.4), tufting (Fig. 14.5), or new braiding techniques allow placement of reinforcing fibers, conducting filaments, electric wires or sensors in the preferred directions (load path), thus saving weight and creating desired anisotropic properties or adaptability to temperature, pressure/tension or electromagnetic radiations.

Fig. 14.4 Sensor integration by Open Reed Weaving

Fig. 14.5 Sensor integration by Tufting

Another textile technique to create complex fibrous structures is 3D-weaving. In cooperation with Prof. James Nebelsick from the University of Tübingen and ITV Denkendorf, biomimetic 3D composite structures derived from the role model of sand dollar are developed. Sand dollars (Fig. 14.6) have a sophisticated composition of their exoskeleton resulting in high stiffness & strength combined with low weight.

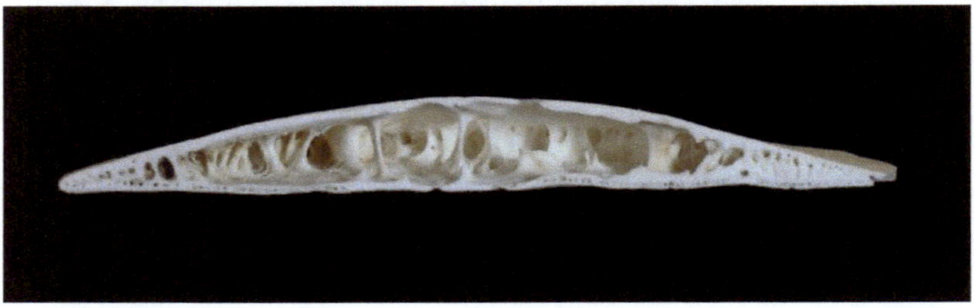

Fig. 14.6 Sand dollar

Fig. 14.7 shows a new 3D-woven, lentil-like, textile structure inspired from the sand dollar made on a specialized ITV spacer weaving machine. The channels in the structures could be used for transportation of materials or fluids. Additionally this structure improves noise- and heat-insulation.

Fig. 14.7 3D woven lentil like structures inspired by sand dollar [H.J. Bauder, ITV]

14.4 Light-Weight-Structures for Robots

One of the main advantages of fibrous materials is the possibility to provide high stiffness and strength by an anisotropic fiber lay-up. When the fibers are arranged along the stress paths, the stress concentration, and thus, the amount of material, is minimized. Nature is highly effective in utilizing this principle to exploit the material in full, so the same principle should be adopted in the manufacture of fiber reinforced composites. Fig. 14.8 shows a hard robot, whose metallic arm was replaced by a load path optimized carbon fiber reinforced composite arm. This makes the robot lighter, increases the payload, reduces energy consumption, as well as increases positional accuracy and repeatability of the robot actions.

Fig. 14.8 Kuka - Robot with carbon composite arm

Composites, inherently, have better vibration damping properties compared to isotropic metallic materials. This makes composite materials one of the best choices to be used in robotics. The vibration damping can be further enhanced by optimization of fiber directions. The so-called technical plant stem invented by ITV Denkendorf and Plant Biomechanics Group Freiburg is inspired by the morphology of giant reed and horse tail plant stems (Fig. 14.9).

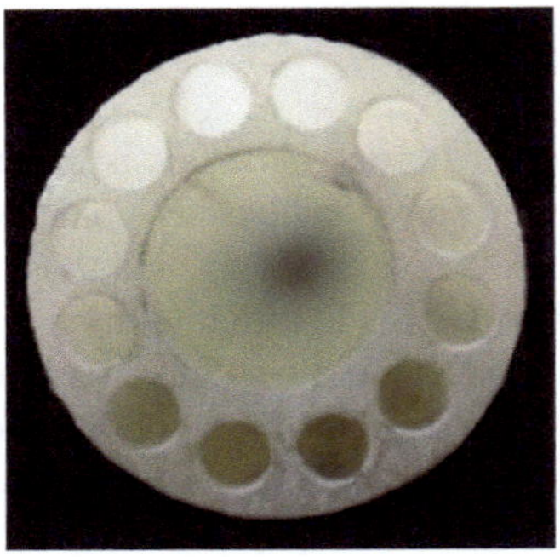

Fig. 14.9 Technical plant stem

The technical plant stem is a good example for an enhanced vibration damping structure by structural optimization. It exhibits a four-fold increase in damping compared to single wall struts/tubes with a monolithic wall design (Fig. 14.10). The results were verified with glass fibers and carbon fibers, both with Polyurethane matrix (PU) and Vinyl ester/Epoxide matrix (VE/EP).

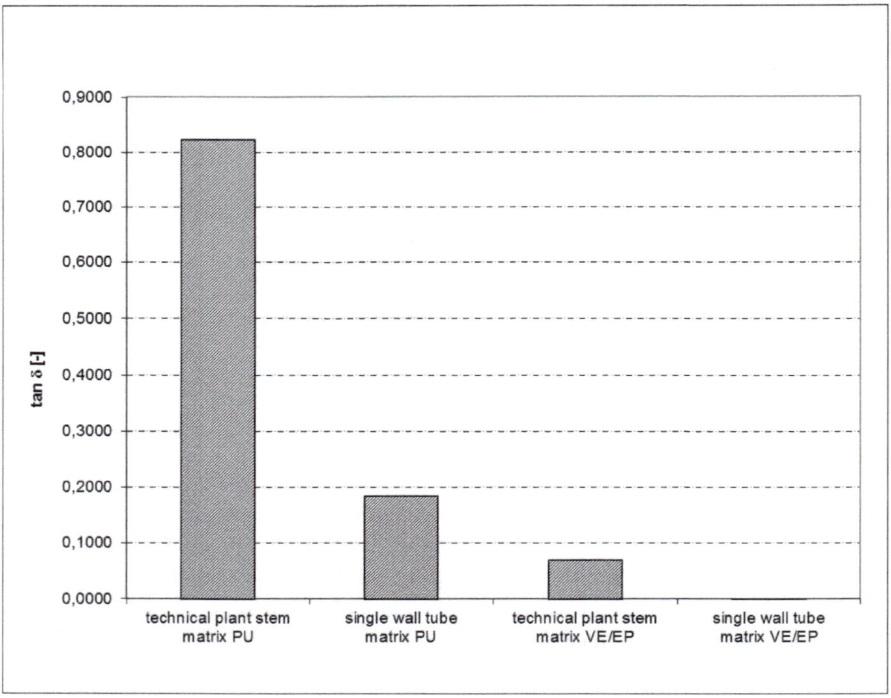

Fig. 14.10 Comparison of the damping behavior "tan δ" of single wall tube and technical plant stem

Fig. 14.11 shows an isogrid structure of German company CirComp, which is an example for ultra-light weight structures with multi-directional stiffness properties. The isogrid can be used in many applications like bicycle frames, robotic arms, or power poles.

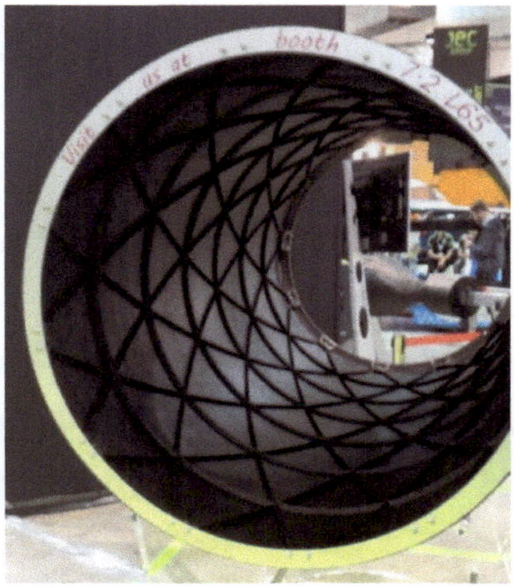

Fig. 14.11 Isogrid structure of German company CirComp

A further optimization of isogrid structural principles was done within the international project "PlanktonTech" which is a Helmholtz Virtual Institute founded in 2008 at the Alfred Wegener Institute for Polar and Marine Science by the Helmholtz Society. It led to extremely light weight fiber reinforced composite structures with pentagon/hexagon structures on different hierarchical levels, exhibiting very high strength and stiffness (Fig. 14.12).

Fig. 14.12 Hierarchical structure inspired from plankton [S. Küppers, ITV]

Plankton, the biological model, has optimized their protective hard shell over millions of years. As a reaction to the ever increasing strength of the jaws of their predators the shells got ever harder and stiffer. At the same time the shell could not increase in mass, otherwise it could not stay near the ocean surface to get sunlight for photosynthesis. This astonishing light-weight design, learnt from plankton, was applied in the manufacture of extremely light-weight robot arms. The thin struts make the robot arm structure compliant when loaded with sideward bending forces created by contact with other objects (Fig. 14.13).

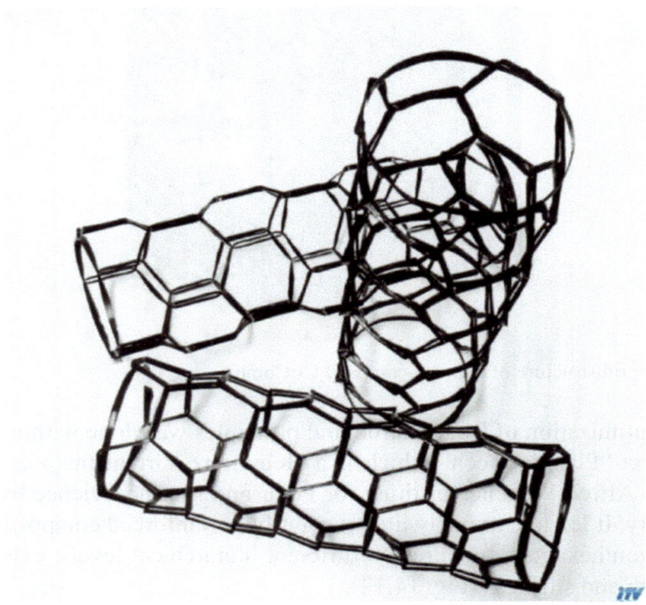

Fig. 14.13 Plankton inspired robot arm [S. K. Selvarayan, ITV]

Another role model for light-weight construction is bamboo, which is a classical fibrous composite material exhibiting good stiffness, because of the presence of nodules, along its structure. With significant support of Amann Corporation, ITV developed a technical process to apply such nodes within a composite tubular structure, enhancing the cross-sectional stiffness properties of technical composite materials (Fig. 14.14).

Fig. 14.14 Bamboo inspired nodal composite structure

This principle, applied in the construction of robot arms and combined with the use of lightweight fiber reinforced materials, increases robot usability, and safety, due to more compliance because of possible thinner skins. The enhanced vibration damping leads to positional accuracy and repeatability of robot movements.

14.5 Adaptive and Intelligent Structures

Unlike the existing hard robots, whose control remains only in their links, the realization of soft robots is complete only when the entire robot becomes intelligent and adaptable; mimicking the multi-functionality of natural systems. In order to realize such highly sophisticated robot systems, functionalized fibrous materials can be processed into intelligent and adaptive structures for sensing, actuation, active vibration damping, energy storage or others. Figure 14.15 shows a surf board made by Hydroflex Corp. which is a sandwich structure consisting of thin glass fiber deck sheets and polystyrene foam core. The stiffness of the surf board can be changed by varying the air pressure inside the core material. This principle is derived from pressure stiffened plant systems and has a big potential to be transferred into many ultra-light-weight constructions.

Fig. 14.15 Adaptive surf board with pressurized core of company Hydroflex

Figure 14.16 shows a classic sun shade with hinges. Hinges are prone to failure and need a lot of maintenance.

Fig. 14.16 Classic folding mechanism with hinges [Photo: J. Lienhard]

A newly developed hinge-less facade shading system called Flectofin© was developed by the Plant Biomimetic Group from the University of Freiburg, the Institute of Building Structures and Structural Design from the University of Stuttgart and ITV Denkendorf and uses the movement principles learned from the bird-of-paradise flower (Fig. 14.17). Stability, and directional force transfers within these structures, were made possible only by the use of anisotropic behavior of the glass fiber composite.

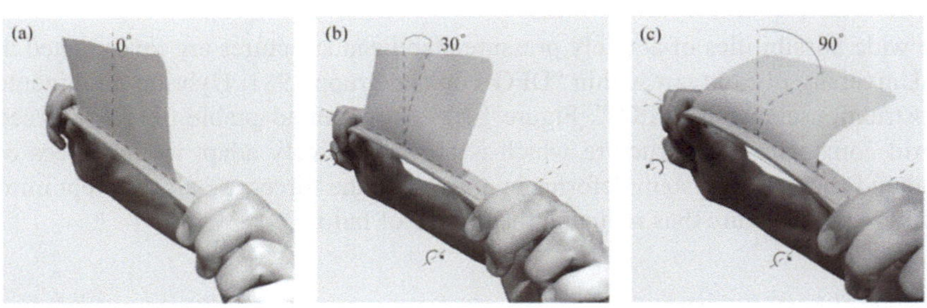

Fig. 14.17 Principle of operation of Flectofin© façade [Photo: J. Lienhard]

Figure 14.18 shows a "soft" actuation system based on textile pressure bags controlled by air pressure. The textile actuators drive precisely a drumstick in a vertical and horizontal motion. Such a system can be used as a low weight and high force actuator for handling materials by controlling the pressure in each chamber of the actuator arm. Thus, this system could be seen in itself as a soft robot, because it delivers both structure and strength without hard elements. The advantages of such pressure activated and pressure stabilized robots are that they can be operated in highly critical environments and also that they work in confined spaces not possible for rigid robot systems.

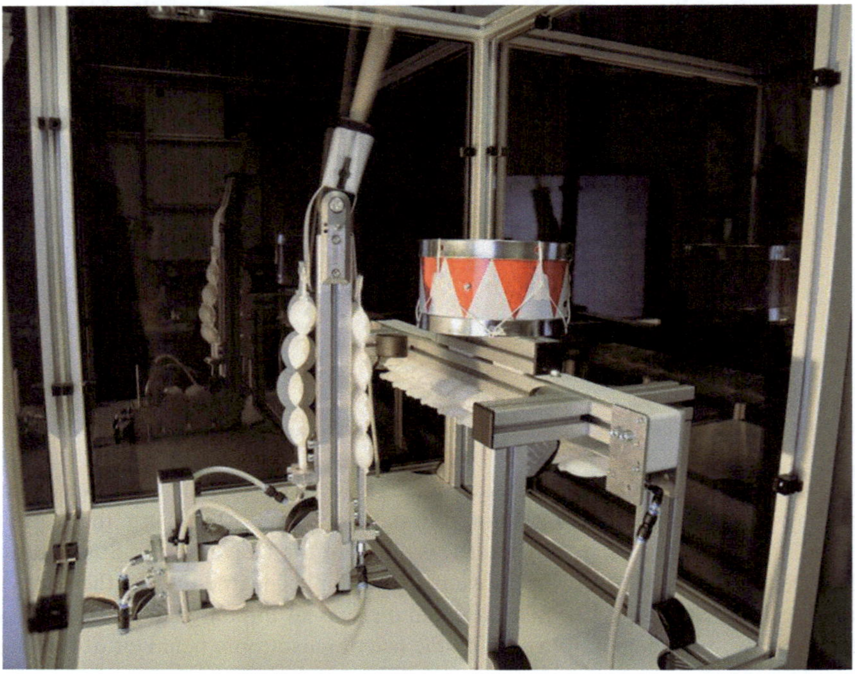

Fig. 14.18 Pneumatic actuation system [C. Riethmüller, ITV]

The wide possibilities of actively pressure-stabilized structures are investigated at the University of Stuttgart within "DFG-Forschergruppe 981: Hybride Intelligente Konstruktionselemente–HIKE". Figure 14.19 shows an adaptable fiber reinforced hybrid composite roof structure which is able to actively adapt to the forces of wind and snow. The system deforms in a way that the forces are always kept minimum on the structure thus reducing the chance of failure.

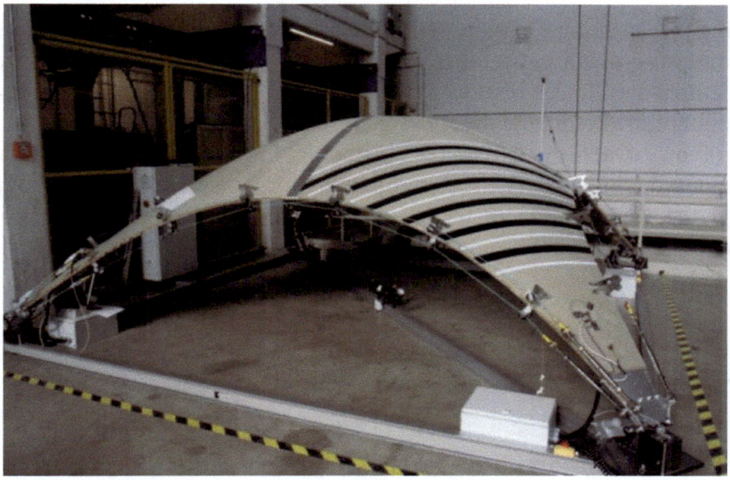

Fig. 14.19 Adaptive hybrid structure [C. Riethmüller, ITV/HIKE Stuttgart]

14.6 Soft Robot Surface Design and Surface Functionalization

Textiles can be functionalized in many ways. A textile can provide a basis so that the robot is aesthetically pleasing. Textiles can be soft and compliant, which are able to create a pleasant touch and atmosphere. So, the inhibition threshold to work closely together with, or be supported by, a robot is decreased. ITV Denkendorf has developed a functional fiber which both emits light in different colors and has tactile functions. Thus, it can be used to show information / entertainment or display a pleasant surface. Additionally, a textile surface equipped with those fibers is tactile and responds to changes in capacitance and pressure (Fig. 14.20). When approached, the sensors give a first signal in form of a voltage output or a color change, (II), when the sensors have contact with human skin they deliver a higher voltage output and change color again (III).

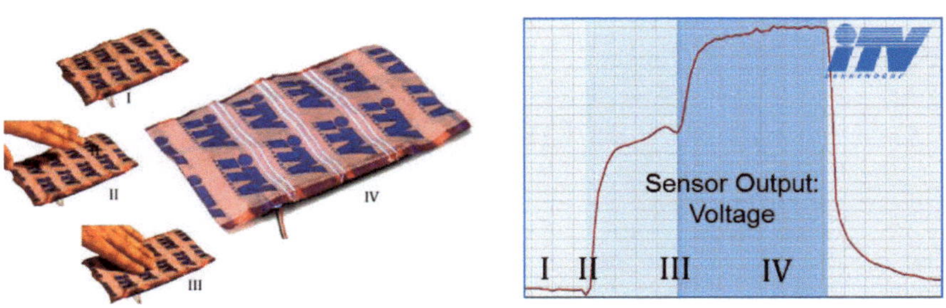

Fig. 14.20 Textile tactile sensor [C. Riethmüller, ITV]

Deploying these approximation sensor-functions into textile robot-skins, robot arms can ensure much better operational safety in human or sensitive surroundings. The light-weight 3D-textiles exhibit compliance and therefore facilitate reaction/stopping time before the inner robot skeleton hits other matter.

Integration of actuators, like piezo-ceramic fibers into the textile materials, makes it possible to actively measure the vibration, to actively induce damping or to change the robot arm geometry to lessen impact damage. An example of using piezo-ceramic fibers can be found in helicopter rotor blades to actively reduce vibration and noise.

14.7 Conclusion

The further advancements of soft robotics rely on the development of novel materials, structures, sensors and actuators. Advanced robot design and optimization combines mobility and "soft" aspects of strength/stiffness. In this regard, the fibrous materials have an enormous potential to contribute in the continued development of soft robotics. Many possibilities that have been discussed in this chapter are still in the lab scale. The commercialization of these technologies may need highly interdisciplinary collaboration. Although, special challenges will remain: the ability to optimally use, and process, fibrous materials will allow far more sophisticated design and cost-effective fabrication of soft robots. Still, a lot of work has to be done to understand the behavior of fibrous materials when transferred into soft robot structures, especially with respect to simulation and visualization. This will lead to improved, and even more advanced, fiber path planning as well as more accurate and reliable control algorithms.

Acknowledgments The authors like to thank the German funding bodies "BMBF, BMWi, AiF and DFG" and the ministries of state Baden Württemberg for the financial support extended for successful completion of the projects.

15 Opportunities and Challenges for the Design of Inherently Safe Robots

Annika Raatz, Sebastian Blankemeyer, Gundula Runge, Christopher Bruns, Gunnar Borchert

Leibniz University Hannover

Abstract An approach for solving the challenges that arise from the increased complexity of modern assembly tasks is believed to be human robot co-operation. In these hybrid workplaces humans and robots do not only work on the same task or interact during certain assembly steps, but also have overlapping workspaces. Therefore, 'safe robots' should be developed that do not harm workers in case of a collision. In this chapter, an overview of methods for designing a hardware based soft robot that is inherently safe in human-machine interaction is given. Recent projects show that robots could be soft enough for interaction but they are not able to resist forces that occur in the assembly process. Current solutions show that the designer of such robots must face a trade-off between softness and dexterity on the one hand and rigidity and load carrying capabilities on the other hand. A promising approach is to integrate variable stiffness elements in the robotic system. The chapter classifies two main design rules to achieve stiffness variability, the tuning of material properties and geometric parameters. Existing solutions are described and four concepts are presented to show how different mechanisms and materials could be combined to design safe assembly robots with a variable stiffness structure.

15.1 Introduction

Due to more and more complex assembly tasks and the high diversity of product variants in production environments, it is necessary to improve hybrid workplaces. This kind of workplace combines the advantages of human beings, such as their high flexibility and sensitivity, with industrial robots that can operate continuously with high precision [1]. Because of this, safety plays a crucial role in human-robot interaction to protect workers from injury. Hence, inherently safe systems utilize passive non-actuated mechanisms that can swerve to avoid a collision, for example with the human body.

Kim et. al. [2] distinguish between a software and hardware based safety approach. For example, established systems for human-robot-interactions are the KUKA LWR or the ABB FRIDA. The LWR has an optimized weight-to-payload ratio and additional torque sensors in its joints, allowing for a programmable compliance and with this a direct human-robot interaction [3]. The FRIDA achieves safety through hardware, such as utilizing motors with small torque and a crush protection structure to prevent danger [4]. A desirable solution would be a passive hardware mechanism that is inherently safe at its instant of impact without superior activation control. From this mechanical viewpoint, Festo's bionic handling assistant is structurally soft, but it lacks precision and ability to withstand process forces [5]. As indicated by the above examples, designers of robots face a trade-off between softness and dexterity on the one hand and rigidity and load carrying capabilities on the other hand.

A desirable approach for creating a soft and, thus, inherently safe robot for human-robot interaction is still an evolving challenge. In this context research done in the field of 'Soft Robotics', dealing with soft and deformable structures, offers high potential to develop new solutions. Robots made of soft materials with tunable characteristic properties can actively vary their stiffness, allowing high deformations and dexterity as well as sufficient stiffness and load capacity. Such robots are deemed, passively i.e. inherently safe because their compliance matches that of human bodies.

The main challenge will be to identify new design principles for the integration of these soft, tunable materials into the robot structure. This is to make use of their chemical or physical effects to absorb a large proportion of the kinetic energy or to yield in the event of a collision.

15.2 State of the Art in Soft Robotics

This chapter will give a short review of characteristic forms of soft robots that have been developed in recent years. The evaluation of published solutions is used to determine practical design principles, so the focus is laid on robots built from soft materials that can vary their stiffnesses and undergo a high degree of deformation.

Trivedi et. al. assert that soft robots can deform because of their infinite degrees of kinematic freedom [1] and the soft materials they are made of. Soft robots are the foundation of continuum robots because they can reach a predefined point in three dimensional space with an infinite number of robot configurations [1]. Because of this high degree of structural flexibility, the mechanical resistance in the event of impact can be reduced.

Inspiration for the development of safe robots can be found in nature, such as octopus arms [2], elephant trunks [1] or earthworms [6]. Such robots could consist of soft and lightweight materials and components [1] actuated by, for example artificial muscles [7, 8], springs [2] or tendons [9]. Robotic researchers show that there are many different fields of applications, such as medical surgery for using the previously described soft robot structures. Tommaso et. al. [8] present a modular, soft, variable stiffness manipulator for minimally invasive surgery. Another possibility to apply soft robots in medical fields is presented by [11] because of its high dexterity, softness, and encapsulated shell. Additionally, robots' ability to squeeze through small holes make them adept at walking or crawling through unknown terrain [2, 10]. These snake-like robots move by creating travelling waves by contracting and expanding their cylindrical bodies [2]. Another approach currently in the early stages of manufacturing is the already mentioned bionic handling assistant [5]. It is a biologically inspired robot arm made of soft plastics and actuated by fluid pressure. Another biologically inspired trunk robot with many segments connected by compliant joints and actuated by shape memory alloy (SMA) is presented in [12]. Majidi [10] asserts that the concepts of soft robotics could be extended. Material compliance matching—the principle that two materials should share the same mechanical rigidity to spread and minimize the acting stress concentrations—could be applied in healthcare, field exploration, and cooperative-human-assistance [10]. For this purpose, the key is to identify a useful measureable property, such as Young's modulus, for adapting the rigidity of a robot. For example, since steel has a modulus greater than 10^9 Pa, it is poor match for humans whose skin and muscle tissue have a modulus of 10^2 Pa-10^6 Pa [10].

But for all their advantages, such as dexterity, softness and safety, soft biological structures have some physical limitations [2]. To support their own weight without a skeleton, they must be very small or if they are large, supported by a surrounding medium such as water or terrain [2].

15.3 Design of Soft Robots with Variable Stiffness

To improve upon existing design approaches in soft robotics, principles that permit the design of robots within the boundaries of desired softness and load-bearing capabilities need be identified. A widespread approach to this problem is given by the notion of variable stiffness. Aside from soft robotics, concepts applied in variable stiffness mechanisms have been studied in disciplines such as in material sciences, aerospace engineering, and human–computer interaction.

From a mechanical viewpoint, stiffness is defined as the measure of resistance of a body to deformation. Different types of stiffness are distinguished as the resistance to different kinds of loads. Bending stiffness S, or flexural rigidity, provides a measure for the resistance to bending, and is mathematically expressed as

$$S = E \cdot I, \tag{1}$$

where E denotes Young's modulus and I the area moment of inertia. Similarly, the torsional stiffness K_t is given by the product of the shear modulus G and the torsional moment of inertia I_t:

$$K_t = G \cdot I_t. \tag{2}$$

The above equations imply that two basic strategies can be identified in variable stiffness mechanisms: in-service adaptation of material properties and alteration of structure geometries. Many of the above examples in soft robotics exploit the former of these strategies. Stiffness variation based on tuning of material properties makes use of intrinsic material effects. An overview of different design elements exploiting material properties for stiffness variation as well as a choice of corresponding design parameters are given in Table 15.1. Furthermore, examples of applications that adopt one or more of these principles are provided in the table. The presented elements are mainly incorporated as passive structure components, however, some of them, e.g. shape memory polymers, are primarily used for actuation.

In soft robotics research, jamming of granular matter has recently received widespread attention as a way of changing the stiffness of passive elements. Jamming is a phenomenon that occurs in many amorphous materials as diverse as glass, colloidal systems, foams, and granular media. It describes the transition from a disordered, fluid-like state to a rigid, solid-like state. Similar to solid bodies, jammed media exhibit a yield stress, where the forces are distributed across chains of particles [13]. The transition from a jammed to an unjammed or liquid state and vice versa is induced by one or more of the following parameters: temperature, density and load. For the purpose of deriving design principles in soft robotics, this transition is considered a change in material properties as the jammed body exhibits a yield stress and can then be treated with similar design guidelines as those for solid bodies. Most applications exploit jamming of granular media by applying a vacuum to a non-porous, flexible membrane filled with particles to remove excess air. Jamming like behavior has also been achieved with layers of overlapping plates that have a sufficiently high frictional coefficient [2]. The stiffness of this so called layer jamming mechanism is also varied by applying a vacuum.

Rheologically complex substances such as magneto- (MR) or electrorheological (ER) fluids have been studied in soft robotics, for example, as tunable stiffness fluids in soft actuators [14] or rigidity-controlled artificial muscles [15]. However, the high electrical and magnetic fields still pose a limit to the use of MR and ER fluids in many robotic applications.

Shape Memory Polymers (SMP) have attracted widespread attention, not only in soft robotics but also in morphing applications such as in aerospace engineering. SMP can reverse their elastic modulus by switching between a lower-temperature glassy state and a higher-temperature rubbery state. Compared to traditional shape memory alloys, SMP feature a higher transduction coefficient but at the cost of a higher response time.

Beyond traditional engineering materials, new classes of gels and colloidal substances are believed to evolve from the field of soft robotics [10], some of which may open new frontiers for researchers seeking to design soft biomimetic machines [16]. Many of these novel fluids and gels exhibit non-Newtonian behavior, i.e. shear-thinning or shear-thickening properties, which can be deliberately exploited for the design of variable stiffness elements for soft robots. In particular, nanocomposite (NC) hydrogels have been shown to have extraordinary mechanical properties compared to conventional chemically crosslinked gels [17]. For NC polymer-clay gels, for example, variations in viscosity of four orders of magnitude have been reported [18].

Table 15.1 Tuning of material properties for variable stiffness mechanisms.

Design principles	Design parameters	Applications
Granular jamming	Porosity, size and shape distribution of grains, interstitial fluid	STIFF-FLOP [8], FormHand [19]
Layer jamming	Number of contact surfaces, frictional coefficient, width and length of flaps	Snake-like manipulator [11]
Magneto-rheological/ electro-rheological fluids	Magnetic material, particle size and concentration	MR fluid rubber actuator for walking robot [14], micropatterned elastomer MR [15]
Shape memory polymers (SMP)	Structure (physical or chemical cross-linked, amorphous), particles (microfibers, micro or nano particles), other ingredients and fillers	Artificial rubber muscle using SMP [20]
Novel fluids and gels, e.g. NC hydrogels, electro-active polymers	Hydrogels: composition of gels and particles, molecular weights	Hydrogel artificial muscles [21]

Variation of geometric parameters, as classified in Table 15.2, seems to have been far less explored in the field of soft robotics than variation of material properties. One reason for this may be that soft robotics draws much of its inspiration from biology, animals and humans in particular, and thus researchers have turned their interest to muscle tensioning as a means of stiffness modulation [2]. Consequently, research has mainly focused on the development and improvement of artificial muscles rather than on the design of soft, variable stiffness structural elements. Few examples in soft robotics exist where stiffness variations are accomplished through tuning of geometric parameters [22]. By studying alternative ways of actively tuning the stiffness of otherwise passive structure elements, researchers may gain inspiration for new designs of inherently soft robots.

One way of tuning the stiffness of an entire structure is to change its geometry. This type of stiffness variation has primarily been studied in morphing applications, e.g. for shape-adaptation of airplane wings. Changing the structure geometry essentially boils down to changing the area moment of inertia. For example, the area moment of a collapsible honeycomb structure can be altered by folding and unfolding it. Such morphing structures have been proposed, e.g. in the form of variable stiffness composite materials (VSM), where the composing cellular elements can take on any gradation of shapes between two different stiffness states [23].

Another method is to alter the effective stiffness of the springs that are contained in the structure. Such variable stiffness mechanisms have been developed, for example, for patient rehabilitation machines [24] and tunable stiffness legs for walking robots [25]. The herein presented designs vary the spring stiffness by changing the effective length of the spring. A similar approach has been presented in [26], which describes a stiffness-adjustable endoskeleton-like structure with alternating compliant and rigid segments. The compliant segments are stiffened by compressing them with an axial force. These segments can be thought of as having an equivalent spring constant. So the stiffness of the entire structure becomes a function of the axial compression.

Finally, shape locking mechanisms can also be employed to switch between flexible and rigid modes in a mechanism. The so-called Dragon Skin structure presented in [26] consists of scale-like elements sheathed in a sealed cover. In contrast to the layer jamming concept described above, the elements are held together through a geometric interlocking mechanism when a vacuum is applied. Another shape locking mechanism that uses a composite of constant stiffness elements and variable stiffness elements is given in McKnight and Henry [23]. In this design, the constant stiffness elements carry the structural load, whereas the variable stiffness elements provide variable connectivity between the rigid elements.

Aside from the methods presented in Table 15.1 and Table 15.2, there are several mechanisms that combine principles from both design strategies simultaneously. In inflatable honeycomb structures [27] intended for morphing of airplane wings for example, the stiffness is varied by changing the cross section of the structure, and thus the area moment of inertia, as well as the pressure inside the cells. A change in pressure then corresponds to a change in the bulk modulus of the composite structure. Other composites which are based on a modulus change are fluidic flexible matrix composites (F^2MC). They are a class of controllable stiffness structures with origins in morphing which may spawn new designs in soft robotics [28]. F^2MC structures consist of fiber wound tubes, where the fibers are arranged in such a way that the tubes exhibit a pronounced anisotropy when pressurized. The stiffness is tuned by controlling the fluid flow in and out of the tube. Embedded in a matrix, multiples of these tubes form building blocks of a variable stiffness composite, e.g. a honeycomb sandwich structure. Related mechanisms have been studied in soft robotics as well. Networks of inflatable chambers have been used for rapid actuation of soft robots [29].

Table 15.2 Tuning of geometry parameters for variable stiffness mechanisms.

Design principles	Design parameters	Applications
Variable geometry structures	Structure material, morphing shape	Deformable variable-stiffness cellular structures [23]
Variable spring stiffness mechanisms, variable stiffness spring-like mechanisms	Spring material, spring geometry, spring length	Variable stiffness mechanism using wire springs [24], stiffness adjustable endoskeleton [22], tunable stiffness composite legs [25]
Shape locking mechanisms	Shape locking material, (laminate material), shape locking geometry	"Dragon skin" shape-locking mechanism [26], variable stiffness materials for reconfigurable surfaces [23]

The classification presented in this section is intended to aid researchers in soft robotics in developing new concepts of tunable stiffness designs to include in their robots. In combining several of the principles outlined above, designs with an even larger range of stiffness variability may be obtained.

15.4 Concepts

The principles stated in Tables 15.1 and 15.2 offer new applications within robotics, but they also raise challenges for the designer. For example, the presented variable stiffness fluids follow different laws than conventional materials. With shear thinning materials, the normal forces which usually occur during an impact must be constructively converted into shearforces since only they will produce the shear thinning effect. In addition, new ideas of stiffness adjustment must be developed to optimize the designs.

The question arises whether these principles and materials are useful for designing soft robots for assembly tasks. As already mentioned, for complex assembly tasks human-robot interaction is a favorable approach to reduce complexity and increase automation where applicable. This entails that the robot must be inherently safe while maintaining its production relevant performance. Depending on the task and the forces which occur during the process, different design types are conceivable. Design type 1 is a robot with a mass-optimized skeleton and an outer soft shell provides limited passive safety protection. In this design, inertial forces and dynamics, depending on the softness and damping capability of the outer shell, should be reduced to achieve desired protection. The advantage of this design is that considerable process forces or loads can be applied. Design type 2 is a highly segmented robot where soft links are combined with small rigid elements. This type is a hybrid structure between type 1 and 3. The potentially softest, and presumably safest structure is type 3, which is a continuum robot built only with soft materials. In addition to the stiffness adaptation to resist forces, continuum robots face more challenges, such as the actuation and kinematics of the robot's position.

The following four design concepts are based on the previously stated principles and are illustrated in Figure 15.1. Concept (a) shows one link that uses the principles of variation of the geometric parameters combined with the mechanism of granular jamming. In concept (b), the principles of layer jamming and an adjustable Young's modulus are implemented in one link. The variation of the geometric parameters and the use of variable stiffness fluids are utilized for the complete robot concept (c). The design idea (d) is based on the principle of variable stiffness fluids.

The first design (Fig. 15.1, a) utilizes pneumatics energy. The body consist of two chambers. The outer chamber is comprised of a solid, granular material that becomes rigid when stressed. The inner chamber is sealed but the air pressure can be regulated by an air valve. When the pressure is increased, the filler in the outer chamber is compressed which enhances the rigidity of the entire system. Compared to the presented soft robots based on granular jamming, this concept uses pressured air so that the link always deforms to the same predefined shape imposed by the outer shell for a given pressure. The use of liquid material for the filler could be an interesting research subject.

The structure of concept (b) is similar to (a). In the outer chamber, plates are embedded in shear thinning material while a vacuum is created in the inner chamber. Due to the adjustable vacuum in the chamber, the plates lie against each other and stick together. The normal forces that occur during a collision are converted into shearforces due to shifting of the diagonally placed plates. An alternative idea is to remove the vacuum while increasing the cross section or the behavior of the shear thinning material to withstand the forces that occur during the process.

The design in Figure 15.1, c shows a concept with alternating rigid and soft segments. The soft segments are made of a hyper elastic hydrogel that preferably matches the stiffness of the human body (compliance-matched). The concept can be actuated by wires that pass through the inside of the robot, and the rigid plates can be clamped to restrain the way the robot moves. By pulling and releasing the wires the effective spring length of the soft segments is changed and therefore the stiffness is adjusted. Further improvement could be made by mixing additives to the matrix to alter the hydrogel properties, e.g. anisotropy or shear thinning behavior. Anisotropy could reinforce the robot's axial direction while keeping the radial direction soft. In other concepts shear thinning behavior could be used to let the rigid plates slip away during a collision.

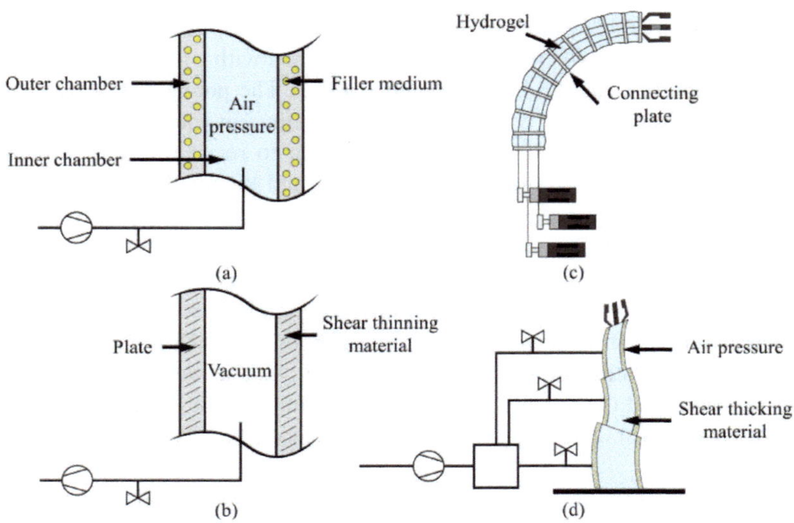

Fig. 15.1 Concepts derived from the design principles

The last concept is based on the shear thickening effect of some materials, e.g. a water-cornstarch suspension. The example in Figure 15.1, d contains three links that are separated by connecting plates. The links' structure is the reverse of concept (a). The inner chamber consists of shear thickening material while the outer chamber works as a pressure actuated, pulsating shell. Because of the pulsating stimulation, the shear thickening material becomes rigid and can withstand forces that occur during the process. These design concepts are a small extract of what could be devised with the stated principles. Obviously, there are other combinations of principles to create a variable stiffness robot.

15.5 Summary and Outlook

Opportunities and challenges for the design of inherently safe robots have been proposed. Existing approaches to safe robots are generally classified into software or hardware based solutions. Hardware based mechanisms, such as soft components with variable stiffness, were assessed to be particularly promising to provide inherent safety. Hence, state of the art designs of soft robots were presented, focused on the trade-off between stiffness and softness as well as stiffness variability. Two main aspects were introduced to achieve stiffness variability: tuning of material properties and geometric parameters. This classification can be used as a basis for the development of variable stiffness robot components and entire robotic systems. As a start, four initial concepts were introduced that rely on the variation of geometric parameters as well as the variation of Young's modulus.

The chapter shows only a small extract of topics which can be addressed within the field of soft robotics. To meet all the challenges, but also to explore all the potentials which arise in the use of soft structures, scientist and engineers from different disciplines need to intensify their collaboration. Extensive studies on soft materials, their fabrication and integration, to develop new tunable material behavior and improve their deployability in robotic applications, seem to be a key point. Once this has been achieved there will be further need for novel solutions in actuation, sensing, modelling, and control.

Acknowledgments The authors would like to thank Prof. H. Menzel (Institute of Technical Chemistry, TU Braunschweig) and Prof. G. Garnweitner (Institute for Particle Technology, TU Braunschweig) for their extensive discussions about new materials.

15.6 References

[1] Trivedi D, Rahn C, Kier W, et al. (2008) Soft robotics: Biological inspiration, state of the art, and future research. Appl. Bionics and Biomechanics 5(3): 99-117

[2] Kim S, Laschi C, Trimmer B (2013) Soft robotics: a bioinspired evolutions in robotics. Trends in Biotechnology 31(5):287-294

[3] Bischoff R, Kurth J, Schreiber G, et al (2010) The KUKA-DLR Lightweight Robot arm - a new reference platform for robotics research and manufacturing. 41st Robotics (ISR) and 6th German Conf. on Robotics (ROBOTIK): 1-8.

[4] Matthias B, Kock S, Jerregard H, et al. (2011) Safety of Collaborative Industrial Robots. IEEE Int. Symp. on Assembly and Manufacturing (ISAM): 1-6

[5] Rolf M, Steil J (2012) Constant curvature continuum kinematics as fast approximate model for the Bionic Handling Assistant. IEEE /RSJ Int. Conf. on Intelligent Robots and Systems: 3440-3446

[6] Seok S, Onal C, Cho K, Wood R (2012) Meshworm: A Peristaltic Soft Robot with Antagonistic Nickel Titanium Coil Actuators. IEEE/ASME Transactions on Mechatronics: 1485-1497

[7] Kang R, Branson D, Zheng T, et al. (2013) Design, modeling and control of a pneumatically actuated manipulator inspired by biological continuum structures. Bioinspiration & Biometrics 8(3)

[8] Ranzani T, Cianchetti M, Gerboni G, et al. (2013) A modular soft manipulator with variable stiffness. 3rd Joint Workshop on New Technologies for Computer/Robot Assisted Surgery

[9] Cheng N, Lobovsky M, Keating S, et al. (2012) Design and Analysis of a Robust, Low cost, Highly Articulated Manipulator Enabled by Jamming of Granular Media. IEEE Int. Conf. on Robotics and Automation: 4328-4333

[10] Majidi C (2013) Soft Robotics: A Perspective – Current Trends and Prospects for the Future. Soft Robotics 1(1):5-11

[11] Kim Y, Cheng S, Kim S, Iagnemma K (2013) A Novel Layer Jamming Mechanism With Tunable Stiffness Capability for Minimally Invasive Surgery. IEEE Transactions on Robotics 29(4): 1031-1042

[12] Schmitt J, Last P, Löchte C, Raatz A (2009) TRoBS – A Biological Inspired Robot. IEEE Int. Conf. on Robotics and Biomimetics: 51-56

[13] Follmer S, Leithinger D, Olwal A, et al. (2012) Jamming User Interfaces: Programmable Particle Stiffness and Sensing for Malleable and Shape-Changing Devices. 25th ACM Symp.on User interface software and technology: 519-528

[14] Taniguchi H, Miyake M, Suzumori K (2010) Development of New Soft Actuator Using Magnetic Intelligent Fluids for Flexible Walking Robot. Int. Conf. on Control, Automation and Systems, Oct., 2010: 1797-1801

[15] Majidi C, Wood R J (2010) Tunable elastic stiffness with microconfined magnetorheological domains at low magnetic field. Applied Physics Letters 97: 164104-1-164104-4

[16] Calvert P (2009) Hydrogels for Soft Machines. Adv. Materials 21: 743-756

[17] Haraguchi K (2007) Nanocomposite hydrogels. Current Opinion in Solid State Materials and Science 11:47-54

[18] Loizou E, Butler P, Porcar L, et al. (2006) Dynamic Responses in Nanocomposite Hydrogels. Macromolecules 39:1614-1619

[19] Löchte C, Kunz H, Schnurr R, Dietrich F, Raatz A, Dilger K, Dröder K. (2013) Form-Flexible Handling Technology for Automated Preforming. 19th Int. Conf. on Composite Materials (ICCM), Montreal, Canada, e-proc

[20] Takashima K, Noritsugu T, Rossiter J, et al. (2011). Development of Curved Type Pneumatic Artificial Rubber Muscle Using Shape-memory Polymer. SICE Annual Conference, Tokyo, Japan: 1691 - 1695

[21] Rogers J A (2013) A Clear Advance in Soft Actuators. Material Science 341:968-969

[22] Hu T M, Park Y-J, Cho K-J (2012) Design and Analysis of a Stiffness Adjustable Structure Using an Endoskeleton. Precision Engineering and Manufacturing 13(7): 1255-1258

[23] Mc Knight G P, Henry C P (2010) Deformable variable-stiffness cellular structures. US Patent 7,678,440

[24] Hayashibara Y (2008) Study on Variable Stiffness Mechanism Using Wire Spring. Robotics and Mechatronics 20(2): 296-301
[25] Galloway K C, Clark J E, Koditschek D E (2009) Design of a tunable stiffness composite leg for dynamic locomotion. Proc. of IDETC/CIE 2009: 1-8
[26] Zuo S, Iijima K, Tokumiya T, Masamune K (2013) Variable stiffness outer sheath with "Dragon skin" structure and negative pneumatic shape-locking mechanism. Computer Assisted Radiology and Surgery 2014: 1-9
[27] Vos R, Barret R, Romkes A (2011) Mechanics of pressure-adaptive honeycomb. Intelligent Material Systems and Structures 22(10): 1041-1055
[28] Shan Y, Lotfi A, Philen M (2008) Fluidic Flexible Matrix Composites for Autonomous Structural Tailoring. Proc. of SPIE 6525: 652517-1-652517-14
[29] Shepherd et al. (2011) Multigait soft robots. Proc. of the National Academy of Science 108(51):20400-20403

16 Aspects of Human Engineering – Bio-optimized Design of Wearable Machines

C. Hochberg, O. Schwarz, U. Schneider

Fraunhofer Institute for Manufacturing Engineering and Automation IPA, Stuttgart

Abstract This chapter outlines important factors for the design process of wearable robots. First, the challenges are discussed and possible user groups are detailed and categories of devices given. Then, major differences of classical design methods from the field of robotics are illustrated. This is due to linking between the machine and the user and challenges of user intention detection. Finally, some design approaches, guidelines and best practices for the development of wearable devices are discussed.

16.1 Introduction

16.1.1 The Challenge

Currently, physical injuries and diseases leading to loss or impediment of parts of the musculoskeletal system can only be rehabilitated or alleviated by high tech devices.

This includes training, or substitution, supported by complex technical systems. These systems are becoming ever more closely linked to the human body. Thus, design approaches have to be determined, which allow the development of safe and effective devices focusing on all relevant aspects.

The challenge is to bring design approaches and solutions form robotics, computer science, health care, physiology, psychology and many more fields together to create an integrated, user-friendly, and safe product.

Traditionally, robotic design approaches take the human only into moderate account, as the machine and the operator are usually apart from each other and should even be shielded by fencing. Developing machines which are to be worn by a human pose very different challenges. The design process has to take human anatomy, psychology, and other needs of a varying, unpredictable and emotional operator into account.

16.1.2 Prevalence

Stroke Patients

A stroke is a sudden disease of the brain, which can lead to failure of brain functions. It is caused by a critical disruption of the blood flow inside the brain. This leads to an abrupt reduction of oxygen and other substrates in the nerve cells [1]. Epidemiology of strokes in Germany:

- Stroke due to too low blood circulation: 80% of all incidents and 150-240 per 100.000 people [2, 3]
- Stroke due to bleeding inside the brain: 20% of all incidents and 24 per 100,000 people [2, 3]
- 196000 first strokes and 55,000 reoccurring strokes in Germany each year (2008) [4]
- Stroke is 3rd most frequent cause of death in Germany with 63,000 fatalities a year (2008) [4]
- Strokes are the 2nd most frequent cause of death worldwide [5]
- Strokes are one of the frequent causes for disabilities [6]

Prevention of Bodily Injury

A number of professions perform lifting activities that lead to musculoskeletal overuse. It is not the mass of the load alone being lifted and carried, but the static position in which the spine is loaded. Especially, high loads to the lumbar backbone are affected [7]. Professions in logistics, assembly, and the military (HULC) are in the target market of exoskeleton developers.

In the nursing profession, for example diseases of the spine, muscles and joints are a serious problem. These represent 35% of all diseases leading to disability. This is the main cause of early retirement [8]. 85% of women who are suspected of such a work-related disease (BK 2108, [9]) are working in the health sector. The reason is that these women spent 25% of their work day in biomechanically harmful postures. The CUELA [8] study examined this by studying the inclinations of health care workers' bodies during a single work day. The study counted 1541 waist inclinations with an angle of more than 20 ° per shift on average [10, 11]. Thus, the nursing staff spent a total average of 2 hours of their working day in an unhealthy position.

Exoskeletons can support transporting heavy loads and avoid the frequent declination, twisting and lateral bending of the upper body. The challenge will be to combine the functions of force enhancement and physical prevention in a light, portable, exoskeleton with high wearing comfort and user-friendly design.

16.2 Designing a Wearable Robot: State of the Art

16.2.1 Different Types of Exoskeletons

To gain an overview of the state of the art of wearable robots they are categorized by their different design goals and intended applications. These categories with respective examples are listed in Table 16.1 and pictures of examples are given in Fig. 16.1.

Functional Replacement describes devices that help patients who have lost functions of their musculoskeletal system such as in paraplegia. They replace the lost function by technical means.

Devices in the category *Therapy and Rehabilitation* are designed for training of patients who have lost, or impeded, musculoskeletal functions. The goal is a partial, or full, recovery of the patient. Furthermore the design focuses on the training task rather than assisting with other tasks such as household chores. It is supposed to become obsolete after the training was accomplished.

Physical Prevention and Force Assistance represents a category of devices which are worn by mostly healthy users or people with minor injuries or signs of fatigue due to physical stress. Such devices are mostly design for work environments and can be applied in fields such as logistics, manufacturing, and healthcare. The user is usually not a patient, but has the option of wearing the device.

The intention is to reduce or prevent physical pain such as chronic back pain, herniated disks, or tennis arms created by stress through heavy lifting, long term stress, or physiologically non-optimal movements. This is mostly the case in routine manual work. As a result of wearing the device the worker is supposed to have a longer, and overall healthier, work life with less absences and more motivation.

The category *Auxiliary Assistance* refers to vision technology for the field of wearable robots, where a wearable device gives the user extra limbs, such as arms, to be able to perform more complex tasks alone, without the help of other people. However, as this robot is body worn, it is not independent from the wearer and has a lesser degree of autonomy as stand-alone assistant robots. Its possibilities will be discussed at the end of this chapter.

Table 16.1 Classification of different types of exoskeletons

Class of exoskeleton	Example of exoskeletons
Functional Replacement	ReWalk, Mina, Active knee prosthetics
Therapy and Rehabilitation	Ekso, Mina, Active arm orthesis
Physical Prevention and Force Assistance	Light Exoskeletons for logistics, production or health care, Bodyweight Support Assist,
Auxiliary Assistance	Body worn Cobots with extra limbs (vision)

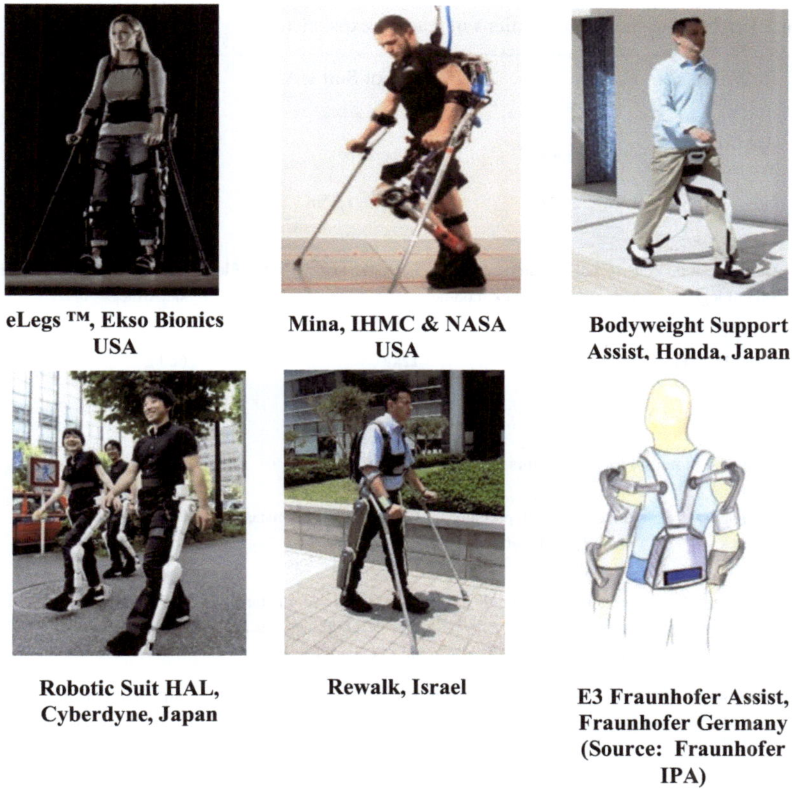

Fig. 16.1 Different Exoskeletons for different purposes

16.2.2 Power and Drives

As shown in Table 16.2, all leading exoskeletons use one or two pairs of electric motors as their source of power. However, depending on the setup and application, the operating times vary.

Table 16.2 Performance and specifications of leading exoskeletons

Product	ReWalk	eLegs	Robot Suit HAL	Bodyweight Support Assist
Company	Argo Medical Technologies	Ekso Bionics	Cyberdyne	Honda
Body Height	N/A	1,50m – 1,90m	1,45m – 1,85m	1,55m – 1,90m
Type	Bi-pedal, at outer side of legs,	Bi-pedal, at outer side of legs, with backpack module	Bi-pedal, at outer side of legs	Bi-pedal, chair-like at inner side of legs
Weight	about 15-18kg	23kg	12kg	6,5kg
Operating Time	N/A	about 6h	about 1h	about 2h
Power Unit	4 Electric Motors	4 Electric Motors	4 Electric Motors	2 Electric Motors
Interface	Remote-control, joystick	Remote-control, pressure sensor for detecting forward leaning	Electrodes at skin measure muscle activity	Force/torque sensors at shoes and motors
Application	Functional Replacement of paraplegic patients	Therapy of paraplegic patients	Rehabilitation for patients with cerebral, nervous and muscle disorders	Physical Prevention and Force Assistance
Source	http://rewalk.com/	http://www.eksobionics.com/ekso/faq	http://www.cyberdyne.jp/english/customer/index_4.html	http://corporate.honda.com/innovation/walk-assist/

16.2.3 Detection of User Intention

Depending on the task, or the application, a variety of control methods and signals can be used to determine the wearer's intentions. These can be sorted from simple signals, to complex control schemes. As the machine is physically linked to the user, even a control scheme can be regarded as an interface and a crucial safety aspect at the same time.

It has to adhere to fundamental human engineering concepts and psychological rules. Therefore, great emphasis has to be put on the mental abilities of the users to understand the intent of their input. Thus, the interface and control scheme has to allow the wearer to generate a mental model of reactions of the machine for it to remain predictable. A simple example is the interaction of a bicycle and its rider. Pushing the pedal leads to acceleration forward. This is easily comprehended by any rider.

Additionally, several aspects of human anatomy and self-perception might interfere with the machine's control scheme. One is the users' subconscious urge for maintaining balance. Powerful, spontaneous acceleration and highly unpredictable torque from the exoskeleton can lead to uncontrollable counter reactions by the user, which could pose a safety risk.

Furthermore, the interface has to take the user's proprioception into account. This allows the user to understand, without seeing where their body parts are positioned, their pose and what forces they apply. Thus, high permanent forces from the exoskeleton to the wearer might lead to mental confusion. Furthermore, unintended movements of inactive extremities of the user might also lead to mental distraction.

Allowing the user to generate a mental model of the machine's reaction includes considering the user's motion planning. As the wearers are autonomous themselves but still linked physically to the control scheme of the exoskeleton, most parts of the motion planning have to be done by the user. Furthermore, systems have to comply with the user's impetus to always execute the simplest or the most direct motion paths.

As most of these effects are subconscious, and related to the user's sense of safety, it is important that they are carefully considered during system design. Otherwise, the system might violate the user's feeling of personal autonomy and safety or simply seem to interfere with, and dominate, the user. This might lead to quick rejection and difficulties convincing the user of wearing other exoskeletons in the future.

The goal is to choose a fast and reliable interface method for the task at hand. Therefore several interfaces and modes of user intent detection are listed below, in order of their complexity:

Pre-Programmed: An algorithm follows a pre-programmed movement with little user interaction.

Example: an active arm orthesis for stroke rehabilitation.

Applicable in: Therapy and Rehabilitation

Design Considerations: Only On-Off switch and optimally a mode selection for speeds or rhythm variations. Only simple movements are possible.

Push-Button: Pushing of a button executes a movement instantaneously.

Example: Leg exoskeleton for paraplegic patients

Applicable in: Functional Replacement, Therapy and Rehabilitation,

Design Considerations: Direct detection of user intent. No complex movements preferable, movement variation is limited due to limited button space, no automatic adaption to different situations or terrains.

Pressure Shift: The exoskeleton detects a shift or change of pressure between the user and the exoskeleton.

Example: paraplegic users shift their weight to control a leg exoskeleton

Applicable in: Functional Replacement, Therapy and Rehabilitation, Physical Prevention and Force Assistance

Design Considerations: Rather slow detection of user intent, noisy signal, Fuzzy Control possible, interface might interfere with sense of balance,

EMG-Deduction: EMG electrodes are in contact with the user's skin and measure the muscle tonus.

Example: A rehabilitation device for patient with cerebral, nervous, and muscle disorders

Applicable in: Functional Replacement, Therapy and Rehabilitation, Physical Prevention and Force Assistance, Auxiliary Assistance

Design Considerations: The sensor has to touch the bare skin of the user. This might lead to hygiene or disinfection problems. EMG has a limited accuracy and high noise, which might lead to safety risks. Only 96% accuracy is reachable when combining EMG with other sensors [12]. As other movement might interfere with the interface, the user intention might not be clearly detectable. EMG is only applicable if the user groups are expected to have some measureable muscle tone.

Sensor Fusion: Using a variety of the above, with extra sensors such as IMUs, radar, or laser, many different signals generate a probability of the user's intent or a situational change.

Example: Change of gait in an active knee prosthesis when climbing stairs

Applicable in: Functional Replacement, Therapy and Rehabilitation, Physical Prevention and Force Assistance, Auxiliary Assistance

Design Considerations: The intent is only indirectly detected. This might be too vague and lead to misinterpretations or long interpretation cycles, which in turn might confuse the wearer and lead to safety risks.

Torque Control: The torque at the joints of the exoskeleton is measured and different control algorithms derive a reaction of the exoskeleton from these measurements.

Example: An exoskeleton supporting a worker in an assembly line

Applicable in: Therapy and Rehabilitation, Physical Prevention and Force Assistance, Auxiliary Assistance

Design Considerations: Different control algorithms such as Admittance control or Zero-Torque control are possible depending on the task. It might be difficult to interpret the user's intention when an outside force is applied through an extra load for example. The reaction time might be too slow.

16.2.4 Human Anatomy

The human anatomy, biomechanical properties, as well as their limitations and variations are a key aspect for designing a wearable device. These are the crucial differences from the classical design constraints for robotics. The most important aspects, and brief conclusions, for their implications on the design process are outlined below.

Complexity Difference of Human and Machine Joints

In most cases the human body's joints do not just have one fixed axis of rotation like in a mechanical joint. Rather, their movement consists of a combination of rotation and sliding. This is called a pole curve [13].

To emphasise this problem, the human knee is taken as an example. The knee as mentioned above consists of a roll-gliding motion. At the beginning of leg flexion the parts of the knee called the femoral condyles are mostly gliding. Then,during the main movement the motion transforms into a stationary rotational gliding which is held in place by the cruciate ligaments [14, 15].

The pole curve of each human is individual and can be measured by a Goniometer. A technical and automatic measurement of the pol curve is described by K. Buttgereit [16]. Some results and the measurement device are illustrated in Fig. 16.2. Designers of wearable machines should take careful considerations when mimicking or supporting human joints. In many cases is important to insert additional passive joints near the human joint, allowing a free rotational-gliding motion and not inducing physiologically harmful movements [17].

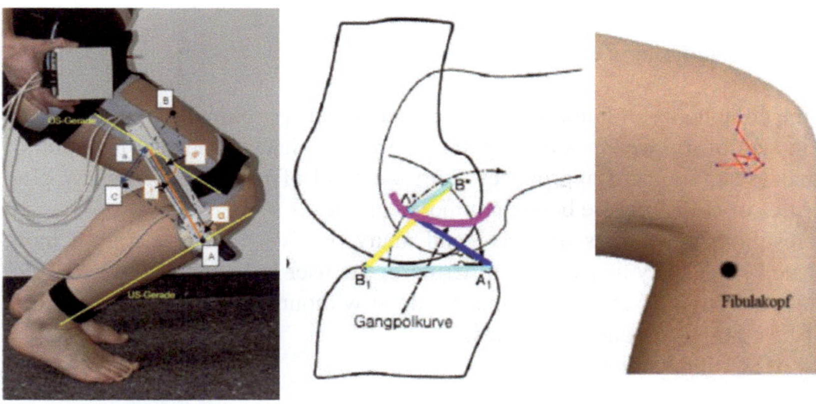

Fig. 16.2 Virtual triangular at the test equipment for measuring pole curves; [16] Middle: model of generating pole curves: the moving centroid curve is purple. Yellow and blue represent the cruciate ligaments and turquoise connecting the dots of each segment is shown. [26] Left: measurement of the knee position with its specific pole curve

Variation of Body Dimension and Motion Ranges

As an exoskeleton is designed for a certain user group, variations in the length, sizes and range, as well as degrees of motion of the prospective users' body parts have to be taken into account. These statistical values can be derived from norms or large population measurements [18, 19].

Here, percentile values are quoted, which relate a certain body part's measurement to the percentage of people whose body part is smaller in size, or whose range of motion is shorter, than the given measurement.

When designing an exoskeleton, some parts close to the body such as legs, backpacks, or arm supports have to be adaptable to different user's sizes and motion ranges. The reason for this is to avoid physical long-term harm to the wearer and to allow the most freedom of motion by the user without reaching a singularity or colliding with the user's body [20].

This is achieved by defining a range within which the sizes, or degrees of motion of certain parts of the exoskeleton, are easily adjustable to fit each user. The most important values for this are the 5th percentile of women, as this indicates the minimum size. Furthermore, the 95th percentile of men best specifies the maximum value of such a range. Designing the exoskeleton to fit the rest of the population is in most cases not cost efficient and only necessary in some high risk applications.

Linked Motion and Singularities

Due to the complexity of human joints, their technical approximation, and the variations in size of the wearer's body parts, as well as varying ranges of motion of individual users, standard approaches from the field of robotics for determining singularities or collisions are becoming more complex.

The resulting kinematics of active and many passive joints generate a more complicated singularity analyses. Furthermore, the user is linked to the exoskeleton. Thus, this kind of analysis cannot be done without considering the wearer's singularities, properties and limitations.

Some of the implications are:

- The point of origin is relative to the user, as the user can move freely around. Some points of origin might even be shifted as the kinematic structure of the exoskeleton moves around the user.
- Passive joints, not only close to the human's joints, but also at other positions around the wearer's body, have to be inserted to accommodate for the exoskeleton movement around the user and the complex kinematics of the human body.
- The additional passive joints have to be considered in all singularity and collision analyses,.

- Human motion axes are variable and move during rotational movement. This creates more complex singularities.
- The motion space, within which the free movement of the operator is guaranteed or necessary, has to be confirmed through a kinematics analysis.
- *Secondary Singularities* deriving from the relative movement of the exoskeleton around the user have to be contemplated. The user and the exoskeleton are regarded as one integrated system. This is different from traditional robotics, where the user is considered a distant operator.
- Collision of the human body and the exoskeleton, both near and far from the machine's and the body's singularities, must be examined.
- The motion of the exoskeleton becomes the interface to the user as well as the task of the machine. Therefore, all rules of interface design from human engineering apply to the motion design of an exoskeleton.
- The motion and reaction of the system has to be comprehendible by the user
- The motion has to facilitate the user's senses and urge for balance.
- The motion has to support the user's proprioception: the sense of where all body parts are positioned in space and what forces are applied.
- Mechanical stoppers have to be specified to guard all users' joints from being over stretched or popped out.
- Force limits and motor-gearing properties have to be set to match the users' limitations. Back-drivability by the wearer or automatic decoupling of the actuator during power-off is necessary in an emergency.

For some of these considerations recoding a real user's motion performing the intended tasks via a 3D-Motion analysis with force plates in the ground could provide the necessary information for simulations and other kinematics analyses.

The point cloud generated by the 3D-Motion analysis could be transformed into a simplified human model. These kinematic data can then be fed into the mathematical model of the exoskeleton to perform tests according to the above implications e.g. search for *Secondary Singularities* or collisions with simplified body parts.

This method can be scaled to perform mass evaluations with an entire range of user of different properties such as heights, ages, medical backgrounds, and so on. Furthermore, in some cases it might be possible to predict stress or other forces on the human body wearing the exoskeleton. Thus, this could support the evaluation of the fundamental design concepts to analyse the benefit of a specific exoskeleton. Additionally, health risks due to long-term stress, or over stretching of joints, for user sub groups with specific physical properties, could be identified.

16.3 Therapy and Rehabilitation

In this field of application a wearable robot is designed for moving an injured or impaired body part for training purposes. The goal is the full recovery of motor control of the body part, or partial improvements of the musculoskeletal system as well as reducing the impact of side-effects due to immobility and non-stimulation.

An example for such an exoskeleton is the active arm orthesis CyberRehab by Fraunhofer IPA [21] (Fig. 16.3).

The success of the rehabilitation process is based on the effect of afferent stimulation of the central nervous system by the activation of muscles and the rendering of forces during training. Therefore, high training intensity and task specific training of daily life movement patterns are the major factors affecting the quality of rehabilitation treatment [22]. These perceptions also influenced guidelines for the rehabilitation process after stroke published in the Netherlands [23]. Additionally, a repetitive locomotion therapy on an electromechanical gait trainer in a non-compliant training mode proved effective in restoring the ability of independent gait in non-ambulatory stroke patients [24].

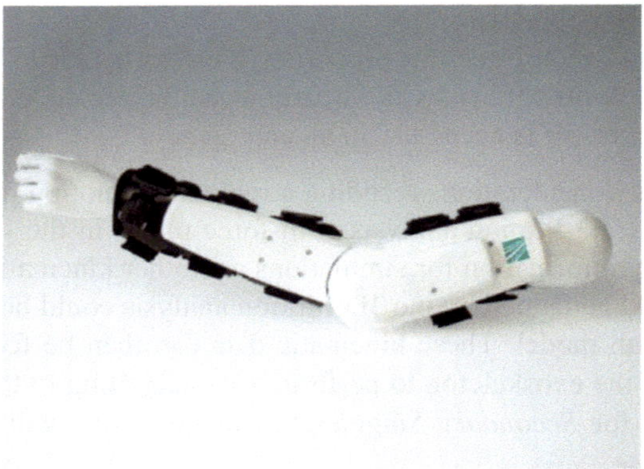

Fig. 16.3 Active Arm Orthesis CyberRehab by Fraunhofer IPA, Stuttgart, Germany

16.4 Physical Prevention and Force Assistance

This category of devices is designed for healthy, or only slightly impaired, operators. The main goal is to prevent future injury, disability, or conic diseases by supporting the biomechanically correct motion and giving power assistance when necessary.

These exoskeletons are applied in fields such as manufacturing and assembly, where the worker is operating on difficult to access parts, or in biomechanically non-ideal posture such as overhead work. Examples are in progress at Fraunhofer IPA.

Moreover, these machines could be employed in logistics, where heavy loads and long term physical stress on the body leads to a large number of health-related absences, motivational decline and work force fluctuation.

Additionally, healthcare workers and care personal for the elderly could be another group of users. As these users often carrier or support patients manually they tend to be affected by back and shoulder pain due to long term overloading.

As these exoskeletons are often applied in the work place, they will have to adhere to the limited technical understanding of the users, quick and stressful working environments and short attention span. They have to be physically robust and easily controllable. A light weight construction, with an emphasis on spinal support as well as flexibility and freedom of motion are important properties for wearable devices in this category.

Though, most importantly, they have to take the users autonomy and motive of personal safety into account. Otherwise, bodily worn devices face a very high rejection rate.

Often exoskeletons tend to be designed for a too heavy load or become very bulky. To prevent this, understanding the Allometric Law for Terrestrial Vertebrates is crucial: $S \sim L^{7/6}$ where S is the weight of the entire skeleton and L is the weight of the maximum acceptable load being lifted [25].

This leads to a trade-off between load and the size of the exoskeleton. Thus, to achieve the best possible outcome of the user's health, and efficient operation, most of the load has to be handled by the operators themselves.

If harmful biomechanical movements, posture, and physical stresses can be reduced by the exoskeleton, it is physiologically preferable to achieve muscle build-up through exercising while working. This might even have a positive effect on the operators' overall well-being in many ways.

16.5 Vision: Auxiliary Assistance

To achieve vision for the wearable robotic field the next step of could be devices for *Auxiliary Assistance*. Such devices are worn similarly to the others described earlier. However, they provide the wearer with extra limbs, machinery or tools that can operate semi-autonomously without being directly attached to a wearer's extremity. One conceptual example is pictured in Fig. 16.4.

Fig. 16.4 Multi-Arm Co-Robot Fraunhofer IPA Design (Source O. Schwarz)

The main differences, however, will be the user intention detection, which might be more heavily relying on image processing, voice interaction and Zero-Torque control or other haptic control for pushed-to-position teaching of the device.

Furthermore, as the device is more separate from the operator and more independent the complexity of *Secondary Singularities* is decreasing and the complexity of collision detection is increasing.

Finally, the user's mental model of the device's movements and reactions shifts from sensing to empathic relationships with the machine. Therefore, high priority has to be given to the user's sense of autonomy, dignity and self-expression. This prevents the device from seeming uncooperative, dominant, or uncaring and raises acceptance levels.

16.6 References

[1] G. F. Hamann, M. Siebler und W. von Scheidt, Schlaganfall: Klinik, Diagnostik, Therapie, Interdisziplinäres Handbuch, Landsberg/Lech: ecomed Verlagsgesellschaft, 2002.

[2] R. K. Institut, „Gesundheit in Deutschland," Gesundheitsberichterstattung des Bundes, pp. 27 - 28, 2006.

[3] Ärztezeitung, „Schlaganfall in Zahlen," 2008. [Online]. Available: http://www.aerztezeitung.de/medizin/krankheiten/herzkreislauf/schlaganfall/article/494572/schlaganfall-zahlen.html. [Zugriff am 16 July 2014].

[4] P. Heuschmann, O. Busse, M. Wagner, M. Endres, A. Villringer, J. Röther, P. Kolominsky-Rabas und K. Berger, „Schlaganfallhäufigkeit und Versorgung von Schlaganfallpatienten in Deutschland," Akt Neurol (Aktuelle Neurologie), Bd. 37, Nr. 07, pp. 333-340, 2010.

[5] WHO, "Factsheets," [Online]. Available: http://www.who.int/mediacentre/factsheets/fs310/en/. [Accessed 16 July 2014].

[6] S. C. Johnston, S. Mendis und C. D. Mathers, „Global variation in stroke burden and mortality: estimates from monitoring, surveillance, and modelling," The Lancet Neurology, Bd. 8, Nr. 4, pp. 301-412, April 2009.

[7] M. Jäger and e. al., "Ermittlung der Belastung der Lendenwirbelsäule bei ausgewählten Pflegetätigkeiten mit Patiententransfer. Dortmunder Lumbalbelastungsstudie 3," Shaker Verlag, Aachen, 2005.

[8] Bundesministerium für Arbeit und Soziales, "Sicherheit und Gesundheit bei der Arbeit 2009," DruckVerlag Kettler GmbH, Bonen, Westfalen, 2011.

[9] S. Freitag and e. al., "Messtechnische Analyse von ungünstigen Körperhaltungen bei Pflegekräften - eine geriatrische Station im Vergleich mit anderen Krankenhausstationen," ErgoMed, vol. 5, no. 5/2007, 2007.

[10] S. Freitag, „CUELA-Studie der BGW - Wirbelsäulenbelastung in der Pflege. Tagungsbericht VIII," in Potsdamer BK-Tage, Berlin, 2011.

[11] Baum, Beck und e. Al., Leitfaden – Prävention von Rückenbeschwerden in der stationären Altenpflege, Hamburg: Berufsgenossenschaft für Gesundheitsdienst und Wohlfahrtspflege – BGW, 2008.

[12] H. v. Rosenberg, „ Identifikation von Willkürsignalen zur Bewegungskontrolle einer Beinprothese," in Stuttgarter Beiträge zur Produktionsforschung, Bd. 3, Stuttgart, Universität Stuttgart, 2012.

[13] I. A.Kapandji, Funktionale Anatomie der Gelenke, Stuttgart: Georg Thieme Verlag, 2009.

[14] H. Frick und e. al, Spezielle Anatomie, 4 Hrsg., Georg Thieme Verlag, 1992.

[15] T. Lanz und W. Wachsmuth, Praktische Anatomie Bein und Statik Ein Lehr- und Hilfsbuch der anatomischen Grundlagen ärztlichen Handelns, Heidelberg: Springer Verlag Berlin , 1972.

[16] K. Buttgereit und O. Schwarz, Neukonzeption und Entwicklung eines Messgeräts zur, Stuttgart: Hochschule Ulm, Fraunhofer IPA, 2011.

[17] A. Schiele, Fundamentals of Ergonomic Exoskeleton Robots, Delft: Technische Universiteit Delft, 2008.

[18] sizeGERMANY Projekt, HUMAN SOLUTIONS GmbH & Hohensteiner Institute, 2008.

[19] H. Schmidtke und I. Jastrzebska-Fraczek, Ergonomie - Daten zur Systemgestaltung und Begriffsbestimmungen, München: Carl Hanser Verlag, 2013.

[20] J. L. Pons und e. al., Wearable Robots - Biomechatronic Exoskeletons, Madrid: John Wiley & Sons. Ltd, 2008.

[21] A. Ebrahimi, D. Minzenmay, B. Budaker und U. Schneider, „Bionic Upper Orthotics with Integrated EMG Sensory," The 23rd IEEE International Symposium on Robot and Human Interactive Communication, 2014.

[22] G. Kwakkel, R. Wagenaar, J. Twisk, G. Lankhorst und J. Koetsier, „Intensity of leg and arm training after middle-cerebral-artery stroke: a randomized trial," Lancet, pp. 191-196, 1999.

[23] R. v. Peppen, G. Kwakkel, B. Harmeling-van der Wel, B. Kollen, J. Hobbelen, J. Buurke und e. al., „KNGF Clinical Practice Guideline for physical therapy in patients with stroke. Review of the evidence. [Translation 2008]," Bd. 114, Nr. 5, p. (Suppl.), 2004.

[24] M. Pohl, C. Werner, M. Holzgraefe, G. Kroczek, J. Mehrholz, I. Wingen-dorf, G. Hoölig, R. Koch and H. S., "Repetitive locomotor training and physio-therapy improve walking and basic activities of daily living after stroke: a single-blind, randomized multi-centre trial," Clinical Rehabilitation, vol. 21, pp. 17-27, 2007.

[25] I. Rechenberg, Bionik auf dem mathematischen Prüfstand - Optimallösungen als Ergebnis der Evolution, TU Berlin, p. 32.

[26] F. Regenfelder, Optimierung eines roboterbasierten Versuchaufbaus zur Bestimmung biomechanischer Eigenschaften des Kniegelenks in 6 Freiheitsgraden, München: TU München, 2006.

17 3D Printed Objects and Components Enabling Next Generation of True Soft Robotics

Andreas Fischer, Steve Rommel, Alexander Verl

Fraunhofer Institute for Manufacturing Engineering and Automation IPA, Stuttgart

Abstract Soft robotics in the content of true softness, with regards to components, parts, or the complete robot, are the next step in the development of tools for humans, especially when used in close proximity. Considering the fact that robots are a multilevel extension of the human body, and that their main purpose should be to help humans perform tasks, then focusing on the development of soft-materials, and product design options allowing for flexibility and softness by design is necessary, for the next development level of the tool "robot". Using additive manufacturing in combination with new materials, design methods, and biomimicry / biomimetics is a key in that development, but also very challenging due to the multi-level complexity. An understanding of the real world tasks required to be performed, and abstracting this information into new applications and robotic designs in the combination mentioned above, is shown in the chapter, functioning as a basis and overview of the state-of-the-art.

17.1 Introduction

Recent developments in the field of additive manufacturing (also called generative manufacturing or 3D-printing) have produced the first examples of new designs and compositions of robots. Purely designing products and objects following the functions required to be performed, tools or robots have received more than a new appearance or outer skin. Soft robotics, in the sense of soft joints, soft shells, soft components in general, with added functionality like embedded components are possible today. One of the first examples is the award-winning innovation called "Bionic Handling Assistant" – a robotic arm inspired by an elephant's trunk, which could only be manufactured using additive manufacturing methods [1].
The possibilities of combining these new technologies with new materials and new design thinking are truly endless and expanding, and are the motivation of the examples and developments described in this chapter.

17.1.1 Additive Manufacturing (AM) as a Manufacturing Technology for Soft-Robotic-Systems

17.1.2 The Production Processes

Additive manufacturing (AM), also called 3D-printing, are new forms of manufacturing technologies to design and create products. Although there is quite a selection of different AM technologies, e.g. Fused Layer Modeling (FLM), Selective Laser Sintering (SLS), Selective Laser Melting (SLM), Digital Light Processing (DLP) just to name a few, the basic principle is always the same. All are based on the fabrication of objects layer by layer.

This characteristic offers design engineers and developers not only undreamed-of shaping possibilities but also new opportunities to adapt functional mechanisms, such as those found in nature. Thus, if combined with biomimicry and biomimetics, as well as new material developments, additive manufacturing technologies offer huge potential for developing many other innovations in the near future.

Two of the technologies, Selective Laser Sintering (SLS) and Fused Layer Modeling (FLM) are described in further detail below.

Selective Laser Sintering (SLS) uses a plastic powder which is first molten or sintered by a laser and then solidified (Fig. 17.1). The laser is hereby reflected by a mirror. Once all the points per layer have been solidified the building platform lowers itself by the predefined amount and the next layer of powder is applied using a wiper system. Once completed the process starts again and is repeated until the complete object is fabricated. Unused powder functions as support material during the printing process. Polyamide 12 (PA) is the material most widely used with a broadening of the material selection being expected over the coming years [1, 2].

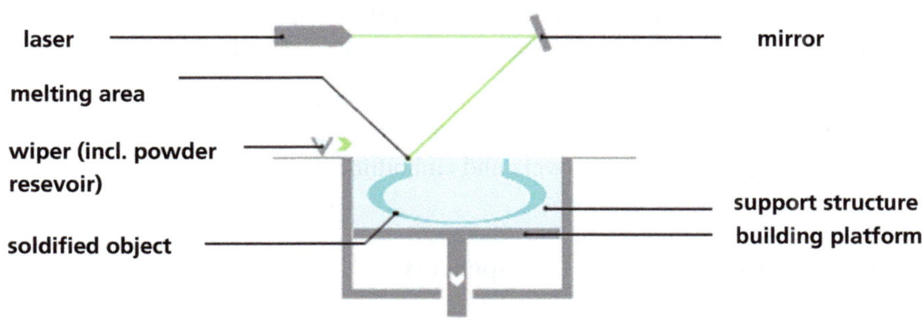

Fig. 17.1 Selective Laser Sintering process principle.

Compared to SLS the FLM-process used a hot nozzle system in order to melt a thermoplastic material string and applying it to a build platform in order to create the specific layers of the object. The classical setup of an FLM system is that of a 3-axis portal machine – x and y movement performed by the print head and the z-axis is the table or building platform. The principle is shown below (Fig. 17.2).

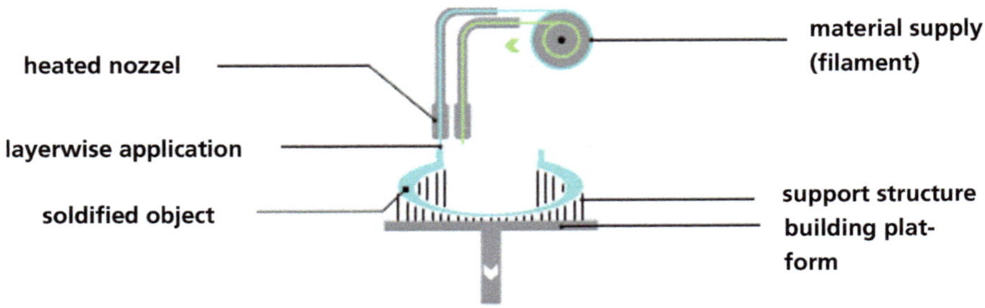

Fig. 17.2 The Fused Layer Modeling (FLM) process.

In general all materials used are thermoplastics of which the most common are ABS, ABSi, PLA, PC, PA, and PPSU. Due to the fast solidification of the molten material, introduction, and embedding, of sensible and also delicate semi-finished products is a real possibility. This process has also been further developed at the Fraunhofer IPA under the name of Pack-FLM (PFLM).

17.1.3 The Term Robot and its Newly Added Additive Components

The Robot Institute of America (RIA) defines the term „robot" as a programmable multi-functional handling tool for moving materials, products, tools, or other specialty equipment. By allowing for freely programmable movements the robot can be used in a versatile way. Following this definition, and expanding it by using additive manufacturing (AM) as well as biomimicry and biomimetics for the design of products, robotic components were redefined and produced. Some examples already designed, built, and in use are presented in this chapter and include:

- Integrated functions and functional components
- Soft actuator systems
- Flexible light-weight covers and embodiments

17.1.4 Integrated Functional Components

With respect to integrating functions and functional components some of the key characteristics of the FLM process are the relatively low process temperature as well as the possibility to pause the fabrication process in a defined matter. This start-stop possibility allows for the previously mentioned PFLM process, and therefore, the integration of semi-finished objects and sub-components like hard metal objects, as well as sensible electronics e.g. sensors. Sensors for instance, can then also be encapsulated expanding the use cases of these objects. Sensors, as well as the other example of encapsulating an electroluminescence foil, then represent added functionality of these objects (Fig. 17.3).

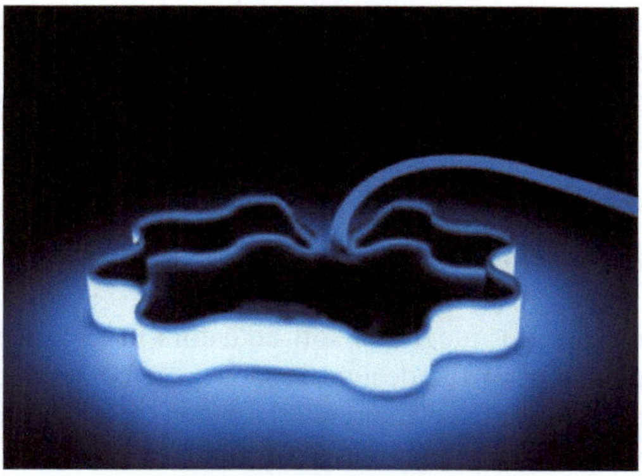

Fig. 17.3 PFLM – integrated electroluminescence (EL) foil inclusive electronics and wiring. Combination of hard ABS plastic on the outside and flexible foil on the inside.

When selecting a specific material with the integrated component or semi-finished object in mind it is possible to create a multi-functional object. Ionic actuators, combining actuator and sensory functions in one component, have to be integrated using a flexible outer skin in order to be able to change shaped in the actuator modus. Additionally, it is advantageous to coat the delicate sensor. This flexible protective skin can be applied using the FLM process and the developed thermoplastic polyurethane (TPU) in combination.

Having the work environment in mind these flexible and stretchable rubber-like materials need to be resilient at the same time to even be considered for use. With these requirements standard off-the-shelf FLM systems are not able to handle these TPU materials. A multilevel specific development for the material supply system and a continuous material curing system made it possible to fabricate TPU parts with the FLM technology. In the current state-of-the-art there are 6 TPU materials of the Desmopan family ready to be used:

- Desmopan 385 S
- Desmopan 9873DU
- Desmopan 3660DU
- Desmopan 3695AU
- Desmopan 3360A
- Desmopan 3070A

The difference of these TPU materials lies within the technical characteristics as well as within the various shore-hardness, allowing for a combination of graded material hardness within one 3D-printed object. Additionally, multi-material objects are possible beyond the TPU material combination by combining TPU with PC for instance. Such a material combination requires object specific calculations and setups within the virtual 3D-CAD world as well as the use of additives during the extrusion process of the real fabrication world.

In general the mechanical stability of FLM objects can be safely sat at approximately 80-85% of that of injection molded objects. This can be vastly improved when utilizing the strengths of additive manufacturing in the design process of objects and therefore creating objects which are hard or even impossible to be manufactured in a conventional way [11, 12].

17.1.5 Soft Actuator Systems

One possibility to generate integrated soft actuators in a defined way, within an object, is the combination of the design freedom offered by additive manufacturing and the maximization of the material characteristics. These properties of the material can be utilized to integrate very stiff, and very flexible, areas in one single part without any assembly like screwing or gluing. Such structures have been demonstrated by the Fraunhofer IPA for a few years. One example is the reinterpretation of the Golem project by Hod Lipson (Fig. 17.4) with a bellow as actuator combined with a solid framework and tube system for the compressed air, all built as a single piece.

Fig. 17.4 Autonomous robot system with flexible and solid areas built in one single piece. The flexible zones is the bellow which is powered by a small air-pump.

The technology used is Selective Laser Sintering (SLS) and its model material Polyamide (PA) 12. This combination allows for very thin walls, which are semi-elastic. This development was also applied at the mentioned Bionische Handlingassistent (BHA) in cooperation with the Festo AG. The setup consists of up to 12 individual air chambers equipped with compressed air regulators as well as rope sensors in order to automatically control the movements. A mathematical model was used to predefine the position of the BHA in space by converting the Cartesian coordinates into joint coordinates of the BHA. The expansion and contraction of the individual air chambers is then responsible of the actual movement and therefore position in space.

One application for this setup is the film industry. The requirements of a film team were divided into the main functional groups: movement, camera, recording, sound combination, and assembly. The most important aspects of the movement are a controlled and soft movement, upscaling of the system of up to 3 m, and an intuitive human-machine-interaction. The result of this development is the CamBot (Fig. 17.5).

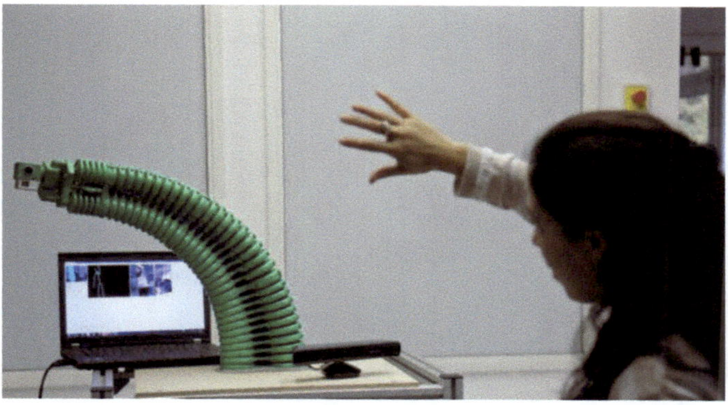

Fig. 17.5 Interactive CamBot with soft-actuator made of PA 12 (SLS).

The CamBot is equipped with a GoPro HD camera and can be used in direct proximity to humans due to its 3D-printed soft-actuators. By combining the CamBot with a Microsoft Kinect-Sensor the system can be used as an interactive system to communicate and react to human actions. One challenge in this constellation is the noise level of a conventional compressor system. This is solved by using gas cylinders [1].

Another option of producing, and using, soft-actuators is by using the FLM technology and a material combination as already mentioned above. An example is the development of the octopus siphon actuator (OSA). The goal was to mimic nature, in this case an octopus, in order to create a new silent underwater propulsion system. This development was only possible through the parallel work of developing TPU materials for the FLM technology. The combination of additive manufacturing with biology and engineering was the only way to truly mimic nature. The movement of the biological octopus, without ground contact, is achieved by contracting muscles of a syphon which then in turn expels the sea water in a jetting matter. This rebound effect allows for the octopus to move in the desired direction. The octopus can also contract the funnel, at will, to accelerate and steer at different speeds.

Using a biomimetic top-down-development process, the octopus principle was analyzed and abstracted to be technically reproduced. The first prototype consisted of four OSA soft-actuators made of Desmopan 385 and being attached to a support structure containing the required mechanical components like hydraulic cylinders etc. The funnel, as well as the water intake system, was fabricated using Macrolon, a polycarbonate (PC) material (Fig. 17.7). The OSA was constructed of an outer shell and inner rope kinematics, all printed by the same fabrication process (Fig. 17.6).

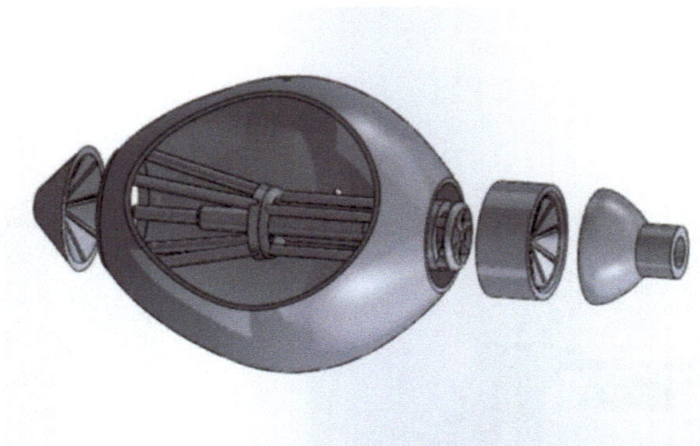

Fig. 17.6 Illustration of the inner structure of the OSA.

The retraction of the OSA is passive, and therefore, material specific, allowing for the OSA to be refilled with water in order to start the cycle again. Due to this setup the OSA is expected to be suitable for all sea level depths as well providing some ground work for possible underwater soft robots [8, 9, 10].

Fig. 17.7 FLM printed OSA with integrated hydraulic system.

17.1.6 Fabrication of Soft Objects Including Endless Fibers

Matrix-composites are used more and more widely across all industrial branches with aerospace, energy harvesting, and automotive being in the forefront with the goal of achieving significant weight reduction. The same focus is also found in the automation and machine building sectors or the medical application industry, e.g. prostheses. Matrix-composites offer, amongst other advantages, a good weight to function ratio.

Additive manufacturing and matrix-composites and their manufacturing processes have, at first glance, very little in common. One reason for this is the partial incompatibility of the two. In order to overcome this issue, and expand the additive manufacturing capabilities even further, the idea of the 3D Fiber Print (3DFR) was born at Fraunhofer IPA. The 3D Fiber Printer is the integration of endless fibers within the thermoplastic material during the build process of FLM thus resulting in an additive manufactured matrix-composite object. The specification involves the typical characteristics of additive manufactured products, being a layer-wise build up. In order to fabricate such an object a, patent pending, print head was developed which is constructed in a way which facilitates the right mix-ratio of fibers to thermoplastic material by guiding the fibers or fiber packs through the nozzle to coat the fibers with the respective material. The principle is similar to that of the water-jet pump. The molten thermoplastic material is forming a composite with the introduced fibers and carries them through the remainder of the nozzle (Fig. 17.8).

Fig. 17.8 1st print test 3D Fibre Printer using a 3K carbon-roving and TPU matrix.

Expanding the typical 3-axis system setup of the FLM process, a 6-axis robot can be used. The robot requires CNC capabilities, and specialty programming commands, in order to function as a 3D-printer. The Fraunhofer IPA implementation is shown below (Fig. 17.9).

Fig. 17.9 FLM based robotic system: 3D Fibre Printer with build volume of 2m x 1m x 1m.

This setup allows for several advantages, like an excellent scalability with regards to object size, a utilization of more than 3-axis, material combination, or using several materials in the same setup via print head quantity.
The first test run was completed using a 3K carbon fiber-roving in combination with also newly developed polyurethane (TPU) of the Desmopan family 385 S (Fig. 17.10).

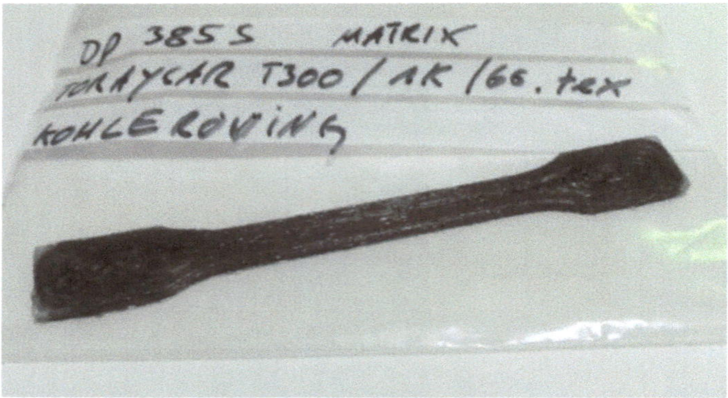

Fig. 17.10 ISO test object made of 1K carbon-fiber-roving and Desmopan 385 S, nozzle exit diameter 1mm (x,y resolution).

The fiber to thermoplastic ratio is currently at approx. 10% for a 1K carbon-fiber-roving. Due to the specific design of the nozzle it is possible to integrate the fibers where required within the object. This accounts not only for an optimized utilization of the fibers following, for example, an FEA (Finite Elements Analysis) –

where the strength needed, and how much, can be determined, and determine the maximum economic, and ecological, utilization of the fibers.

Combining the fiber implementation process with thermoplastic polyurethane (TPU) creates a very interesting combination between a very strong, and stiff, material and a very flexible material. The potential use within the physical machine composition of soft robotics becomes now a very real and promising future application. In addition to the mechanical advantages further advantages could be realized, like sending electrical current or data through [3, 4, 5, 6, 7].

17.2 Discussion and Outlook

Selecting the right manufacturing method and the right material combination as well as applying a new design philosophy, focusing on a "no limits approach", soft robotics can be developed into a new way of robotic appearance and application. The safe use in close proximity to humans will enable new human-machine-interactions and use cases beyond the service field. In addition, new developments, like the combination of materials or the integration of objects, will increase their functionality while decreasing the complexity of the assembly system. Using additive manufacturing also allows for a production of components, or robots, as well as their replacement parts on-demand and on-site, potentially increasing the OEE and use time of a robot.

Weight reduction (potentially up to 40%) results in more load capacity, longer use times when run on autonomous energy supplies, and a reduced risk of injury to humans.

Summarizing these benefits; the manufacturing of complex robotic systems, which consist of hard and soft components, integrated sensors, electronics and functions and complex shapes, allow the best performance and are currently possible to manufacture. For a standardization process, however, further development work is however needed in the areas of:

- Accuracy
- Speed
- Material selection
- Part size
- Training of engineers

in order to fully realize their potential.

Acknowledgments The presented work in the field of 3D printed TPU has been made possible by the support of Bayer Material Science AG (BMS). Special thanks to Jürgen Hättig and Eckhard Foltin of BMS for their support and future vision.

17.3 References

[1] Breuninger et al (2013) Generative Fertigung mit Kunststoffen. Springer-Verlag, Berlin-Heidelberg

[2] Fahrer (2002) Ganzheitliche Optimierung des Laser-Sinterprozess. Herbert Utz Verlag GmbH, München

[3] Flemming et al (2012) Faserverbundbauweisen, Fasern und Matrices. Springer-Verlag, Berlin-Heidelberg

[4] Flemming et al (1996) Faserverbundbauweisen, Halbzeuge und Bauweisen. Springer-Verlag, Berlin-Heidelberg.

[5] Flemming et al (1999) Faserverbundbauweisen, Fertigungsverfahren mit duroplastischer Matrix. Springer-Verlag, Berlin-Heidelberg

[6] Grießbach (2009) Praxis Rapid Technologien, Handbuch für Produktentwickler, Techniker, Kaufleute. Vonroth & Bode KG Verlag

[7] Jäger et al (2010) Carbonfasern und ihre Verbund-werkstoffe: Herstellungsprozesse, Anwendungen und Marktentwicklung. Die Bibliothek der Technik, Band 326. Verlag Moderne Industrie

[8] Marguerre (1991) Bionik – von der Natur lernen. Siemens, Berlin

[9] Lickfeld (1993) BIONIK – Patente der Natur. Pro-Futura-Verlag, München

[10] Nachtigall (1997) Vorbild Natur: Bionik-Design für funktionelles Gestalten. Springer-Verlag, Berlin-Heidelberg

[11] VDI-Richtlinie (2009) Generative Fertigungsverfahren – Rapid-Technologien (Rapid Prototyping) – Grundlagen, Begriffe, Qualitätskenngrößen, Liefervereinbarungen. VDI, Düsseldorf

[12] Zäh (2006) Wirtschaftliche Fertigung mit Rapid-Technologien: Anwender-Leitfaden zur Auswahl geeigneter Verfahren. Carl Hanser Verlag, München-Wien

Part V Soft Robotic Applications

18 Soft Hands for Reliable Grasping Strategies

Raphael Deimel and Oliver Brock

Technische Universität Berlin

Abstract Recent insights into human grasping show that humans exploit constraints to reduce uncertainty and reject disturbances during grasping. We propose to transfer this principle to robots and build robust and reliable grasping strategies from interactions with environmental constraints. To make implementation easy, hand hardware has to provide compliance, low inertia, low reaction delays and robustness to collision. Pneumatic continuum actuators such as PneuFlex actuators provide these properties. Additionally they are easy to customize and cheap to manufacture. We present an anthropomorphic hand built with PneuFlex actuators and demonstrate the ease of implementing a robust multi-stage grasping strategy relying on environmental constraints.

18.1 Introduction

Humans are very proficient graspers. In fact, humans grasp so reliably and robustly that experimenters usually assess grasp difficulty by execution speed instead of error rate. For autonomous robots, on the other hand, comparably robust and reliable grasping and manipulation remains an open challenge, despite established theories for assessing the quality of a grasp [15,9]. Recent studies of human grasping indicate a plausible reason for the difference in human and robot performance Deimel and Brock [7] and Kazemi et al. [12] showed that humans interact more with the environment if their vision is impaired (occluded or blurred). To maintain grasp reliability, humans seek contact to counteract uncertainty. We believe that they use available constraints to guide the motion of their hands and fingers to make the execution reliable and robust. The interactions can be terminated by sensing simple events, such as a stop of motion, or sensing contact. By concatenating those interactions, humans can draw from a rich repertoire of reliable grasping strategies for each situation. Our hypothesis shares many ideas with sensorless manipulation proposed by Mason and Erdmann [8,13], indeed our hypothesis can be seen as a continuation of this work.

To be able to transfer the principle of exploiting constraints successfully to robots, the hardware needs to facilitate easy implementation. Accurately enacting joints torques or angles is less important than making collisions simple, stable, fast and safe. While classical hand designs (see Controzzi et al. [3] for an overview) can provide the required behavior to some degree, hands such as the Pisa/IIT Soft

Hand [2], or the iHY hand [14] are suited much better to this task. These hands provide desirable features with elastomer based or dislocatable joints, but are otherwise based on rigid links. The earliest gripper design with a large number of compliant joints is the soft gripper by Hirose and Umetani [10], which was able to grasp prismatic objects with widely varying shapes. With the RBO Hand [4] and its successor RBO Hand 2 [5], we go one step further and investigate the opportunities and limits of a literally soft hand. To explore the design space, we developed a method for creating customizable, soft continuum actuators, inspired by the work of Ilievski et al. [11]. The positive pressure gripper [1] uses a balloon filled with granular material to provide ultimate adaptability to object shape, but its homogeneous structure limits the availability of diverse grasping strategies.

In the following section we will present a selection of interactions that can be used as building blocks for constraint exploiting grasping strategies. From those interactions we will extract desired hardware features that simplify their implementation and present the PneuFlex actuators and the RBO Hand 2 that realizes these features. We will demonstrate the feasibility of our approach by implementing an example grasping strategy using the RBO Hand 2.

18.2 Exploiting Constraints

Our main hypothesis is that competent grasping and manipulation is enabled by actions that exploit the ability of environmental constraints to reduce uncertainty about certain state variables, specifically those that are relevant for the success of subsequent manipulation actions. For example, contact with a surface can reduce uncertainty about distances and orientations between two or more objects. This enables us to execute sequences of actions with a reliable outcome. In many typical situations, the table surface provides such a constraint. The actions that exploit constraints relate to interactions between hand, object and environmental constraints. Hand morphology and end effector control also determine which interactions are possible. Therefore, hand design should be guided by a description of desirable actions on commonly encountered constraints and should facilitate easy implementation of control. We will now present three examples of interactions between hand, constraint and object to grasp. As we will see in the subsequent sections, implementing control of these interactions is easy with a soft, compliant hand.

Compliant Collision with a Surface

The first interaction we consider is to collide with a surface (see Figure 18.1). The collision path can e.g. be guided by visual servoing. This interaction makes the distance between the two colliding objects very certain. To be able to implement this interaction easily, the hand must limit impact forces and be able to react to disturbances quickly.

In a special case, the object to grasp itself provides the constraining surface. Compliant actuation of each joint then makes it easy to establish many contact points on the object and to balance contact forces to achieve a grasp [4,5,7].

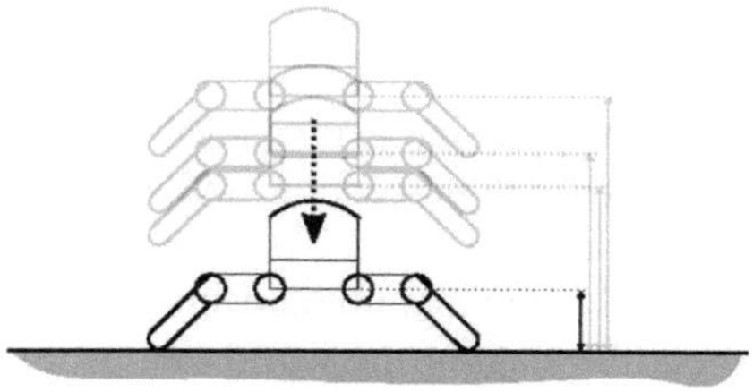

Fig. 18.1 Two objects can be positioned relative to each other by using compliant collision

The second interaction we consider is to slide fingers and object along a surface and is shown in Fig. 18.2. This interaction fixates motion along the surface normal and rotation out of plane, which in turn simplifies the control of finger position and hence object position. For most reliable execution, the fingers should always stay in contact with the constraint during the sliding motion.

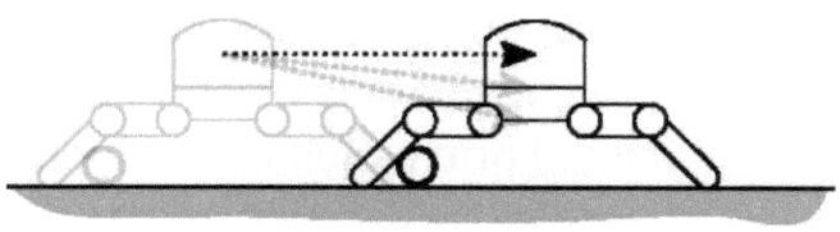

Fig. 18.2 Staying in contact simplifies motion control

The third interaction we consider is closing a cage around an object that was previously formed with the hand and a surface (see Fig. 18.3). By sliding the contacts between surface and fingertips, the cage gets smaller while at the same time an enclosed object cannot escape the cage. That effectively reduces the uncertainty about object location. For maintaining the cage, the fingers should continuously stay in contact with the surface. Additionally, the cage itself can also reject disturbances that concurrent manipulation, such as sliding, may introduce otherwise. This makes caging a very useful interaction in a grasping strategy.

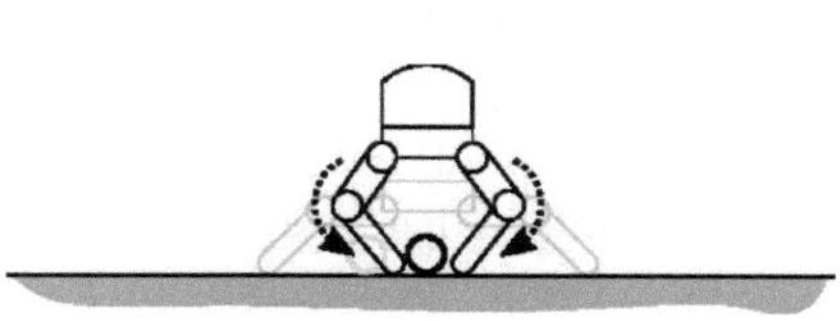

Fig. 18.3 Reducing uncertainty by closing a cage

The list of possible constraint exploiting interactions here is not comprehensive. Grasping strategies may also include intermittent servoing to be able to concatenate interactions, but should be avoided as they lower reliability of execution.

18.3 Requirements to Hardware

Guidance by constraints – which was explained in the previous section – can be used to reject uncertainties from perception and reject anticipated disturbances arising from actuation. In the best case, this results in a fixed sequence of interactions that yields a predictable outcome in many situations. While better perception reduces uncertainty about the state of the world too, using constraints also simplifies planning. A planner that constructs these strategies will greatly benefit from a large and diverse set of interactions to choose from, therefore manipulator hardware should facilitate as many different interactions as possible. Additionally, this approach frees up perceptual resources for other tasks. From this problem description, we can extract a set of goals for hand design:

Low Inertia

Movable parts of the robot should exhibit a low apparent inertia. It enables reaction to fast changes in contact location and limits the energy transferred upon impact. The former makes it easy to maintain contact, while the latter limits the increase of uncertainty to position and orientation of the contacted object.

No Reaction Delay

Actuation should provide a very low time delay for reactive motion to stay compliant during fast disturbances. This requirement is especially difficult to achieve when actively controlling compliance, e.g. with geared electric motors.

Robustness to Arbitrary Collisions

The hardware has to be robust against arbitrary collisions. The robot needs to contact objects of unknown shape and position frequently and quickly, without having full or accurate knowledge of the world. Errors will happen, and therefore unexpected collisions will occur. A suitable hardware will tolerate these collisions and not break. Robustness can be accomplished by providing compliance in every direction and about many rotation axes.

Safe for the Environment

To a lesser extent, it is also desirable for the manipulator to generally not break or injure objects. If safety to the environment can be ensured by passive, mechanical means, more actions can be tried without risking catastrophic damage. This requirement also facilitates autonomous learning.

18.4 PneuFlex Actuators

To build literally soft hands, we developed a process to create customizable, mechanically compliant, and pneumatically actuated continuum actuators. An example actuator is shown in Fig. 18.4. These so called PneuFlex actuators bend with an approximately constant ratio of curvature per pressure [4], and have a fixed stiffness.

The actuator is made of silicone rubber and forms a closed air chamber. A thin silicone tube is inserted into the actuator at a convenient position to inflate and deflate the actuator. The actuator is restricted from expanding radially by the thread wound helically around it. Additionally, the bottom side of the actuator embeds a flexible mesh, making it inextensible. Polyester (PET) is used as the fiber material throughout, as it is readily available and easy to handle. The manufacturing method is openly documented[9] to facilitate reuse and application by independent research.

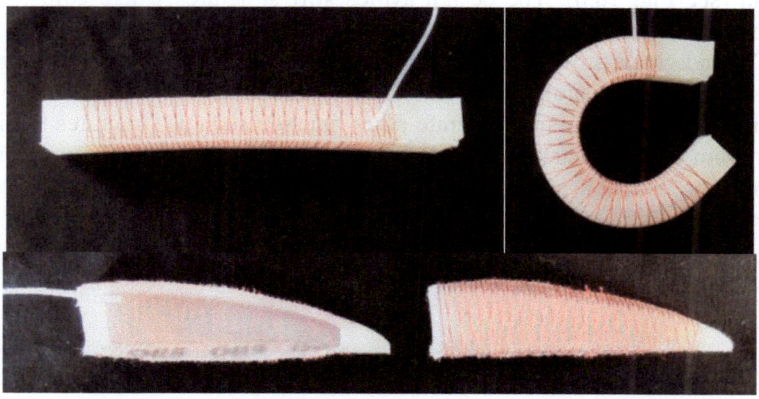

Fig. 18.4 A PneuFlex actuator in deflated and inflated state, and a cross section of an actuator used as a finger, revealing its inner structure

[9] http://www.robotics.tu-berlin.de/index.php?id=pneuflex_tutorial

The actuation ratio (curvature with respect to applied pressure) and actuator stiffness (change of curvature with respect to change of moment) can be customized with the cross section geometry. PneuFlex actuators are manufactured using printed molds, which greatly simplifies customizing and replicating actuators. The required materials are cheap, encouraging a Rapid Prototyping work flow for exploring design space.

The PneuFlex actuators enable us to build hands that have the properties we require for exploiting constraints. The actuation method ensures very low inertia, which is complemented by local deformation of the rubber body. The fingers provide high quality compliant actuation, and are able to comply to collision forces from any direction. The rubber used (SmoothOn DragonSkin brand) offers high tear strength and large strains, making it very robust. The attainable contact pressures are limited, which makes the hand passively safe for direct interaction with humans.

18.5 Anthropomorphic Soft Hand Prototype

The RBO Hand 2 (see Fig. 18.5) is the latest in a series of experimental prototypes and the first anthropomorphic soft hand and is capable of diverse grasp postures [5]. The hand is built from seven actuators, five for each finger including thumb, and two forming the palm and providing a dexterous thumb. The compliance and robustness of its PneuFlex actuators is complemented by the flexible polyamide scaffold (see Fig. 18.6). The design avoids stiff structures where collisions are probable while providing a rigid connection to the wrist of a conventional robot arm. The individual struts are stabilized by a flexible palmar sheet connecting the fingers and palm actuators. As the scaffold is manufactured with selective laser sintering, we can also easily integrate other function such as structures to distribute air from control channels to individual actuators. The pervasive compliance, robustness to collision and limited contact pressures of the RBO Hand 2 make contact with constraints easy to control.

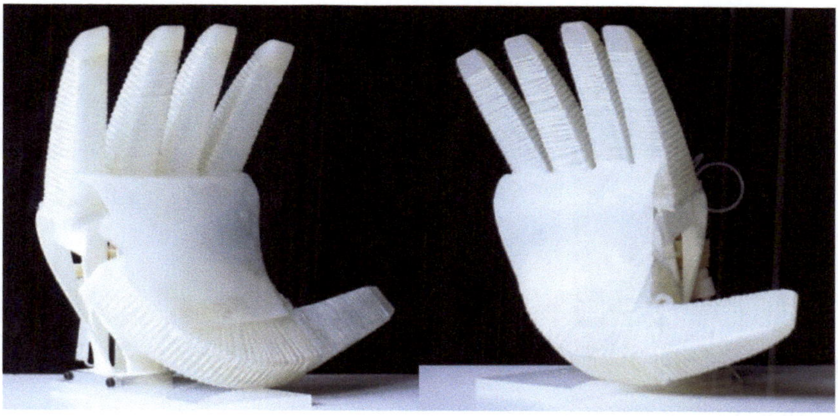

Fig. 18.5 First prototype of a soft, anthropomorphic hand for exploring the capabilities of soft hands

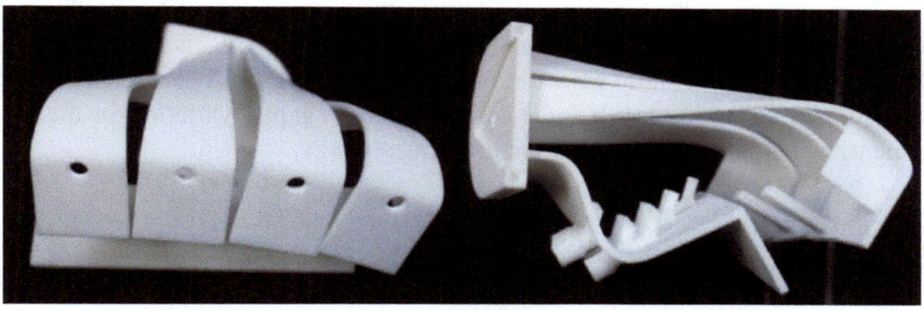

Fig. 18.6 Top and side view of the printed polyamide scaffold

18.6 Example Implementation of a Grasping Strategy

To demonstrate the simplicity of implementing interactions with constraints using the RBO Hand 2, we explain the creation of an example grasping strategy. Fig. 18.7 shows the execution of a strategy we will refer to as slide-to-wall-grasp. The implementation uses an RBO Hand 2 and a Meka robot arm. The environment provides a horizontal surface and a wall whose inclination can be modified. First, the robot moves the hand until the fingertips contact the table surface. It then slides the fingers across the horizontal surface towards the corner. In the corner, the robot rotates the fingers around their tips and slides them under the object to grasp. After that, the hand again is rotated around the fingertips while slightly flexing the fingers to cage the object against the wall, and to bring the fingers into the position for the final step: The object is grasped by flexing the fingers while dragging them upwards along the wall.

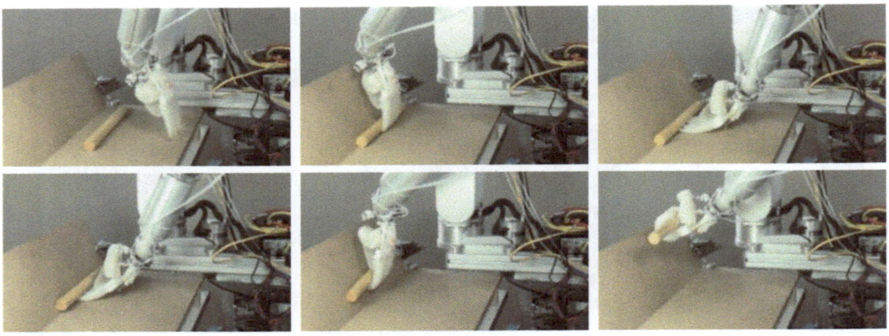

Fig. 18.7 Slide-to-wall grasping strategy

Fig. 18.8 shows the probability of success when picking a cylinder with 22 mm diameter. The experiment tested at 40°, 45°, 50°, 60°, 70°, 80°, 85°, and 90° wall inclination. Angles were first tested in 10° increments, intermediate angles where success changes rapidly were tested additionally to increase resolution. The wall inclination was varied from the initial configuration of 60°, each angle was tested 10 times. The grasp reliably works at 60°, but the strategy still succeeds when deviating up to 15°. This result indicates robustness against variation in the environment, which is a stated goal of the grasping strategy.

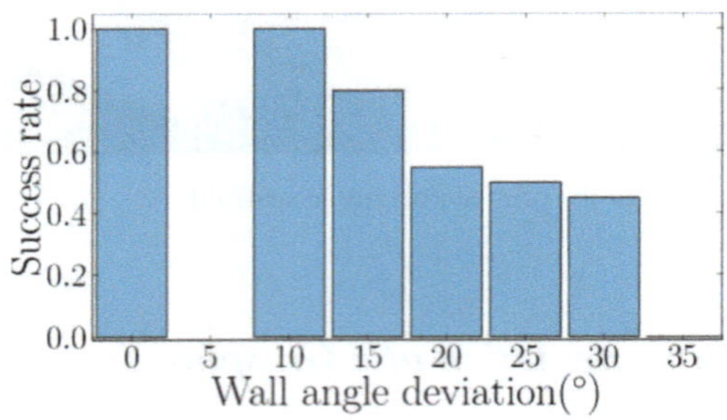

Fig. 18.8 Data indicating the robustness of the grasping strategy with respect to changes of the environment

18.7 Used Interactions

The grasping strategy uses several interactions at various phases to reduce uncertainty or reject disturbances. Here, we will restrict ourselves to analyzing an example for each of the interactions explained in a previous section.

Fig. 18.9 shows the hand sliding along the table surface. In the first phase (first two images), the compliant fingertips are used to contact the horizontal table surface. As the fingertips are compliant, the arm does not need to stop in an accurate position. Also, approach direction is not critical, as long as it is well within the friction cone of the finger contact. The collision can be done relatively quickly too, as the arm's inertia is decoupled from the contact by the compliant, soft fingers. This makes implementation of this interaction simple.

For reliably sliding small objects across the table, as illustrated by third and fourth image in Fig. 18.9, we have to ensure that the fingertips move as low as possible, which can be done by keeping them in contact with the table. This is accomplished by the compliant fingers and greatly reduces the accuracy requirements for the wrist trajectory compared to a hand with few or stiff joints.

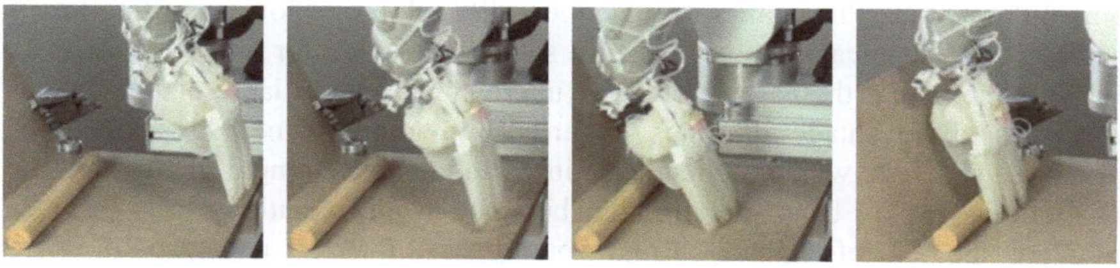

Fig. 18.9 Using the flat surface to vertically align fingertips prior to and during sliding a cylinder into the corner

Closing of a cage is done in the final stage of the grasp, where the object is first caged against the wall (see Fig. 18.10). The cage ensures that the cylinder reliably ends up between palm and closing fingers, while at the same time it also rejects disturbances in the cylinder's orientation that are caused by not grasping it at exactly its center of mass. The compliance of the fingers – and the palm – is enough to handle a deviation of wall inclination of at least 15°, as shown in the experimental results in Fig. 10.8. Accurate knowledge of wall orientation is therefore not necessary for reliable execution.

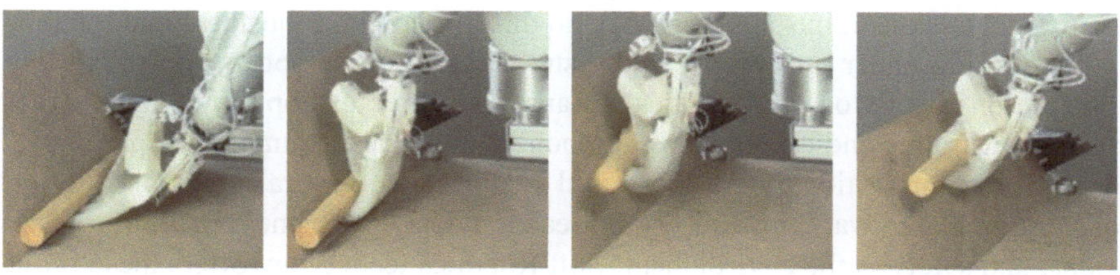

Fig. 18.10 Caging during of the slide-to-wall grasp strategy.

The example shows that constraints can be used for creating robust, multi-stage grasping strategies, and that the interactions can readily be implemented with simple joint controllers of limited accuracy when errors are compensated by the compliant end effector.

18.8 Limitations

Soft Robotics turns out to be a well suited technology to implement grasping strategies that utilize the environment. It is difficult today though, to create soft mechanisms that are as sophisticated as conventional rigid-bodied robots. This is due to the lack of established best practices for integration and off-the-shelf solutions for sensing, actuation, control, and modeling. The missing integration of sensing and actuation is especially unfortunate because soft structures may provide several of these function at once and this size and cost advantage is currently is not exploited to its full potential. Pneumatics are also more difficult to use for controlling forces than electromechanical systems. This is a severe restriction for many applications, but we believe that the ease of creating interactions with reliable outcomes outweighs this disadvantage for hands. A current limitation of PneuFlex actuators is their fixed stiffness, which indirectly limits the attainable strength of a grasp: The actuators can easily be made stronger, but they would simultaneously get less compliant too. Therefore, hand design would benefit from an actuator with variable stiffness. Finally, needles and sharp edges are able to damage the actuator. This disadvantage could be remedied by using cut-resistant gloves, or by following work safety rules designed for human manipulators. Also, the robot currently does not adapt grasping strategies or plan new ones to accommodate for a large variety of situations. Integration of a suitable perception, representation and planning with the actuation principle is an open issue requiring further research.

18.9 Discussion

Recent research on human grasping indicates that humans intentionally exploit environmental constraints, and that they do this to improve robustness and reliability of grasping under uncertainty and disturbance. We attempt to implement this principle on robots too, and for this we analyzed three example interactions that exploit constraints and can serve as components of robust example grasping strategies. These interactions were then used to formulate several beneficial design goals for hand hardware: low inertia, no reaction delay, robustness to arbitrary collisions, and safety to the environment. Soft Robotics technology offers these properties as we demonstrated by building the RBO Hand 2 and implementing a grasping strategy with it. In turn, grasping seems to be a promising reference application to drive the development of Soft Robotics, as it strongly benefits from compliance and many passive joints. The manufacturing process developed to rapidly prototype soft hands may also help in other research areas.

18.10 References

[1] Amend JR, Brown EM, Rodenberg M, Jaeger HM, and Lipson H (2012) A positive pressure universal gripper based on the jamming of granular material. IEEE Transactions on Robotics, 28(2):341–350.

[2] Catalano M, Grioli G, Farnioli E, Serio A, Piazza C, and Bicchi A. (2014) Adaptive synergies for the design and control of the Pisa/IIT SoftHand. Int. J. of Robotics Research, 33(5):768–782.

[3] Controzzi M, Cipriani C, and Carozza MC (2014) Design of artificial hands: A review. In: The Human Hand as an Inspiration for Robot Hand Development, volume 95 of Springer Tracts in Advanced Robotics, Springer.

[4] Deimel R and Brock O (2013) A compliant hand based on a novel pneumatic actuator. IEEE Int. Conf. on Robotics and Automation (ICRA), 2047–2053.

[5] Deimel R and Brock O (2014) A novel type of compliant, underactuated robotic hand for dexterous grasping. Robotics: Science and Systems (RSS).

[6] Deimel R, Eppner C, Alvarez-Ruiz J, Maertens M, and Brock O (2014) Exploitation of environmental constraints in human and robotic grasping. Int. J. of Robotics Research. To appear.

[7] Dollar AM and Howe RD (2010) The highly adaptive SDM hand: Design and performance evaluation. Int. J. of Robotics Research, 29(5):585–597.

[8] Erdmann MA and Mason MT (1988) An exploration of sensorless manipulation. Int. J. of Robotics Research, 4(4):369–379.

[9] Gabiccini M, Farnioli E, and Bicchi A (2013) Grasp analysis tools for synergistic underactuated robotic hands. International Journal of Robotics Research, 32(13):1553–1576.

[10] Hirose S and Umetani Y (1978) The development of soft gripper for the versatile robot hand. Mechanism and Machine Theory, 13(3):351–359.

[11] Ilievski F, Mazzeo A, Shepherd RF, Chen X, and Whitesides GM (2011) Soft robotics for chemists. Angewandte Chemie Int. Edition, 50(8):1890–1895.

[12] Kazemi M, Valois J, Bagnell J, and Pollard N (2014) Human-inspired force compliant grasping primitives. Autonomous Robots, 37(2):209–225.

[13] Mason MT (1985) The mechanics of manipulation. IEEE Int. Conf. on Robotics and Automation (ICRA), volume 2, 544–548.

[14] Odhner L, Jentoft L, Claffee M, Corson N, Tenzer Y, Ma R, Buehler M, Kohout R, Howe RD, and Dollar AM (2014) A compliant, underactuated hand for robust manipulation. Int. J. of Robotics Research, 33(5):736–753.

[15] Prattichizzo D and Trinkle J (2008) Grasping. In: Springer Handbook of Robotics, Springer.

19 Task-specific Design of Tubular Continuum Robots for Surgical Applications

Jessica Burgner-Kahrs

Leibniz University Hannover

Abstract Tubular continuum robots are the smallest among continuum robots. They are composed of multiple, precurved, superelastic tubes. The design space for tubular continuum robots is infinite: each one of the component tubes can be individually parameterized in terms of its length, segmental curvatures, diameter, and material properties. Ad-hoc selection of those parameters is extremely challenging, since the elastic coupling of concentrically arranged and actuated tubes is hard to predict with common sense, especially under the presence of workspace constraints. In this chapter, an overview of the design process is given and the current state of the art in task-specific design of tubular continuum robots is reviewed.

19.1 Introduction

Continuum robots do not consist of discrete joints when compared to conventional robot types, such as rigid-link serial or parallel manipulators. They are characterized by a curvilinear structure and are inherently flexible. Often, continuum robots can extend along their structure. A review on continuum robots can be found in [1]. In the scope of this chapter, we will consider continuum robots with a tubular structure, which is the smallest type among continuum robots in terms of their diameter.

Concentric tube continuum robots have been first proposed in 2006 from both [2] and [3]. The kinematic structure of these robots is continuously tubular, i.e. composed of several, superelastic, precurved tubes that are nested inside of each other. Actuation is achieved by telescoping the tubes, thus each tube can be controlled with two degrees of freedom: translation and rotation at its base. Elastic interaction of the tubes causes them to bend and twist, which results in a continuous curvilinear manipulator as a net result. An overview of the historical development of this category of continuum robots and a summary of the state of the art can be found in [4].

The inherent flexibility of a tubular continuum robot and its small size (superelastic tubes are available with diameters down to tenths of a millimeter) make them favorable for surgical applications in constrained environments. As of now, tubular continuum robots have been exclusively considered for surgical applications, mainly for transnasal surgery [5], cardiac surgery [6], and neurosurgery [7]. In the medical domain, tubular continuum robots can be applied as manipulators - either teleoperated or operated (semi-) automatically - through natural orifices or small incisions, or as steerable needles within tissue applied percutaneously.

The design space for tubular continuum robots is infinite: each component tube can be individually parameterized in terms of its length, segmental curvatures, diameter, and material properties. Ad-hoc selection of those parameters is extremely challenging, since the elastic coupling of concentrically arranged and actuated tubes, especially under the presence of workspace constraints is hard to predict with common sense. This motivated the development of design heuristics and computational optimization methods in recent years, which are reviewed in the following.

19.2 Continuum Robots with Tubular Structure

19.2.1 Kinematic Structure

Continuum robots with a tubular kinematic structure are composed of multiple precurved, superelastic, and concentric tubes. By telescoping all component tubes, i.e. translating and rotating each tube at its base, and by constraining the tubes at an outlet with equaling diameter to the outermost tube, a tentacle-like overall manipulator motion is generated. The tubes interact elastically and conform to a common, curvilinear, continuous shape. Fig. 19.1 illustrates the kinematic structure and actuation principle of a tubular continuum robot composed of three tubes.

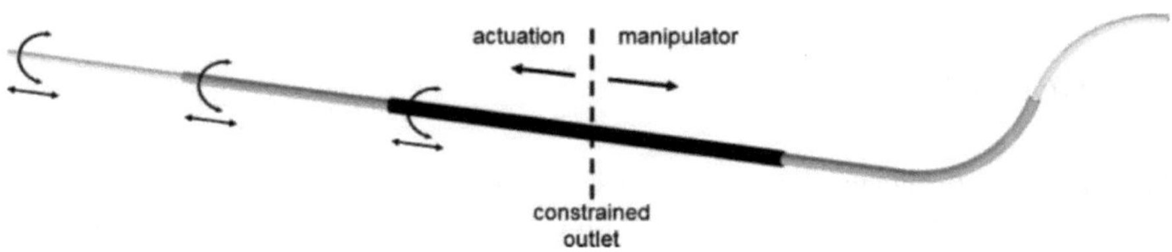

Fig. 19.1 Continuum robot with a tubular structure.

19.2.2 Kinematic Modelling

State of the art kinematic modelling applies Cosserat rod theory to each component tube as a continuum undergoing bending and torsion [8, 9]. Each tube is expressed by a Cosserat equation. Concentricity with all other tubes is enforced by requiring conformation to the same curvature as a function of arc-length. The tubes are left free to rotate axially with respect to one another. Boundary conditions for the resulting set of ordinary differential equations are the axial angles of the tubes at the constrained outlet and the vanishing internal moments at the tip. Integration along the tubes determines the space curve describing the robot. The model considers external loads applied to the robot structure [10, 9]. Further, the manipulator's Jacobian and compliance matrix can be determined [11], e.g. to allow for differential kinematic mapping [5, 12].

19.2.3 Component Tube Parameters

The characteristics of a tubular continuum robot are dependent on the number and composition of its tubes. Each tube can be parameterized by its properties:

- Length
- Diameter
- Wall thickness
- Segmental precurvature, and
- Material.

The individual tube parameters are deciding factors for the resulting manipulator in terms of its workspace, dexterity and manipulability, workload, stiffness etc. Usually tubular continuum robots are made from nitinol, a superelastic shape-memory alloy from nickel and titanium. The shape-memory property of nitinol allows to prebend a tube into segmental, arbitrary curvatures. Given the number of parameters per tube, the composition of multiple, individually parameterized tubes into a tubular manipulator, and the elastic interaction of all tubes during actuation, makes the design space for a tubular continuum robot diverse.

19.3 Task-specific Design

The task-specific design process is illustrated in Fig. 19.2. Surgical applications imply that the robot has to perform certain task(s) under the anatomical and pathological constrains which are individual for each patient. Thus, the surgical application imposes requirements onto the tube selection, tube prebending, and tubular manipulator. Further, the tube selection imposes constraints on the tube prebending, e.g. the recoverable strain defines the maximum curvature of a tube. Specifications on the tubular manipulator also influence tube prebending and tube selection. Hence, tube selection, tube prebending, and the tubular manipulator are interdependent.

The tubular continuum robot structure allows the component tubes to be disposable, individually parameterized for a specific patient and surgical application, while the actuation mechanism can be reused. This allows for task- and patient-specific tubular continuum manipulator design. Here, we only consider the design of tubular continuum robots in terms of their tube properties and composition. Research in the scope of tube actuation and end-effector design is summarized in [4].

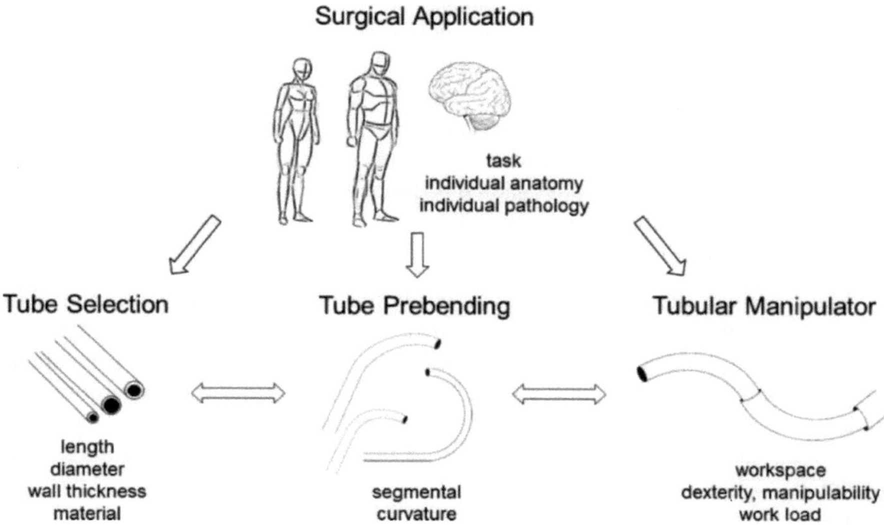

Fig. 19.2 The surgical application imposes constraints and requirements onto the design of tubular manipulators.

19.3.1 Design Heuristics

Along with proposing the general idea of tubular continuum robots, Webster et al. [2] and Sears and Dupont [3] provided tube parameter selection recommendations. For instance, tube curvatures are constrained by the material properties, i.e. the recoverable strain limit for superelastic materials. The maximum precurvature for a single tube such that it can fully straightened without plastic deformation is

$$\kappa = \frac{2\varepsilon}{D(1+\varepsilon)}, \tag{1}$$

where κ is the tube curvature, ε is the material strain, and D the diameter of the tube.

Dupont et al. proposed a design strategy, which distinguishes two desirable types of a manipulator [8]: The first type with the ability to move the tip of the manipulator without motion of the shaft, and the second type with the ability to maneuver through narrow curved environments without exerting high forces to the tissue. Dupont et al. suggests a design heuristic with four rules [8]:

- Telescoping Dominant Stiffness:
  Each tube should be selected such that their stiffness dominates all tubes extending from it.
- Fixed and Variable Curvature Sections:
  Each telescoping section is either a dominating (fixed curvature, if the outer tube dominates all inner tubes) or balanced (variable curvature, if the outer tube pair dominates all inner tubes) tube pair.
- Piecewise-Constant Tube Precurvatures:
  Extension of the tubes must proceed from the outer tube to the most inner tube to avoid lateral motion. Tubes should be prebend into one constant curvature.
- Increasing Curvatures from Base to Tip:
  Curvatures should increase form the outer to the most inner tube.

Limitations of these design rules lie in the idealization of the telescoping precurved, superelastic tubes principle. Dominant stiffness tube pairs are intuitively appealing and practical for some applications, but the heuristic is not generalizable. Especially, since the number of tube pairs is not infinite when regarding feasible actuation mechanisms. In fact, all existing tubular continuum robot prototypes today utilize designs with two to four tubes, whereas three tubes are most common.

While the heuristic differentiates between two desirable types of manipulators, ad-hoc selection of tube parameters is almost impossible when specific requirements on the tubular continuum robot are imposed. The diverse design space for the component tubes motivated the development of computational design optimization methods.

19.4 Computational Design Optimization

Anor et al. proposed the first tube design optimization method, which ensures the reachability of a discrete set of target positions by a tubular manipulator [13]. This method simplifies the manipulator kinematics to a pure geometric model such that the tubes are considered as constant curvature arcs (i.e. no mechanics based model). Given a discrete number of desired target positions, the pattern search optimization algorithm determines a sequence of constant curvature arcs, specified by their length and curvature, which guarantee reachability. The methodology has been applied to a neurosurgical scenario which requires reaching of specific positions within the ventricles [13].

Bedell et al. proposed a computational design method which finds tube parameters to explicitly reach discrete poses [14]. The design space is reduced to those component tubes with constant circular curvature. Bedell et al. use a kinematic model which neglects torsional compliance [8]. A generalized pattern search algorithm is applied to minimize an objective function, which evaluates the reachability of the target poses within the given anatomical constraints by a set of component tubes. Anatomical constraints are represented as a surface model in which boundaries the manipulator has to reach the target poses. Application of this method has been cardiac surgery [14].

Torres et al. introduced a task-based design method that searches for a set of component tubes that enables a collision-free path to a desired region [15]. The method combines random exploration of the design space with a sampling based motion planner in the robot's configuration space using rapidly exploring random trees. The used kinematic model takes torsional compliance into account [16], but since collision free paths are desired, external forces and moments acting on the robot are not considered. A lung application has been considered where deployment of the robot through the bronchi is required. In this scenario the number of tubes has been restricted to three tubes with an initial straight section followed by a precurved section with constant circular curvature [15].

Burgner et al. proposed a computational design optimization algorithm which determines the optimal set of tubes with the objective to fully cover a predefined working volume [17]. The desired surgical working volume is specified using combinations of geometric primitives, such as spheres, cones, cylinders, and ellipses. This allows the surgeon to define the respective parameters, such as height, diameter, and length of those geometric primitives, upfront in medical images of the patient. Fig. 19.3 illustrates the surgical workspace required for transnasal skull base surgery and its approximation by an ellipse to describe the tumor and the frustum of a cone to represent the access through the sphenoid sinus. The objective function which is minimized using the Nelder-Mead simplex algorithm, determines the percentage of the surgical volume which cannot be reached by a set of component tubes.

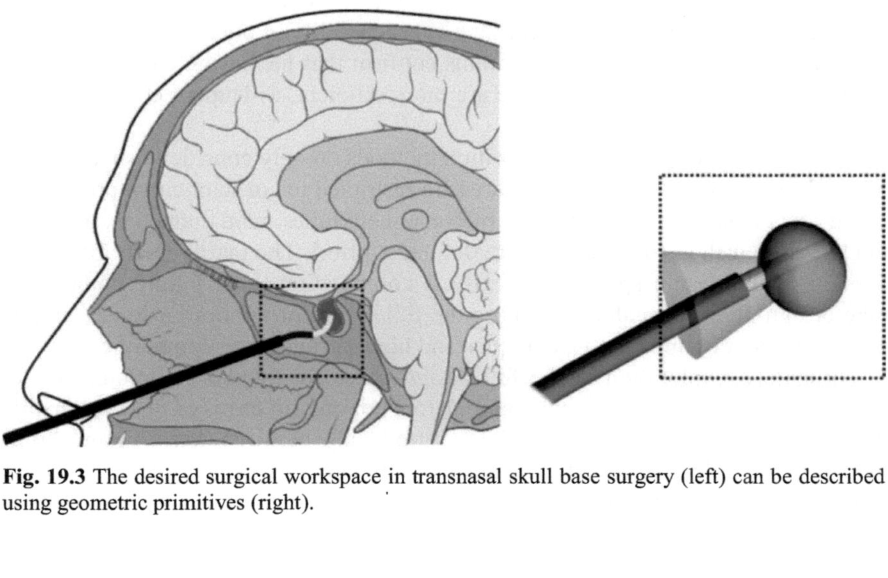

Fig. 19.3 The desired surgical workspace in transnasal skull base surgery (left) can be described using geometric primitives (right).

To reduce the complexity of the tube design space, Burgner et al. only considers manipulators made of three component tubes, with each tube having an initial straight section followed by a precurved section with circular constant curvature [17]. The kinematic model used is the state of the art model described earlier [9]. The volume-based objective function has been applied to transnasal skull-base surgery [17] and neurosurgery [7]. The volumetric representation has further motivated efficient workspace computation methods [18]. These four computational design algorithms have in common, that they seek in optimizing the component tubes of a tubular manipulator in regard of one objective: Either to reach a discrete set of positions required by the surgical application, to reach a defined region, or to cover a desired surgical volume.

While the earlier approaches by Anor et al. [13] and Bedell et al. [14] used simplified kinematic models and reduced the computation expense, Torres et al. [15] and Burgner et al. [17] used state of the art kinematic models. Hence, the results of their methods ensure feasibility for real tubular continuum robots.

19.5 Discussion and Outlook

Selecting the parameters of component tubes for tubular manipulators for a specific task is challenging. The design space is diverse, such that heuristics on the tube precurvatures and composition are useful, but yet ad-hoc selection of tube parameters is almost impossible. Burgner et al. have shown that tube parameters selected by experienced engineers for a particular surgical application did in fact cover the desired surgical workspace only by 25% [17].

The design algorithms proposed so far indicate the effectiveness of optimization methods for task-based criteria such as reachability of poses or volume coverage. The results of the design algorithms confirm that increasing curvatures from the most outer to the innermost tube are more efficient, as Dupont et al. proposed in their design heuristics [8].

However, existing design optimization methods do only consider a small subset of the design space. The tube curvatures are restricted to constant curvatures of the curved sections. Further the number of component tubes, tube diameters, and wall thickness, as well as tube material are defined a-priori. The main reason for the reduction of the design space complexity is computation time. Future component tube design algorithms should explore the full extent of the design space, by applying efficient optimization techniques and high-performance computing.

Ultimately, exploration of the full design space for tubular continuum robots will enable task-specific optimal manipulators not only in terms of pose reachability or surgical workspace coverage. By introducing multi-objective optimization methods, multiple task-specific design requirements can be considered, such as robot stiffness, manipulability at the surgical point of interest, or optimal motion paths by minimizing energy bifurcations.

As of today, task-specific design optimization for tubular continuum robots cannot be regarded as a solved problem. However, the proposed objective functions will be part of future multi-objective design optimization methods.

Acknowledgments This work was funded within the German Research Foundation's Emmy Noether Programme under award number BU2935/1-1. The author thanks Carolin Fellmann for generating Fig. 19.1.

19.6 References

[1] Walker ID (2013) Continuous Backbone "Continuum" Robot Manipulators. ISRN Robot 2013:1–19. doi: 10.5402/2013/726506

[2] Webster R, Okamura A, Cowan N (2006) Toward Active Cannulas: Miniature Snake-Like Surgical Robots. 2006 IEEE/RSJ Int. Conf. Intell. Robot. Syst. pp 2857–2863

[3] Sears P, Dupont P (2006) A Steerable Needle Technology Using Curved Concentric Tubes. 2006 IEEE/RSJ Int. Conf. Intell. Robot. Syst. pp 2850–2856

[4] Gilbert HB, Rucker DC, Webster III RJ (2013) Concentric Tube Robots: The State of the Art and Future Directions. Int. Symp. Robot. Res.

[5] Burgner J, Rucker DC, Gilbert HB, Swaney PJ, Russell PT, Weaver KD, Webster III RJ (2014) A Telerobotic System for Transnasal Surgery. IEEE/ASME Trans Mechatronics 19:996–1006. doi: 10.1109/TMECH.2013.2265804

[6] Gosline AH, Vasilyev N V., Butler EJ, Folk C, Cohen A, Chen R, Lang N, Del Nido PJ, Dupont PE (2012) Percutaneous intracardiac beating-heart surgery using metal MEMS tissue approximation tools. Int J Rob Res 31:1081–1093. doi: 10.1177/0278364912443718

[7] Burgner J, Swaney PJ, Lathrop RA, Weaver KD, Webster RJ (2013) Debulking from within: a robotic steerable cannula for intracerebral hemorrhage evacuation. IEEE Trans Biomed Eng 60:2567–75. doi: 10.1109/TBME.2013.2260860

[8] Dupont PE, Lock J, Itkowitz B, Butler E (2010) Design and Control of Concentric-Tube Robots. IEEE Trans Robot 26:209–225. doi: 10.1109/TRO.2009.2035740

[9] Rucker DC, Jones BA, Webster III RJ (2010) A Geometrically Exact Model for Externally Loaded Concentric-Tube Continuum Robots. IEEE Trans Robot 26:769–780. doi: 10.1109/TRO.2010.2062570

[10] Lock J, Laing G, Mahvash M, Dupont PE (2010) Quasistatic Modeling of Concentric Tube Robots with External Loads. IEEE/RSJ Int. Conf. Intell. Robot. Syst. pp 2325–2332

[11] Rucker DC, Webster RJ (2011) Computing Jacobians and compliance matrices for externally loaded continuum robots. IEEE Int. Conf. Robot. Autom. pp 945–950

[12] Xu R, Asadian A, Naidu AS, Patel R V. (2013) Position control of concentric-tube continuum robots using a modified Jacobian-based approach. IEEE Int. Conf. Robot. Autom. pp 5813–5818

[13] Anor T, Madsen JR, Dupont P (2011) Algorithms for Design of Continuum Robots Using the Concentric Tubes Approach: A Neurosurgical Example. IEEE Int. Conf. Robot. Autom. pp 667–673

[14] Bedell C, Lock J, Gosline A, Dupont PE (2011) Design Optimization of Concentric Tube Robots Based on Task and Anatomical Constraints. IEEE Int. Conf. Robot. Autom. pp 398–403

[15] Torres LG, Webster III RJ, Alterovitz R (2012) Task-oriented Design of Concentric Tube Robots using Mechanics-based Models. IEEE/RSJ Int. Conf. Intellegent Robot. Syst.

[16] Rucker DC, Webster III RJ, Chirikjian GS, Cowan NJ (2010) Equilibrium Conformations of Concentric-Tube Continuum Robots. Int J Rob Res 29:1263–1280. doi: 10.1177/0278364910367543

[17] Burgner J, Gilbert HB, Webster III RJ (2013) On the Computational Design of Concentric Tube Robots: Incorporating Volume-Based Objectives. IEEE Int. Conf. Robot. Autom. pp 1185–1190

[18] Granna J, Burgner J (2014) Characterizing the Workspace of Concentric Tube Continuum Robots. 45th Int. Symp. Robot. pp 1–7

20 Soft Robotics with Variable Stiffness Actuators: Tough Robots for Soft Human Robot Interaction

Sebastian Wolf, Thomas Bahls, Maxime Chalon, Werner Friedl, Markus Grebenstein, Hannes Höppner, Markus Kühne*, Dominic Lakatos, Nico Mansfeld, Mehmet Can Özparpucu, Florian Petit, Jens Reinecke, Roman Weitschat and Alin Albu-Schäffer

German Aerospace Center (DLR)

Abstract Robots that are not only robust, dynamic, and gentle in the human robot interaction, but are also able to perform precise and repeatable movements, need accurate dynamics modeling and a high-performance closed-loop control. As a technological basis we propose robots with intrinsically compliant joints, a stiff link structure, and a soft shell. The flexible joints are driven by Variable Stiffness Actuators (VSA) with a mechanical spring coupling between the motor and the actuator output and the ability to change the mechanical stiffness of the spring coupling. Several model based and model free control approaches have been developed for this technology, e.g. Cartesian stiffness control, optimal control, reactions, reflexes, and cyclic motion control.

20.1 Introduction

Robots interacting with humans in direct physical contact or even acting in place of a human in given situations are of high interest for current research. Manipulating objects in direct contact with the human or in unstructured environment likely results in contacts and collisions that are unpredictable. Furthermore, the desired robot skills include fine manipulation as well as highly dynamic and powerful movements. This results in special demands on the robots capabilities:

- Robustness to fast impacts to reduce the risk of robot damage
- Precision for fine manipulation
- Sensitivity for gentle interaction with the environment
- High dynamics for fast and controlled movements

If the intended application demands the combination of the four aspects, namely to be robust, precise, sensitive, and dynamic at the same time, robots with stiff structures and stiff drive-trains come to their limits. Robots with stiff structure, but inherent compliance in the drive-train, promise to overcome the restrictions of classical stiff robots, being able to combine the four required aspects. The technology of compliant actuators has been intensively investigated in the last decade, which gained a great variety of implementations [1]. It can be distinguished between a Variable Impedance Actuator (VIA) with physical variable stiffness and damping, a Variable Stiffness Actuator (VSA) with physical variable stiffness, and a Variable Damping Actuator (VDA) with physical variable damping only.

This article will first introduce the concept of compliant actuation on the example of the DLR Hand Arm System [2]. Then we discuss strategies for control of the hand, Cartesian stiffness control, and optimal control. Furthermore, we show reactions and reflexes to protect the hardware and the human, and address cyclic motion control.

20.2 Compliant Actuation

The concept of soft robotics at DLR is based on stiff link structures and compliant elements in the actuator drive-train (see one-dimensional model in Fig. 20.1). A mechanism with a mechanical spring decouples the motor inertia from the output and the link mass, which makes the actuator more robust and enables energy storage in the spring (see Sec. 20.8, [3], [4], [5]). The difference between motor position θ and link position q is the passive spring deflection ϕ. Joint torques are calculated by position measurement of the passive spring deflection. The joint torque information is used in control, e.g. impedance control, where the control adds active impedance to the passive compliance of the spring mechanism (see Sec. 20.6.2). There is no passive damping element in the current mechanics, so the actuator types are VSAs. Damping is realized by active control [6], [7]. Most of the link structure is equipped with flexible and padded covers, which give the robot a soft tactile touch and reduce local contact forces.

The anthropomorphic DLR Hand Arm System was built as an experimental platform to evaluate the VSA concept on 3 different types of VSAs [2]. It has a total of 26 DoF with 7 in the arm and 19 in the hand (Fig. 20.2). The flexibility is solely located in the actuators, which are connected with stiff aluminum link structures. The housings are padded with soft foam and textile.

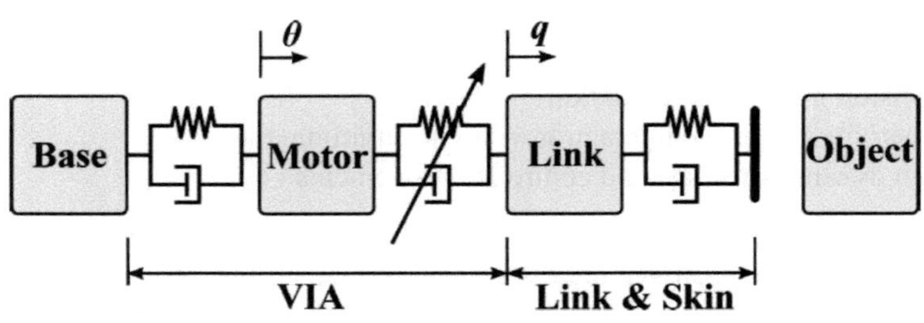

Fig. 20.1 1-DoF mass model of a Variable Impedance Actuator (VIA), which has physical variable stiffness and damping, interacting with an external object.

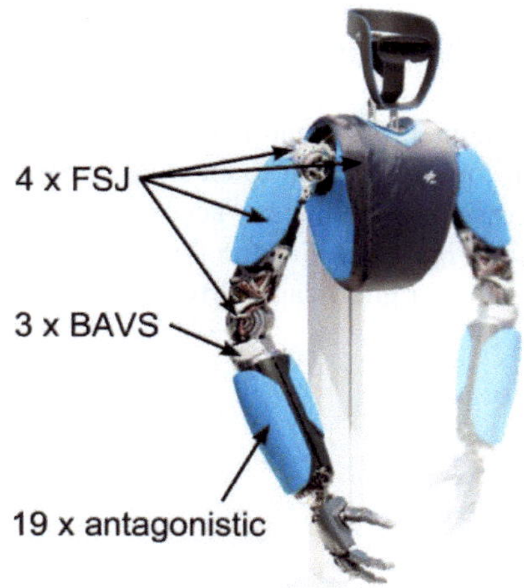

Fig. 20.2 The DLR Hand Arm System is equipped with 3 different types of Variable Stiffness Actuators (VSA).

20.2.1 Floating Spring Joint (FSJ)

The mechanical principle of the DLR FSJ (see Fig. 20.3) is a VSA module designed for the first 4 axes of the DLR Hand Arm System [4]. For this purpose, the joints have to be extremely compact to fit into the arm. At the same time they require a high power density in order to approximate the human arm skills. The DLR FSJ is designed to have energy efficient components and low friction. The potential energy of the spring is used to a maximal extend in order to have a high energy capacity to weight ratio. The DLR FSJ has one big motor to change the output position of the actuator and one small motor to change the stiffness preset of the spring mechanism. In this setup only one motor has to move to change the link position of the robot.

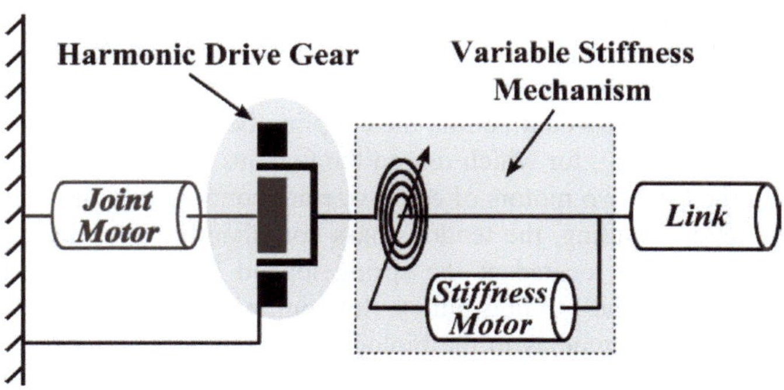

Fig. 20.3 Schematic of the DLR FSJ mechanism. [4]

If the small motor is kept in a fixed position, the actuator behaves like a Serial Elastic Actuator (SEA). The nonlinear spring mechanism is in series between the main gear and the output.

The torque is generated by a rotational cam disk and roller system which transmits the rotational joint deflection to an axial compression of a linear spring. The shape of the cam disks can be chosen according to the desired torque vs. displacement behavior (see Fig. 20.4). For the DLR Hand Arm System the cam disk shapes were chosen to have a good capability of stiffness variation under all load conditions.

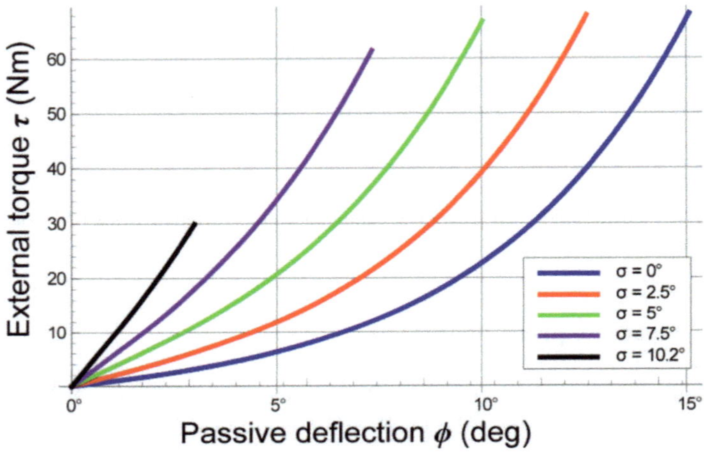

Fig. 20.4 The elastic torque characteristics of the DLR FSJ is parameterized by stiffness setup σ. The stiffness setup is set by the small motor. [4]

20.2.2 Flexible Antagonistic Spring Element (FAS)

The FAS is motivated by the antagonistic arrangement of the human muscular system. Herein, two muscles act in opposing directions to the joint. To change the joint position, the muscles generate asymmetric forces. The stiffness can be varied by co-contraction of the muscles, which generate internal forces. In order to change the stiffness by co-contraction, the coupling between joint and muscles has to be a non-linear spring, for which reason the tendons are exponential spring elements [8]. In the FAS two motors of equal size are connected to the output by tendons. In the tendon routing, the tendons pass compliantly supported pulleys (see Fig. 20.5). The pulley located at the spring loaded lever (right pulley) rotates around the center of the guiding pulley (left pulley). The principle to move or stiffen the joint is the same as in the biological muscle arrangement [5]. It is depicted in Fig. 20.5a) in a low mechanical stiffness due to small α and in Fig. 20.5b) in a high stiffness due to a large α.

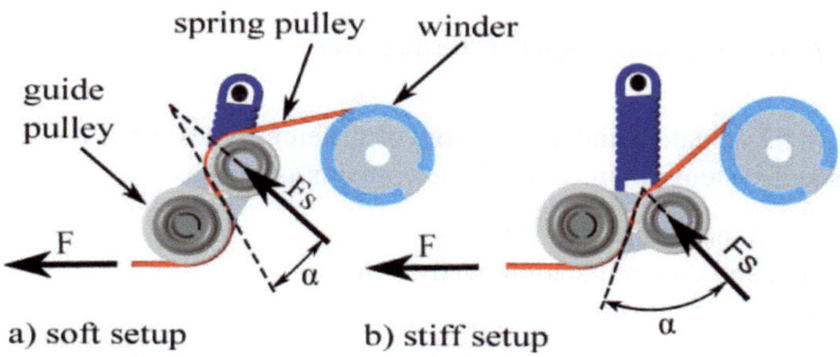

Fig. 20.5 Antagonistic drive compliance mechanism ("tan α mechanism") used for the FAS drive in a) a low mechanical stiffness, and b) a high stiffness. [5]

20.2.3 Bidirectional Antagonism with Variable Stiffness (BAVS)

The forearm rotation (pronation and supination) and both rotations of the wrist (radial and ulnar deviation; flexion and extension) are based on the principle of a Bidirectional Antagonism with Variable Stiffness (BAVS) [9], [10], [11]. A normal antagonistic mechanism has the drawback that only the power of one motor is available at the drive side of the joint. With the principle of BAVS we overcome this drawback by allowing both motors to push and pull and thus to assist each other in both directions. For this reason we also use the term helping antagonism (see Fig. 20.6). In the case of equally sized motors and gears, the output torque is twice the maximum individual motor torque. In the BAVS of the DLR Hand Arm System each motor is connected to a harmonic drive gear with the circular spline bedded in a ball bearing instead of fixing it to the base frame. In order to change the position of the VSA, both motors rotate in the same direction. In order to change the VSA stiffness, the motors move in the opposite direction, which causes a pretension in the non-linear springs, but keeps the output position constant. The non-linear spring characteristic is realized by a linear spring with a non-linear cam disc mechanism. The cam disks can be easily changed in order to test different stiffness-deflection relations.

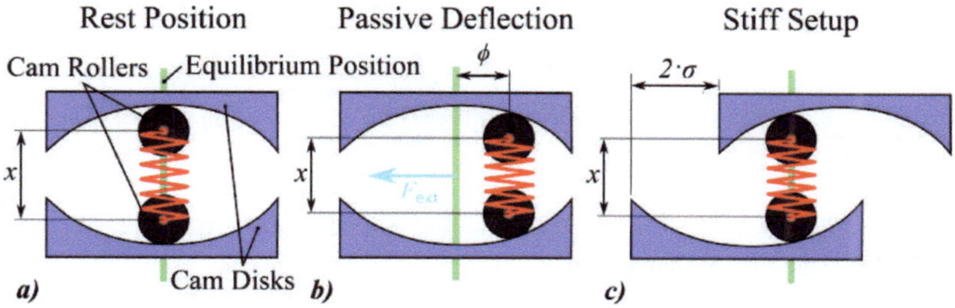

Fig. 20.6 BAVS drive: in a) the mechanism is in rest position, in b) the joint is deflected by an external torque, and in c) an equilibrium position in a stiff configuration is depicted (adapted from [2]).

20.3 Electronics and System Architecture

Dealing with 52 actuators and 430 sensors of different types poses a challenge for electronics and systems architecture. Hence, major aspects as

- reliability and maintainability
- computation and communication bandwidth
- power density and integration level

have to be taken into account during the development and design process [2]. Therefore the system is designed modularly and hierarchically. Electronics as well as the computation and communication architecture follow this approach. Highly integrated electronic subcomponents which are independent and scalable guarantee the required flexibility, maintainability and reliability. This also is mirrored in a well-structured communication concept described in [2], [12], [13].

Fig. 20.7 shows the hierarchical topology of the DLR Hand Arm System. SpaceWire [14] is used as the communication backbone of the system. It is a standardized packet based bus with a bandwidth up to 1 Gbit/s and is deterministic for a given topology. Actuators and sensors are connected via the industry standard BiSS [15]. BiSS is a master / slave bidirectional serial bus with a typical data rate of 10 Mbit/s [2].The physical interfaces in turn are connected via their dedicated interfaces (I2C, SPI, PWM, etc.) to the sensor and actuator modules. Due to this hierarchical design the complexity of the system can be hidden from the application designer and on the other hand a good performance can be achieved with only a minimal execution overhead.

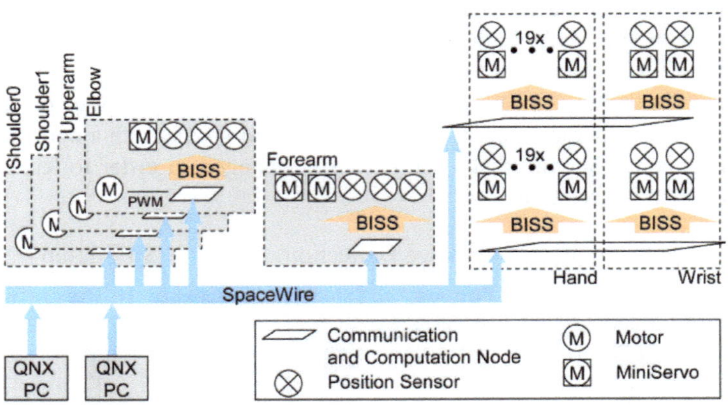

Fig. 20.7 DLR Hand Arm System topology. [2]

20.4 Hand Design and Control

The hand is the most exposed part of the robot, although it has the smallest force capabilities. The required features are large dexterity, manipulation capabilities, and robustness against unwanted collisions 251[16], [17]. As the human hand is highly under-actuated and uses several muscle/tendon synergies [18], which are not technically realizable, a plain copy of the human hand is not feasible. The hand design has to be based on an abstraction of the fundamental functionalities of the human hand. To reduce the number of drives, DoFs which have relatively low impact on grasping abilities should be omitted. These missing DoFs have to be compensated functionally by the kinematics design of the hand [19].

The kinematics of the hand (see Fig. 20.9) is closely adapted to the human hand on a functional basis [19]. It consists of 19 independent DoFs[10] in order to reduce the number of necessary drives. Like human fingers, the index and middle finger have 4 DoFs. The 2nd (PIP) and 3rd (DIP) finger joint[11] of the ring and fifth finger are coupled to reduce the number of necessary actuators. The 5th DoF of the human thumb turned out to be of low relevance [20], [19] and have been omitted. To ensure proper opposition of the thumb and the 5th finger, an antagonistically driven 4 bar mechanism was designed to mimic the motion of the 5th finger metacarpal bone[12]. The structure of the finger is designed as an endoskeleton with "bionic joints" [21]. The finger base (metacarpal) joint is a hyperbolically shaped saddle joint because the human condyloid joint type cannot be replicated technically[13]. The finger (interphalangeal) joints, on the other hand, are designed as hinge joints. All joints allow dislocation without damage in case of overload[14]. In addition to robustness due to short-term energy storage, the use of antagonistic actuation (see Sec. 20.2) enables to cope with tendon slackening or overstretching, which is one of the major problems of nowadays tendon-driven hands having inevitably constant tendon length. In contrast, antagonistic actuation is able to compensate unaligned pulley axes, and other geometrical errors via the elastic elements of the drive train. Therefore no explicit tendon tensioner is needed [22].

The forearm is composed of 3 major parts: 1-DoF forearm rotation and 2-DoF wrist both with BAVS actuation and a 19-DoF hand with antagonistic actuation. The 38 motors and their corresponding nonlinear compliance mechanisms are tightly integrated into the forearm (see Fig. 20.8). The fingers are actuated via the ServoModules. A multiturn winder transfers the rotational gear motion to the ten-

---

[10] Thumb 4 DoF, index finger 4 DoFs, middle finger 4 DoFs, ringfinger 3 DoFs, 5th finger 4 DoFs

[11] starting from the fingers base

[12] the first bone of the finger located within the palm

[13] in a meaningful way

[14] the elongation of the tendons in case of dislocation is compensated by the elastic elements of the antagonistic drives

dons. The compliance mechanism is similar to the one described in [22]. One difference is that the winder also acts as the first pulley of the "tan α mechanism" (see Fig. 20.5). This reduces the number of required pulleys. Each compliance element is adapted to the different finger and joint characteristics. In order to facilitate maintenance, the finger actuators are placed on two almost identical halves. This configuration allows to replace ServoModules without dismounting the tendons. If a tendon replacement is required, the forearm can be opened to grant access to the compliance mechanisms as well as the winders.

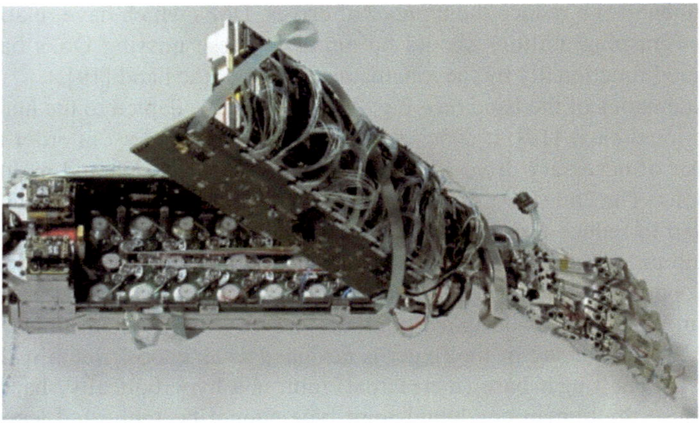

Fig. 20.8 Opened forearm with the ServoModules. The ServoModules are located on the outside of each half of the forearm. The tendons and the elastic elements are located in the middle layer between both halves.

The compact ServoModules are used for both the wrist and forearm rotation since the requirements are almost identical. The major difference is a higher required torque for the forearm rotation and wrist actuation than for the finger actuation. This requirement motivated the use of the newly developed BAVS actuation (see Sec. 20.2).

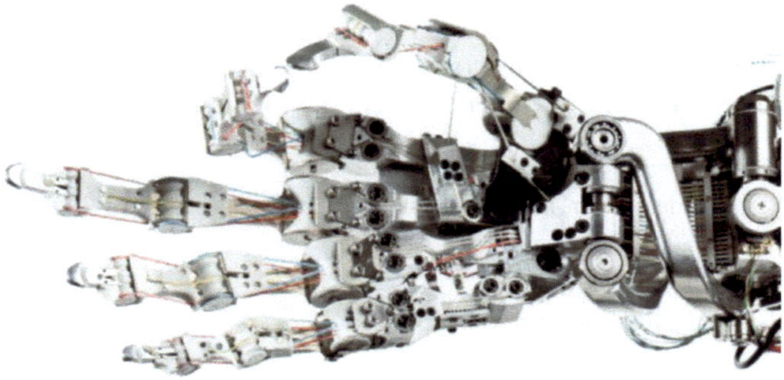

Fig. 20.9 Hand and wrist of the DLR Hand Arm System.

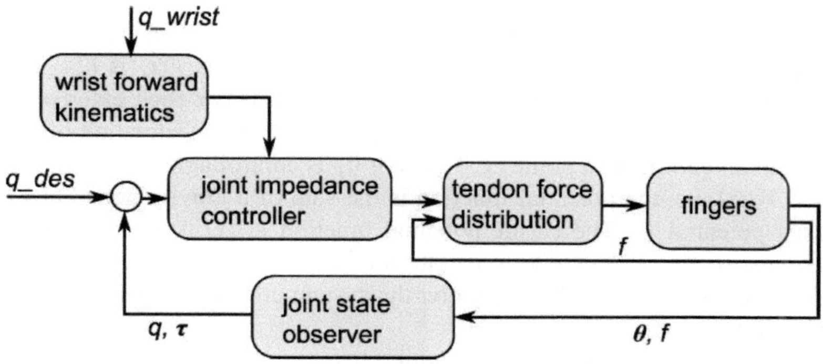

Fig. 20.10 Hand controller structure. q and θ represent the joint and motor positions. f and τ represent the tendon forces and the joint torques.

The hand controller implements a joint level impedance behavior to facilitate interactions with the environment. The top level control scheme implemented on the system is depicted in Fig. 20.10. The joint state observer provides an estimate of the joint position and the joint torque based on the measured motor displacement and the FAS sensors. An inner loop ensures that the tendon pulling constraints are holding and that the maximal tendon forces are not exceeded. The joint impedance controller is using a back-stepping structure in order to deal with the flexibility of the joints. A kinematic model of the wrist is used to compensate for the tendon displacement due to wrist motion.

20.5 Modeling Soft Robots

A soft robot with lumped elasticities such as the DLR Hand Arm System can be imagined as a set of directly actuated rigid bodies with configuration $\theta \in \mathbb{R}^m$, connected to the indirectly actuated rigid bodies (with configuration vector $\theta \in \mathbb{R}^n$) through visco-elastic forces [23]. The entire configuration space of the system is denoted by $x = (\theta, q)$, $x \in \mathbb{R}^{(n+m)}$. A quite general abstraction of a compliant robot, which can be used for the generic design of controllers, is given by

$$M(x)\ddot{x} + c(x,\dot{x}) + \frac{\partial V(x)}{\partial x} + d(\dot{x}) = \begin{bmatrix} \tau_m \\ \tau_{\text{ext}} \end{bmatrix} \tag{1}$$

with $M(x)$ being the inertia matrix, $c(x,\dot{x})$ the Coriolis and centrifugal vector, V the potential energy of the elastic element and of the gravity forces, $\tau_m \in \mathbb{R}^m$ the actuator generalized forces acting as control inputs and $\tau_{ext} \in \mathbb{R}^n$ the external torques acting on the robot as a disturbance. The most relevant property of this structure is its under-actuation, meaning that the system has less control inputs (m) than its configuration space dimension $(n+m)$. However, in contrast to other purely inertially coupled under-actuated systems[15], for the considered robots $V(x)$ is [24] positive definite, implying that a unique equilibrium point exists for each external torque with actuators in a fixed configuration $\theta = \theta_0$ and that the linearization of the system around an equilibrium point $\{x = x_0; \dot{x} = 0\}$ is controllable. Typically, $V(x) = V_G(x) + V_\tau(x)$, i.e. the potential function is the sum of a gravity potential V_G and an elastic potential V_τ. The elastic potential function $V_\tau(x)$ is a convex function, increasing strongly enough to compensate for the destabilizing effects of V_G, such that $V(x)$ is convex as well. Furthermore, the system contains in general a dissipative friction force $d(\dot{x})$ with $\dot{x}^T d(\dot{x}) \leq 0$.

If there is no substantial inertial coupling between motors and links and under some further mild simplification assumptions, the model can be written in a form resembling more the classical flexible joint robots:

$$\begin{bmatrix} M(q) & 0 \\ 0 & B \end{bmatrix}\begin{bmatrix} \ddot{q} \\ \ddot{\theta} \end{bmatrix} + \begin{bmatrix} c(q,\dot{q}) \\ 0 \end{bmatrix} + \begin{bmatrix} g(q) \\ 0 \end{bmatrix} + \begin{bmatrix} \tau \\ -\tau \end{bmatrix} + d(\dot{x}) = \begin{bmatrix} \tau_{\text{ext}} \\ \tau_m \end{bmatrix}. \quad (2)$$

20.6 Cartesian Stiffness Control

20.6.1 Cartesian Impedance Control

While the mechanical stiffness of VIA robots can be adjusted in a wide range, not any task impedance can be realized by shaping the intrinsic mechanical properties. This is due to intrinsic limitation in minimal and maximal mechanical stiffness as well due to the fact that so far, most robots do not have coupled stiffness elements which would require multi-joint actuators [25]. This is in contrast to biological systems, which have a much larger number of muscles and also contain a large number of multi-articular joints. However, one can design a stiffness controller on motor side, which acts in serial interconnection with the intrinsic mechanical stiffness. Following a central paradigm of VIA, the mechanical robot parameters need to be tuned such that the desired behavior is naturally achieved on a mechanism level to the largest extent possible.

The stiffness behavior is described by a constant stiffness matrix $K_C = \frac{\partial f}{\partial x} \in \mathbb{R}^{m\mathrm{x}m}$ as the relation between the Cartesian wrench f and the Cartesian displacement x. The n passive and adjustable joint stiffness components provide the diagonal[16] matrix $K_J = \frac{\partial \tau}{\partial q} \in \mathbb{R}^{n\mathrm{x}n}$ with the joint torques τ and the joint positions q. The

---

[15] as for example multiple pendulums such as the Acrobot [27]

[16] Decoupled joint compliance is assumed.

mapping from the Cartesian stiffness space to the joint stiffness space is given by $\tau : K_J = T(K_C)$.

This transformation can be written as

$$\begin{aligned} K_J &= \frac{\partial \tau}{\partial q} = \frac{\partial (J(q)^T K_C \Delta x)}{\partial q} \\ &= J(q)^T K_C J(q) - \frac{\partial J(q)^T K_C \Delta x}{\partial q}. \end{aligned} \tag{3}$$

$J(q) = \frac{\partial f(q)}{\partial q}$ is the manipulator Jacobian, where $f(q)$ is the forward kinematics mapping. $\Delta x = x_d - x$ is the Cartesian position error between the desired and the actual position. The first part of (3) reflects the stiffness around the equilibrium point. The second part of (3) is due to the change of the Jacobian, see [26].

The Cartesian stiffness a the equilibrium position ($\Delta x = 0$), resulting from a specific joint stiffness can be obtained by solving the inverse problem of (3), $K_C = \tau^{-1}(K_J)$[17]. Using compliance matrices $C_\cdot = K_\cdot^{-1}$, it results from

$$C_C = J(q)C_J J(q)^T) \tag{4}$$

that

$$K_C = (J(q)K_J^{-1} J(q)^T)^{-1}. \tag{5}$$

Active control adapts the stiffness in a wider range, and the elastic elements are capable of absorbing impacts and increase the energy efficiency. By combining the two concepts, one can benefit from the individual advantages. The serial interconnection of the active stiffness K_{active} and the passive one $K_{passive}$ results in an overall stiffness K_{res} (see Fig. 20.11):

$$K_{\mathrm{res}}^{-1} = K_{\mathrm{active}}^{-1} + K_{\mathrm{passive}}^{-1} \tag{6}$$

To compute the active and passive stiffness components, a two-step optimization algorithm can be used [28]. It achieves first a passive stiffness as close as possible to the desired one and secondly designs the active stiffness to minimize the residual.

[17] Note that at the equilibrium position the second term in 3 vanishes

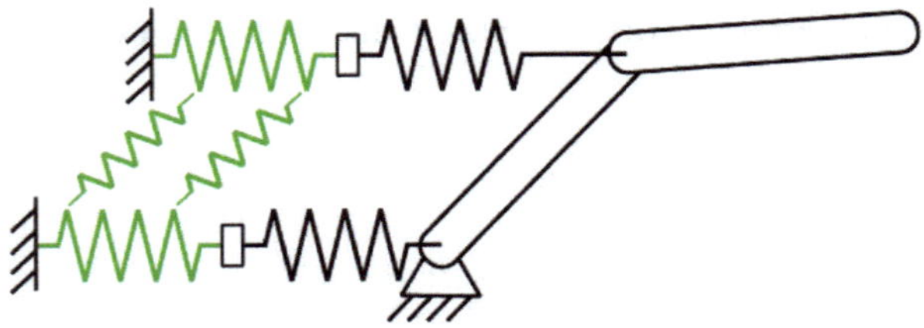

Fig. 20.11 Combination of active impedance controller with passive compliance to achieve a desired Cartesian stiffness.

20.6.2 Independent Position and Stiffness Control

The classical robot positioning task has some obvious particularities in the case of VIA systems. One does not only need to control the position of the link, but simultaneously has the chance to control the joint stiffness as well. This is possible due to the two independent actuators per joint. Additional control goals, especially important in the presence of disturbances, are the minimization of deflections and the suppression of vibrations.

The solutions to achieve independent control of link motion and stiffness can be classified in two categories.

- One class of controllers exploits the knowledge of system models. In [29] and [30] feedback linearization approaches are used to transform the robot dynamics into an equivalent model of simpler form. A decoupled chain of integrators can be achieved, as long as system inversion is possible. The simple structure of the equivalent dynamics allows for simultaneous decoupling and accurate tracking of motion and stiffness reference profiles. However, the abstraction of the robot dynamics hinders the implementation of performance criteria.
  Another model based approach aims to achieve a reduced order model [31]. Therefore, separate dynamics are identified in the robot system, namely the arm, the positioning actuators, and the stiffness actuators. The independence of these dynamics is shown by a singular perturbation analysis. A cascaded control structure is based upon this analysis.
  The abstraction that this class of controllers provides, allows for theoretical simplicity and design flexibility. On the other side, robot performance and robustness in the presence of model uncertainty are not guaranteed. High model accuracy and high derivatives of states are often required. Furthermore, to conform with the idea of embodiment additional effort is necessary, as the controllers often enforce a very different robot behavior than the natural one.
- The second class of controllers are energy shaping based controllers.
  One of the first controllers in this category was presented by [32]. The controller acts on motor position and uses a transformation to independently control

the joint position and stiffness. The controller is validated on a 1 DoF VSA joint.

An extension and generalization has been presented in [33]. The control design formulation is valid for a quite general form of underactuated Euler-Lagrange systems including variable impedance robots. Herein, the controller action can be interpreted as shaping of the potential and kinetic robot energy ensuring system passivity.

A general task is described as to control k independent output variables given by $q = h(x)$ to desired constant values $q_d \in R^k$. $\in R^n$ is the vector of generalized coordinates, where $n = 2k$ is the usual case for VSA robots. Given the structure of VSA robots (2), a new variable $\bar{q}$ can be found. This is a collocated (directly actuated) variable, which is statically equivalent to the noncollocated (indirectly actuated) q. Using this collocated variable for a passive feedback ensures stability and can be interpreted as shaping the potential energy of the system.

The variable $\bar{q}$ is achieved by solving the static solution of the link side equation

$$u = g(\bar{q}) - J_{\bar{q}}^T(\theta)K_p(\bar{q} - q_d) - K_d\dot{\theta} \tag{7}$$

for q. Except for very simple cases, this equation has to be solved numerically. Due to the convex nature, this is a fast and numerically robust task in practice. Consequently, the controller

$$\tau + g(q) = 0 \tag{8}$$

is stabilizing the desired position q_d. $g(\bar{q})$ is a feed forward term, compensating for the gravity. The use of $\bar{q}$ enables arbitrarily low controller gains even for large displacements from the equilibrium. $J_{\bar{q}}$ is a Jacobian mapping the collocated variable on the statically equivalent $\bar{q}$. Global asymptotic stability based on La Salle's invariance theorem can be shown.

The approach can be extended in order to include also feedback of torque and torque derivative, with the effect of reducing the apparent actuator inertia and friction. This allows improving the transient performance while remaining within the passivity framework.

The energy shaping approach provides excellent performance in the static case and for well damped systems. However, some joints show low intrinsic damping to enable joint torque estimation and energy efficiency. In this case it is desirable to add additional damping via control or to include a physical variable damping element. Several damping control structures can be considered. A simple gain scheduling approach for a one DoF system has been presented in [34]. Local linear sub-problems are identified on which a LQR state feedback controller is designed. The nonlinearity of the robot dynamics requires adapting the controller poles dependent on the system state, which is especially hard for the multi joint robots.

A physically motivated state feedback control approach for multi-DoF VSA robots has been presented in [6] using an eigenvalue based modal decomposition.

20.7 Optimal Control

Humans are capable of highly dynamic motions such as throwing or kicking. A major feature that presumably enables them to perform such tasks is their ability to optimally store and release elastic energy in the compliant elements of the musculoskeletal system, in combination with inertial energy transfer between the rigid parts of the body. Based on this feature, using elastic elements in robot joints, and finding control strategies which exploit this elasticity optimally has recently drawn significant attention.

The question of how to make use of the (adjustable) potential energy stored in elastic elements of a robotic system in the best possible way is not a trivial one, especially when considering the increased complexity of robotic systems with additional elasticity. A powerful mathematical tool for dealing with this question is Optimal Control (OC) Theory [35], which provides conditions for the control to minimize a given cost functional. Generally, a cost functional can be written as a sum of two terms

$$J(u) = \vartheta(x(t_f), t_f) + \int_{t_0}^{t_f} L(x(t), u(t), t)\mathrm{d}t, \tag{9}$$

where the first term, the terminal cost, considers the final state of the state and the terminal time, whereas the second term, the integral cost, takes the whole trajectory into account. By choosing an appropriate cost functional, the optimal cost can then be used as an indicator for the performance one can gain from elastic joints. Unfortunately, the conditions which OC theory provides, cannot always directly be formulated as control laws, especially when one wants to analyse a complex robotic system with multiple degrees of freedom and various nonlinearities. Nevertheless, under some simplifying assumptions, it is in some cases also possible to find analytical solutions, which are useful in understanding the best control strategies for these novel devices. Consequently, two main approaches are being followed for the analysis of these systems.

The first one is to work on simplified models such that analytical results can be obtained ([36][37][38][39][40][41][42][43][44][45]). These results are useful to reveal the relation between the system's parameters such as eigenfrequency, damping ratio and the maximum attainable performance regarding for instance the system's peak velocity during a throwing motion. The second approach is the use of powerful numerical tools such as pseudospectral methods [46], which have been successfully applied to various robotic systems including the DLR Hand Arm System ([37], [2], [48], [49]).

Even though one can compute the optimal control strategy for complicated systems following the second approach, this computation may require a significant amount of time depending on the complexity of the used model. Consequently, for applications where the optimal control strategy is needed in real-time, these methods need to be further developed. In [50], the problem of generating optimal motions in real-time was addressed by encoding trajectories via Dynamic Movement Primitives (DMPs). The optimal trajectories learned via DMPs have the advantage of being reconstructable in real-time with high precision. Furthermore, the extrapolation to other tasks show near-optimal behavior, which has been illustrated in [50] by realizing a ball throwing motion with the DLR Hand Arm System, when the goal position is varying.

Comparing the existing results regarding the optimal control of elastic joints in literature, a lack of existing analytical results especially for n-DoF systems is apparent. Consequently, filling this gap and combining the gained insight for more efficient numerical methods and learning algorithms is an ongoing research.

20.8 Collision Detection and Reaction

As mentioned previously, intrinsic joint elasticity improves impact robustness. In addition, the potential energy storage and release capabilities in the joints allow to outperform rigid manipulators by means of achievable peak link velocity. While high link velocities are desirable in terms of performance, they may also increase the robot's level of dangerousness and the risk of self-damage during collisions. In other words we have to consider both, threats for the environment and the robot. Threats for the environment are mainly caused by contact forces with the robot, whereas threats for the robot are dominated by the torque in the drive-train of the robot actuators. Thus, the problem of ensuring safety has to be treated differentiated. In physical human-robot interaction (pHRI) it is therefore important to detect contact with the environment and take appropriate countermeasures by control in case of unwanted collisions.

For detecting and isolating collisions with the DLR Hand Arm System, we use the generalized link momentum observer proposed in [51]. As this observer requires knowledge of the elastic joint torque, we use an estimate obtained by model identification. An alternative scheme described in [30] observes both the motor and link momentum and therefore does require identification of the joint torque. However, this scheme showed poor performance in practice because it is sensitive to unmodeled motor friction. Having detected collisions with the environment, collision reaction schemes can be activated to ensure the physical integrity of human and robot.

20.8.1 Reactions

The most intuitive collision reaction is to stop the robot. For rigid robots or robots with high joint stiffness such as the LWR-III, stopping can simply be achieved by engaging the brakes or commanding the current position to the position controller.

However, when halting the motors of VSA robots the links can still oscillate significantly due to the intrinsic joint elasticity. This can result in large link side velocities and even constraints such as maximum elastic deflection may be violated. The decrease of link velocity then depends on the damping of the joints, which is typically undesired for VSA robots and therefore very low.

Braking of elastic joint robots can be achieved by introducing active damping by control. For this, several schemes have been proposed recently. The state feedback controller described in [6] bases on a eigenmode decoupling approach and selects feedback gains such that dynamics and control behave like damped second order systems. A second decoupling based method described in [52] aims at braking an elastic robot as fast as possible, in other words, introducing as much damping as possible. The controller makes use of optimal control theory to brake each decoupled SISO mass-spring system in minimum time. It is noteworthy, that this optimal control problem can be solved in real-time. Both previously mentioned methods require a dynamic model of the robot. A model-free approach was presented in [7] were the energy storage and release process of every joint is exploited to form a passive damping controller.

An overload in the drive-train of a VSA robot typically results in the situation that the spring deflection limit is approached, e.g. by an external impact. In this case the mechanics or spring is likely damaged. In order to avoid this, an extremely large spring could be utilized, which would be of no use during the normal operations and would increase weight and bulkiness. An alternative is to actively move the VSA position in the direction of the threatening torque, so that the spring is discharged. This for example can be initiated at a load level above a given remaining potential energy capacity, or a remaining passive deflection angle (see red dashed line in Fig. 20.12), [53].

Another possible reaction to avoid spring overload takes advantage of the capability of a VSA to change its stiffness. It uses the same activation criteria as the previous reaction. Depending on the construction and the implementation principle, a VSA features different spring energy capacity or maximum deflection angle at different stiffness presents. If the VSA is set due to application demands to a less advantageous stiffness preset concerning the robustness issue, this can be mitigated by the reaction strategy. As long as the threat of overload is present the VSA changes the stiffness preset to a safer state and returns to the normal state when the critical situation is over. Spring preload type and antagonistic VSAs have a higher energy capacity at lower stiffness presets, because the energy used to preload the spring(s) for a stiffer setup cannot be used for passive deflection (light grey area in Fig. 20.12). In this case the motors release the preload during the reaction so that they do not have to supply additional power and at the same time increase the actuator energy capacity. An additional benefit is that also the maximum passive deflection angle increases with a softer stiffness preset.

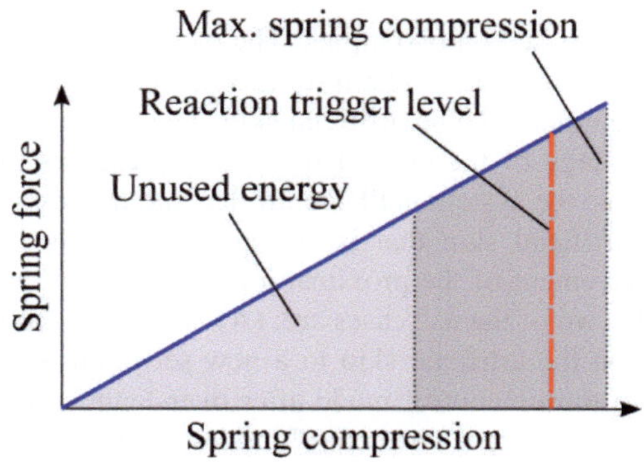

Fig. 20.12 Spring force vs. spring compression with the spring energy (grey area below the blue line) and the reaction trigger level (red line).

20.8.2 Reflexes

A novel approach to increase safety and security of VSA systems is to take the human mimetic approach one step further: The actuator unit consisting of motor and spring can be controlled analogously to the human muscle and tendon, which are protected by reflex actions. Those reflex actions can be interpreted on a control level to protect the mechanical VSA system. But what are those protective reflexes and what do they regulate? Among them is the stretch reflex, commonly known from the tendon jerk experiment: A tap on the tendon of the quadriceps passively stretches the muscle and special receptors, the so called muscle spindles, lead to a contraction of the same muscle. Moreover there is the more complex nociceptive withdrawal reflex that leads to a withdrawal movement elicited by a painful stimulus. The number of activated muscles is scaled with the intensity of the stimulus and might lead to the retreat of only one limb, e.g. after tapping on a needle, up to a reaction of the whole body jumping away from the source of stimulus. A third reflex is the autogenic inhibition that leads to a relaxation of the muscle threatened by overload: If a force is much too strong for a muscle, golgi tendon organs induce its inhibition to protect it from harmful tear.

Those reflexes are very different, but our assumption is that they follow common principles, which are namely: hierarchy, since they operate concurrently in a predefined order, where the central nervous system (CNS) is at the highest level and can suppress any reflex activity by conscious thought; operation on joint level and not necessarily a reaction in Cartesian space; reflex reversal, which means that the same stimulus can lead to different reactions, depending on the situation; irradiation, because the stronger a stimulus the more muscles are activated and the reaction is spread (irradiated) over the joints; the preservation of a status quo, which is e.g. stability or sanity. The reflexes, interpreted on a control level, are a combination of PD control (stretch reflex), force/torque control (autogenic inhibition) and a fast trajectory (nociceptive withdrawal reflex).We propose an activation strategy of the reflexes based on two inputs: The measured deflection of the spring mechanism, directly correlating to the torque, since the stiffness of the spring is known, and the force input from an artificial skin. The activation is complemented with a switching strategy of the control mode that preserves the stability of the system in action. Thus, the system is PD controlled and moves away from a source of stimulus to the artificial skin that is arranged in so called reflex responsive fields, eliciting a movement of the proximal joint away from the source of stimulus (see Fig. 20.13). Two of the use cases are: (A) An evasive trajectory of the motor after an impact on the artificial skin to a new set-point, symbolized by a red dot; (B) A switch to torque control mode after over-lengthening of the spring to reduce the stored elastic energy. The control modes can be used conjointly and the reaction can be spread over multiple proximal joints. After each reflexive reaction a trajectory back to original set-point, symbolized by a green dot, is computed. The stronger the stimulus, the more joints are activated to support the movement. If overload of the spring is detected, the system switches into torque control, hence reducing the energy storage in the spring, and switches back to PD control, after necessary energy and stability conditions are satisfied. All control action takes place on joint level, can be spread over multiple proximal joints as well as suppressed by raising the activation-thresholds. It only steps in when necessary and thus enhances existing control schemes by providing additional safety and security.

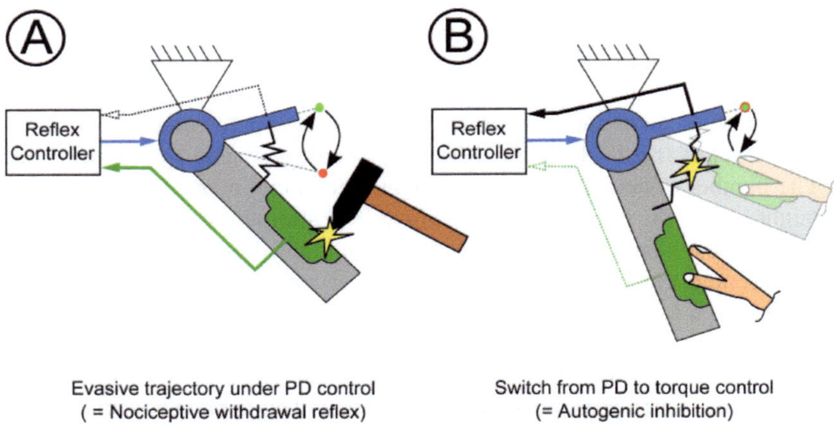

Fig. 20.13 Reflex use cases for a VSA joint, where the motor is depicted in blue, the link in grey, a passive compliance as black spring and the artificial skin in green.

20.9 Cyclic Motion Control

Key properties of compliant actuators in robotic arms are the capability to improve performance and energetic efficiency. Especially in case of cyclic motion tasks, the capability to buffer and release elastic energy may reduce the size and weight of required actuators and save a substantial amount of energy. These properties are even of increased importance for legged robots, which need to wirelessly walk, jump, or run over rough and uneven terrain. The step from rigid towards elastic actuation introduces natural oscillation dynamics into the plant. In our recent works [54], [55], [56], [57], [58], [58], novel concepts to exploit these natural dynamics have been derived.

Considered limit cycle control methods range from classical principles based on the well-known Van der Pol oscillator to novel concepts following the idea of resonance and natural dynamics excitation. The former method is adapted from rigid actuator control, where the generalized force of the second-order dynamics can be considered as control input. An implementation for compliant actuators is proposed in [54]. The latter method is derived from human motor control [56], [58]. This method directly applies to continuous second-order systems[18] of the form

$$m\ddot{q} + d(q,\dot{q}) = -\left.\frac{\partial U(\phi)}{\partial \phi}\right|_{\phi=q-\theta},$$

with positive plant parameters[19] representing a large class of compliantly actuated robotic joints. The limit cycle controller is simply the switching law[20].

$$\theta(q,\theta^-) = \begin{cases} -\mathrm{sign}(q-\theta^-)\hat{\theta} \text{ if } \left|q-\theta^-\right| > \epsilon \\ 0 \qquad\qquad\qquad\qquad \text{otherwise} \end{cases}$$

For the maximally efficient controller parameters $\hat{\theta} \leq 2\epsilon$, this controller generates a globally asymptotically stable limit cycle without any model knowledge, only based upon measurements of the spring deflection [58].

The extension of the above methods from the single to the multiple joint case under the concept of natural dynamics exploitation is derived from the notion of intrinsic mechanical oscillation modes [58]. Exploiting the idea of two dimensional invariant submanifolds, $2n^{th}$-order dynamics of multiple joint robots collapses to second-order dynamics of a single joint. This dimensionality reduction can be achieved using one of the following approaches:

- Considering a locally valid linearization of the nonlinear plant dynamics, linear eigenmodes can be identified. The valid range of these modes can be extended to hold globally by feedback control [54].

---

[18] The system has continuous states $q, \dot{q} \in \mathbb{R}$

[19] The inertia m > 0, the generalized damping force is positive semi-definite in a sense that d(q,q˙) q ≥0 and the elastic potential U(ϕ) is positive definite w.r.t. the spring deflection ϕ.

[20] The finite dynamics is piecewise continuous from the left, i.e. θ^- represents the state of θ before the switching.

- On the basis of the switching control, a directed excitation along an intrinsic mechanical oscillation mode can be achieved by adaptive control [55].
- A method to generate a limit cycle along a predefined oscillation mode has been derived from classical impedance and null-space control [36].

The above methods can be nearly arbitrarily combined to achieve a multiple degree of freedom limit cycle control. One can adapt to one of the following requirements: task adaptability, performance, or efficiency.

20.10 Conclusion

We highlighted in this chapter that intrinsic flexibility in robotic systems is of crucial importance to obtain human level power density and to interact safely with humans and unknown environments. Using classical electro-mechanical actuation in combination with nonlinear elastic joint mechanisms allows realizing high performance actuators, with high motion repeatability and high efficiency. A large amount of control literature on flexible joint robots can be therefore accessed and adapted to the nonlinear, adjustable stiffness case. In this chapter we addressed on the basis of the DLR Hand Arm System all major aspects related to the mechatronic design and control of VSA robots. As a research prototype, the DLR Hand Arm System fully validates the feasibility of the VSA concepts. A major drawback remains the relatively high complexity of the nonlinear elastic joint. New technological approaches to implement the variable elastic element and integrated electronics solutions would therefore support the evolution of VSA systems towards commercial applications.

20.11 References

[1] Vanderborght B, Albu-Schäffer A, Bicchi A, Burdet E, Caldwell D, Carloni R, Catalano M, Eiberger O, Friedl W, Ganesh G, Garabini M, Grebenstein M, Grioli G, Haddadin S, Höppner H, Jafari A, Laffranchi M, Lefeber D, Petit F, Stramigioli S, Tsagarakis N, Damme MV, Ham RV, Visser L, Wolf S (2013) Variable impedance actuators: A review. Robotics and Autonomous Systems 61(12):1601–1614, DOI http://dx.doi.org/10.1016/j.robot.2013.06.009

[2] Grebenstein M, Albu-Schäffer A, Bahls T, Chalon M, Eiberger O, Friedl W, Gruber R, Haddadin S, Hagn U, Haslinger R, Höppner H, Jörg S, Nickl M, Nothhelfer A, Petit F, Reill J, Seitz N, Wimböck T, Wolf S, Wüsthoff T, Hirzinger G (2011) The DLR Hand Arm System. In: IEEE International Conference on Robotics and Automation (ICRA), pp 3175–3182, DOI 10.1109/ ICRA.2011.5980371

[3] Wolf S, Hirzinger G (2008) A new variable stiffness design: Matching requirements of the next robot generation. In: Robotics and Automation, 2008. ICRA 2008. IEEE Interna-

tional Conference on, Pasadena, CA, USA, pp 1741–1746, DOI 10.1109/ROBOT.2008.4543452
[4] Wolf S, Eiberger O, Hirzinger G (2011) The DLR FSJ: Energy based design of a variable stiffness joint. In: Robotics and Automation (ICRA), 2011 IEEE International Conference on, pp 5082–5089, DOI 10.1109/ICRA.2011.5980303
[5] Friedl W, Chalon M, Reinecke J, Grebenstein M (2011) FAS A flexible antagonistic spring element for a high performance over. In: Intelligent Robots and Systems (IROS), 2011 IEEE/RSJ International Conference on, pp 1366–1372, DOI 10.1109/IROS.2011.6094569
[6] Petit F, Albu-Schäffer A (2011) State feedback damping control for a multi dof variable stiffness robot arm. In: IEEE Int. Conf. on Robotics and Automation (ICRA), IEEE, pp 5561–5567
[7] Petit F, Ott C, Albu-Schäffer A (2014) A model-free approach to vibration suppression for intrinsically elastic robots. In: IEEE Int. Conf. on Robotics and Automation (ICRA), IEEE
[8] Cui L, Maas H, Perreault EJ, Sandercock TG (2009) In situ estimation of tendon material properties: Differences between muscles of the feline hindlimb. Journal of Biomechanics 42(6):679 – 685, DOI http://dx.doi.org/10.1016/j. jbiomech.2009.01.022
[9] Friedl W, Höppner H, Petit F, Hirzinger G (2011) Wrist and forearm rotation of the DLR Hand Arm System: Mechanical design, shape analysis and experimental validation. In: Intelligent Robots and Systems (IROS), 2011 IEEE/RSJ International Conference on, pp 1836–1842, DOI 10.1109/IROS. 2011.6094616
[10] Petit F, Chalon M, Friedl W, Grebenstein M, Albu-Schäffer A, Hirzinger G (2010) Bidirectional antagonistic variable stiffness actuation: Analysis, design amp; implementation. In: Robotics and Automation (ICRA), 2010 IEEE International Conference, pp 4189–4196, DOI 10.1109/ROBOT.2010.5509267
[11] Petit F, Friedl W, Höppner H, Grebenstein M (2014) Analysis and synthesis of the bidirectional antagonistic variable stiffness mechanism. Mechatronics, IEEE/ASME Transactions on PP(99):1–12, DOI 10.1109/TMECH.2014. 2321428
[12] Jörg S, Nickl M, Nothhelfer A, Bahls T, Hirzinger G (2011) The computing and communication architecture of the DLR hand arm system. In: Proceedings IEEE/RSJ International Conference on Intelligent Robots and Systems, pp 1055–1062
[13] Nickl M, Jörg S, Bahls T, Nothhelfer A, Strasser S (2011) Spacewire, a backbone for humanoid robotic systems. In: Proceedings of the 4th International SpaceWire Conference, pp 356–359
[14] European Cooperation for Space Standardization (ECSS) (2003) ECSS E-50-12A SpaceWire - Links, nodes routers and networks. http://spacewire.eas.int
[15] IC Haus (2007) BiSS C Interface Protocol (C-Mode). http://www.ichaus.com, c1 edn
[16] Grebenstein M (2014) Approaching Human Performance: The Functionality-Driven Awiwi Robot Hand (Springer Tracts in Advanced Robotics). Springer, Berlin; Heidelberg; New York
[17] Grebenstein M, Chalon M, Friedl W, Haddadin S, Wimböck T, Hirzinger G, Siegwart R (2012) The hand of the DLR Hand Arm System: Designed for interaction. IJRR 31(13):1531–1555
[18] Gray H (1999) Anatomy, descriptive and surgical. Courage Books, Philadelphia
[19] Grebenstein M, Chalon M, Hirzinger G, Siegwart R (2010) A method for hand kinematics designers; 7 billion perfect hands. International Conference on Advances in Bioscience and Bioengineering
[20] Chalon M, Grebenstein M, Wimböck T, Hirzinger G (2010) The thumb: Guidelines for a robotic design. Intelligent Robots and Systems, (2010) IEEE/RSJ International Conference on, pp 2153–2858

[21] Grebenstein M, Chalon M, Hirzinger G, Siegwart R (2010) Antagonistically driven finger design for the anthropomorphic DLR Hand Arm System. Humanoids Robots, IEEE/RAS International Conference, pp 609–616

[22] Grebenstein M, van der Smagt P (2008) Antagonism for a highly anthropomorphic hand-arm system. Advanced Robotics 1(22):39–55, DOI 10.1163/156855308X291836

[23] Albu-Schäffer A, Wolf S, Eiberger O, Haddadin S, Petit F, Chalon M (2010) Dynamic modelling and control of variable stiffness actuators. In: Robotics and Automation (ICRA), 2010 IEEE International Conference on, pp 2155 –2162, DOI 10.1109/ROBOT.2010.5509850

[24] Jafari A, Tsagarakis N, Caldwell D (2013) A novel intrinsically energy efficient development of a novel actuator with adjustable stiffness (awas). IEEE Transactions on Mechatronics 18(1)

[25] Albu-Schäffer A, Fischer M, Schreiber G, Schoeppe F, Hirzinger G (2004) Soft robotics: What cartesian stiffness can we obtain with passively compliant, uncoupled joints? In: Proc. of the IEEE/RSJ Int. Conf. on Intelligent Robots and Systems, pp 3295–3301

[26] Hogan N (1990) Mechanical impedance of single- and multi-articular systems. In: Winters J,Woo SY (eds) Multiple Muscle Systems, Springer New York, pp 149–164

[27] Fantoni I, Lozano R, Spong MW (2000) Energy based control of the pendubot. IEEE Trans on Automatic Control 45(4):725–729

[28] Petit F, Albu-Schäffer A (2011) Cartesian impedance control for a variable stiffness robot arm. In: Proc. of the IEEE/RSJ International Conference on Intelligent Robots and Systems, pp 4180–4186

[29] Palli G, Melchiorri C, Luca AD (2008) On the feedback linearization of robots with variable joint stiffness. In: Proc. IEEE Int. Conf. on Robotics and Automation, pp 1753 – 1759

[30] De Luca A, Flacco F, Bicchi A, Schiavi R (2009) Nonlinear decoupled motion-stiffness control and collision detection/reaction for the vsa-ii variable stiffness device. In: IEEE/RSJ Int. Conf. on Intelligent Robots and Systems (IROS2009), IEEE, pp 5487–5494

[31] Palli G, Melchiorri C (2011) Output-based control of robots with variable stiffness actuation. Journal of Robotics

[32] Tonietti G, Schiavi R, Bicchi A (2005) Design and control of a variable stiffness actuator for safe and fast physical human/robot interaction. In: Proc. IEEE Int. Conf. on Robotics and Automation, pp 528–533

[33] Albu-Schäffer A, Ott C, Petit F (2012) Energy shaping control for a class of underactuated euler-lagrange systems. In: IFAC Symposium on Robot Control

[34] Sardellitti I, Medrano-Cerda G, Tsagarakis NG, Jafari A, Caldwell DG (2012) A position and stiffness control strategy for variable stiffness actuators. In: Proc. IEEE Int. Conf. on Robotics and Automation

[35] Pontryagin L (1987) Mathematical Theory of Optimal Processes. Classics of Soviet Mathematics, Taylor & Francis

[36] Garabini M, Passaglia A, Belo FAW, Salaris P, Bicchi A (2011) Optimality principles in variable stiffness control: the VSA hammer. 2011 IEEE/RSJ International Conference on Intelligent Robots and Systems (IROS2011), San Francisco, USA pp 3770 – 3775

[37] Garabini M, Passaglia A, Belo F, Salaris P, Bicchi A (2012) Optimality principles in stiffness control: The VSA kick. In: Robotics and Automation (ICRA), 2012 IEEE International Conference on, pp 3341–3346

[38] Haddadin S, Weis M, Albu-Schäffer A, Wolf S (2011) Optimal control for maximizing link velocity of robotic variable stiffness joints. In: Proceedings IFAC 2011, World Congress pp 3175–3182

[39] Haddadin S, Krieger K, Mansfeld N, Albu-Schäffer A (2012) On impact decoupling properties of elastic robots and time optimal velocity maximization on joint level. In: In-

telligent Robots and Systems (IROS), 2012 IEEE/RSJ International Conference, pp 5089–5096, DOI 10.1109/IROS.2012.6385913
[40] Haddadin S, Özparpucu M, Albu-Schäffer A (2012) Optimal control for maximizing potential energy in a variable stiffness joint. In: Decision and Control (CDC), 2012 IEEE 51st Annual Conference, pp 1199–1206
[41] Incaini R, Sestini L, Garabini M, Catalano MG, Grioli G, Bicchi A (2013) Optimal control and design guidelines for soft jumping robots: Series elastic actuation and parallel elastic actuation in comparison. In: IEEE International Conference on Robotics and Automation (ICRA2013), pp 2477 – 2484
[42] Özparpucu MC, Albu-Schäffer A (2014, Accepted) Optimal control strategies for maximizing the performance of variable stiiffness joints with nonlinear springs. In: Decision and Control (CDC), 2014 IEEE 53rd Annual Conference on
[43] Özparpucu MC, Haddadin S (2013) Optimal control for maximizing link velocity of visco-elastic joints. In: Intelligent Robots and Systems (IROS), 2013 IEEE/RSJ International Conference on, pp 3035–3042
[44] Özparpucu MC, Haddadin S (2014) Optimal control of elastic joints with variable damping. In: Control Conference (ECC), 2014 European, pp 2526–2533
[45] Özparpucu MC, Haddadin S, Albu-Schaffer A (2014, Accepted) Optimal control of variable stiffness actuators with nonlinear springs. In: Proceedings. IFAC 2014, World Congress
[46] Garg D, Patterson MA, Hager WW, Rao AV, Benson DA, Huntington GT (2010) A unified framework for the numerical solution of optimal control problems using pseudospectral methods. Automatica 46(11):1843–1851
[47] Braun D, Howard M, Vijayakumar S (2011) Exploiting variable stiffness in explosive movement tasks. In: Proceedings of Robotics: Science and Systems (RSS2011), Los Angeles, USA, pp 25–32
[48] Haddadin S, Huber F, Albu-Schäffer A (2012) Optimal control for exploiting the natural dynamics of variable stiffness robots. In: Robotics and Automation (ICRA), 2012 IEEE International Conference, pp 3347–3354, DOI 10.1109/ICRA.2012.6225190
[49] Mettin U, Shiriaev A (2011) Ball-pitching challenge with an underactuated two-link robot arm. Proceedings IFAC 2011, World Congress pp 1–6
[50] Weitschat R, Haddadin S, Huber F, Albu-Schäffer A (2013) Dynamic optimality in real-time: A learning framework for near-optimal robot motions. In: Intelligent Robots and Systems (IROS), 2013 IEEE/RSJ International Conference on, pp 5636–5643
[51] De Luca A, Albu-Schäffer A, Haddadin S, Hirzinger G (2006) Collision detection and safe reaction with the DLR-III lightweight manipulator arm. In: IEEE/RSJ Int. Conf. on Intelligent Robots and Systems (IROS2006), IEEE, pp 1623–1630
[52] Mansfeld N, Haddadin S (2014) Reaching desired states time-optimally from equilibrium and vice versa for visco-elastic joint robots with limited elastic deflection. In: IEEE/RSJ Int. Conf. on Intelligent Robots and Systems (IROS2014), IEEE
[53] Wolf S, Albu-Schäffer A (2013) Towards a robust variable stiffness actuator. In: Intelligent Robots and Systems (IROS), 2013 IEEE/RSJ International Conference on, IEEE/RSJ, Tokyo, Japan, pp 5410–5417
[54] Lakatos D, Garofalo G, Petit F, Ott C, Albu-Schäffer A (2013) Modal limit cycle control for variable stiffness actuated robots. In: Proc. IEEE Int. Conf. on Robotics and Automation, pp 4934–4941
[55] Lakatos D, Görner M, Petit F, Dietrich A, Albu-Schäffer A (2013) A modally adaptive control for multi-contact cyclic motions in compliantly actuated robotic systems. In: Proc. IEEE/RSJ Int. Conf. on Intelligent Robots and Systems, pp 5388–5395
[56] Lakatos D, Petit F, Albu-Schäffer A (2013) Nonlinear oscillations for cyclic movements in variable impedance actuated robotic arms. In: Proc. IEEE Int. Conf. on Robotics and Automation, pp 508–515

[57] Lakatos D, Garofalo G, Dietrich A, Albu-Schäffer A (2014) Jumping control for compliantly actuated multilegged robots. In: Proc. IEEE Int. Conf. on Robotics and Automation

[58] Lakatos D, Petit F, Albu-Schäffer A (2014) Nonlinear oscillations for cyclic movements in human and robotic arms. IEEE Transactions on Robotics pp 865–879

[59] Lakatos D, Albu-Schäffer A (2014) Switching based limit cycle control for compliantly actuated second-order systems. Accepted for publication at the IFAC World Congress

21 Soft Robotics Research, Challenges, and Innovation Potential, Through Showcases

Cecilia Laschi

The BioRobotics Institute, Scuola Superiore Sant'Anna, Pisa

Abstract Soft robotics, intended as the use of soft materials in robotics, is a young yet promising and growing research field. The need for soft robots emerged in robotics, for facing unstructured environments, and in artificial intelligence, too, for implementing the embodied intelligence, or morphological computation, paradigm, which attributes a stronger role to the bodyware and its interaction with the environment. Using soft materials for building robots poses new technological challenges: the technologies for actuating soft materials, for embedding sensors into soft robot parts, for controlling soft robots are among the main ones. Though still in its early stages of development, soft robotics is finding its way in a variety of applications, where safe contact is a main issue, in the biomedical field, as well as in exploration tasks and in the manufacturing industry. Literature in soft robotics is increasingly rich, though scattered in many disciplines. The soft robotics community is growing worldwide and initiatives are being taken, at international level, for consolidating this community and strengthening its potential for disruptive innovation.

21.1 Introduction: The Need for Soft Robots

Soft robotics, intended as the use of soft materials in robotics, is a young yet promising and growing research field [1]. Soft robotics stems from robotics, on one side, and from artificial intelligence, on the other side.

Robotics evolved significantly from its recent birth in the '50s and is today a solid discipline with a good wealth of knowledge and technologies. Though robotics basically evolved in the field of industrial manufacturing, service applications were soon investigated by roboticists [2]. The huge difference in the two application domains can be summarized in being the environment *structured*, in industrial manufacturing, and *unstructured* in service applications, where robots are expected to operate in a variety of scenarios, of our own world. For this reason, service robotics brought to bioinspiration, as investigating the animals that evolved and successfully live in these environments is definitely a good starting point for building like-wise successful robots [3][4]. It is then clear that soft tissues have a dominant role in animals, with respect to their rigid skeletons and exoskeletons. The use of soft deformable and variable stiffness technologies in robotics represents an emerging way to build new classes of robotic systems that are expected to interact more safely with the natural, unstructured, environment and with humans, and that better deal with uncertain and dynamic tasks (i.e. grasping and manipulation of unknown objects, locomotion on rough terrains, physical contacts with human bodies, etc.) [5].

In artificial intelligence, too, one of the modern views is based on the decisive role of the interaction with the environment. This interaction is not just intended as reaction to external forces and perturbations, but especially as control of movements. This means that the morphology of the body and the mechanical properties also play a decisive role in intelligent behaviour [6]. In other words, a part of movement control is not given by computation and neural processes, but by passive adaptation of body parts to the forces borne, in prevailing tasks, in so-called *embodied intelligence* and *morphological computation*.

21.2 The Challenges for Soft Robotics, Through the Octopus Showcase

Look at an octopus with a roboticist's eyes: its arms are soft and deformable, they can bend in any direction, at any point along the arm; however, they can stiffen when needed and they can grasp and pull objects with considerable strength. The octopus is undoubtedly a good model for soft robotics, and an extreme one, considering that it has no rigid structures, of any kind.

The octopus does not have a large brain, yet it can control this huge amount of possible movements and motion parameters [7]. Its soft body seems to simplify control exploiting its rich interaction with environment, which is thought to be at the base of its unexpected intelligence. The octopus then represents an ideal model for morphological computation, too.

Understanding the secrets of the octopus soft dexterity and copying some of the key principles is an effective case-study for facing the different challenges, in different disciplines, related to the development of soft robots [8]. These have been the main objectives of the OCTOPUS project.

21.2.1 Biological Insights

The first big lesson from biology is the muscular hydrostat [9][10]: a muscular structure composed of longitudinal, transverse and oblique muscles, which allows all-direction bending, by selective contractions, elongation and shortening, by contractions of the transverse and longitudinal muscles, respectively, torsion, by contractions of the oblique muscles, and stiffening, by co-contractions.

Among open questions were: how long can an octopus arm stretch? How are nervous fibres arranged inside the arm, not to be stretched and damaged? What force can each arm generate when pulling? How are longitudinal and transverse muscles arranged and anchored along the arm? What are the mechanical properties of the octopus muscular tissue? What is the density of an octopus? How are the fibres of the connective tissue arranged? What are principles of the crawling mechanism?

Biomechanical measurements on arms of specimens of *Octopus vulgaris*, with purposively designed bioengineering tools (see Fig. 21.1), gave evidence of an average 73% elongation and an average 40N force applied with one arm [11][12]. Imaging and biological analysis of the octopus arm tissue (ecography, histology, SEM) gave evidence of the helicoidal arrangement of nervous fibres in the central nerve chord [13], of the radial transverse fibres and of the anchoring of longitudinal fibres at different lengths along the arm. Mechanical tests showed a hyperelastic behaviour of the octopus arm tissue and a density very close to the water density, giving neutral buoyancy. Video analysis of the swimming and of the crawling movements outlined the mechanisms of pulsed jet propulsion and back-arm pushing for crawling [14][15].

The results have been translated into specifications for the robot design [16][17].

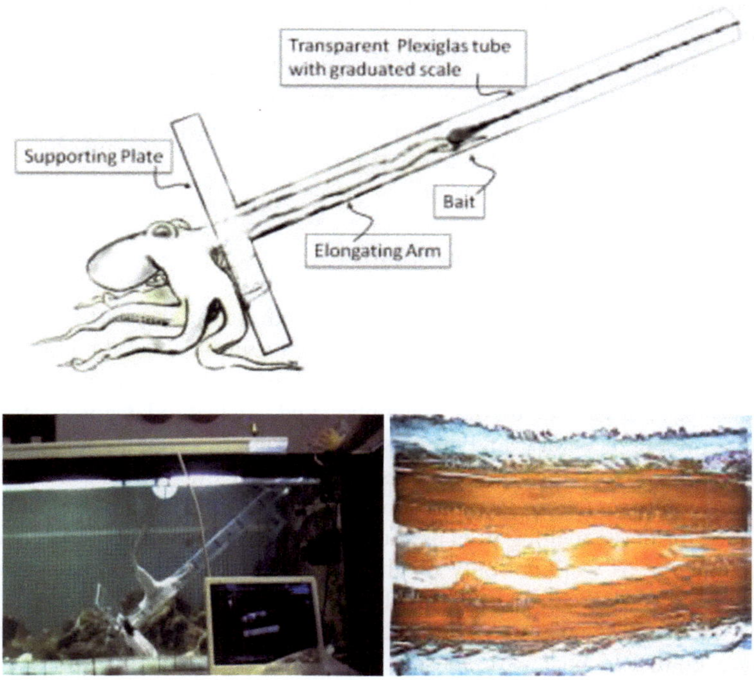

Fig. 21.1 From top to bottom, left to right: The experimental set-up for measurements and for echography of the octopus arm, and an image of corresponding histology, used for comparison.

21.2.2 Soft Actuation Technologies

The actuation of soft bodies, producing the desired deformations and desired forces, is one of the main challenges in soft robotics. In animals, analogous actuation is given by a number of muscle fibres, well distributed in the body. While artificial muscles are still an open objective for engineering, several technologies are currently being investigated to this purpose.

An important technological field, in the quest for artificial muscles, is represented by EAP, ElectrActive Polymers [18][19]. They use the property by which two layers of conductive material tend to attract when powered, by a Coulomb attraction force, thanks to the Maxwell effect, if the medium between them can be compressed. They are then well-suited to stand as soft actuators, as they can be built with soft materials like silicone, though the geometry needs to be carefully designed in order to obtain useful strain [20].

Shape Memory Alloys (SMA) are alloys that deform to an original shape when heated [21]. They are not strictly soft materials, but they are used in wires that can well serve the purpose of actuating soft materials [22][23]. Despite of their well-known drawbacks in slow deformation, difficult controllability, low strain, SMA springs stand as an effective solution for the OCTOPUS front arms, well complying with the specifications of the water environment, slow contractions, and on-off control [24].

Compressed air is a powerful actuation system for soft materials. In addition to the well-known McKibben actuators [25], compressed air has recently been used for deforming soft body parts at lower scales. In the starfish-like robot by [26], networks of channels in elastomers inflate for actuation.

21.2.3 Soft Robot Modeling and Control

New approaches are necessary for modelling and controlling soft-bodied structures. The 50-year history of robotics and of robot control is based on the assumption that robots are kinematic chains of rigid links. Robot control theories and techniques have developed on this assumption and reached today a very high level of solidity, rigour, accuracy, and progress [2]. The use of soft materials in robotics is going to unhinge these fundamentals, as most rules no longer stand. Known techniques for kinematic and dynamic modelling in robotics cannot be used, while techniques for the modelling of continuum structures are needed. Control needs a deep rethinking, as well, not only for the lack of exact kinematics and dynamics models, but also for the increased role of interaction with environment.

Most of the approaches currently in use for the direct model of continuum soft robot are limited to piecewise-constant-curvature approximation [27]. Recently, Jones et al. [28] presented a steady state model of continuous robot neglecting the actuation. In the work by Boyer et al. [29] the distributed force and torque acting on the robot are estimated but no discussion is made concerning on the actuators that could generate them. A continuum geometrically exact approach for tendon-driven continuum robot has been proposed by Renda et al. [30]. It is capable of properly simulating the coupled tendon drive behaviour of non-constant curvature manipulators, because it takes into account the torsion of the robot. The inverse model proposed in literature for controlling continuum soft robot follows different approaches. A modal approach was proposed by Chirikjian et al. [31]. A successful Jacobian method for a non-constant curvature tendon-driven manipulator is proposed and compared with a neural approach [32][33].

21.2.4 Integration and Validation of an Octopus-like Robot

The final OCTOPUS prototype is the first completely soft robot, which integrates a central body with 8 arms extending in radial directions and the main processing units (see Fig. 21.2). The front arms are mainly used for manipulation, elongation, grasping, the others are mainly used for locomotion. To optimize elongation, reaching and manipulation tasks the front arms are based on the SMA actuators, which reproduce the internal anatomical features of the real octopus arm, and thus allow to perform finely controlled and precise movements. The other arms, which are used for crawling, are based on silicone and cables, embedding the features needed to obtain an octopus-inspired locomotion. The robotic octopus works in water and its buoyancy is close to neutral.

21.3 Soft Robots at Work

21.3.1 Biomedical Applications of Soft Robotics: Octopus-derived Technologies in Surgery

Minimally invasive surgery is nowadays widely used in clinical practice and progresses are going on at a good pace. Few limitations in modern laparoscopic and robot-assisted surgical systems are due to restricted access through Trocar ports, lack of haptic feedback, and difficulties with rigid robot tools operating inside a confined space filled with organs.

The STIFF-FLOP[21] project aims at taking inspiration from biological manipulators like the octopus arm and the elephant trunk and at taking advantage of the OCTOPUS research and technologies for developing a highly dexterous soft robotic arm able to locally control its stiffness for both being compliant with the environment and accomplishing surgical tasks.

---

21 www.stiff-flop.eu

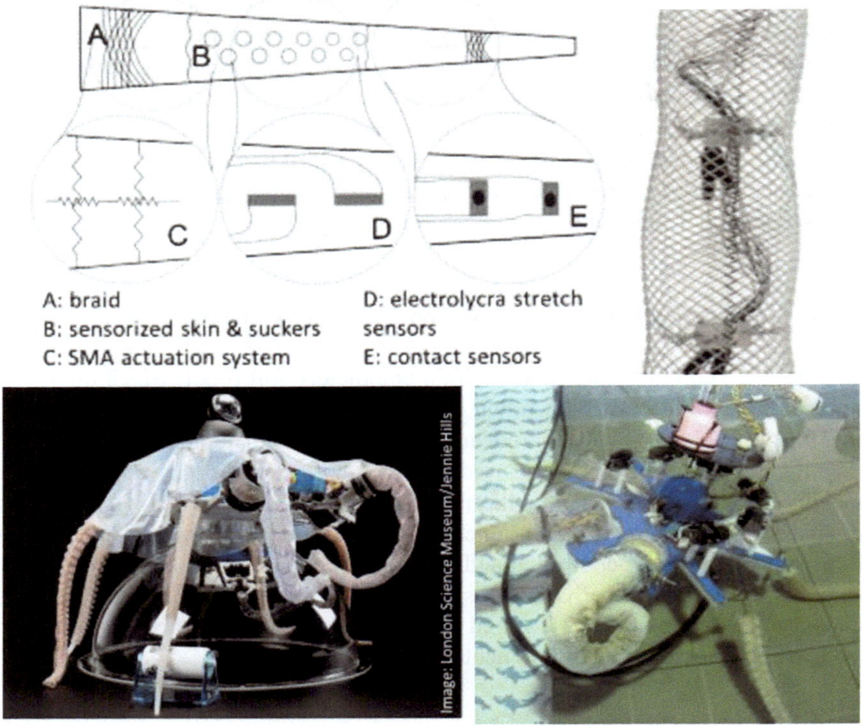

Fig. 21.2 From left to right, top to bottom: Scheme of the OCTOPUS arm components; images of the OCTOPUS arm with SMA actuators in the braided structure (credits to Massimo Brega, The Lighthouse); an image of the OCTOPUS prototype exhibited at the Science Museum in London (credits to Jennie Hills, Science Museum); image of the 8-arm robot in water, with 2 front SMA arms and 6 crawling arms.

A combination of pneumatic actuation and granular jamming led to the first prototype of a soft endoscope with controllable stiffness [34].

21.3.2 Soft Robots in Explorations: An Octopus-like Underwater Robot

Robotics has proved to be an essential tool for underwater operations. A number of tasks are today accomplished by robots, such as Remotely Operated Vehicles (ROVs) and Autonomous Underwater Vehicles (AUVs). Standard working procedures for these kinds of vehicles envisage the robots to work at a safe distance from the sea bottom or the submerged structure upon which operation is being carried out in order to avoid the risk of damage. Instead, the introduction of soft robots in this field brings a disruptive perspective to underwater explorations and operations.

The PoseiDRONE Project aims at developing a soft robot capable of swimming, crawling over irregular and uneven substrates and perform complex manipulation tasks in cramped environments underwater. The capability to perform multigait locomotion in the aquatic environment and manipulation along with an overall highly compliant structure provide this robot with unprecedented assets [35] (see Fig. 21.3). This robot will be applicable in marine operations such as those entailed with coastal and offshore engineering, petroleum and drilling technology as well as underwater archeology and environmental protection.

Fig. 21.3 Images of the PoseiDRONE prototype, from left to right: view of the 4-arm prototype in a salt water tank, the prototype in sea water, detail of the pulsed-jet propulsion with the fluorescein dyed vortex ring (credits to Massimo Brega, The Lighthouse).

21.3.3 Soft Grippers for Manufacturing

While robotics has contributed fundamental technologies for manufacturing processes, there are still few industrial tasks that cannot be performed with current robotic grippers, requiring higher flexibility and adaptability to different shapes. For those tasks, soft robotics can be effectively applied, by producing soft grippers than can intrinsically adapt to grasp different shapes.

The SMART-e[22] Marie Curie Action aims at investigating this application of soft robotics through a European network of PhD research programmes. Specific topics are soft robotics and morphological computation, octopus-based technologies for manipulation in manufacturing, soft robotic grippers for industrial manufacturing.

21.4 Conclusions

The many challenges and the many potential applications of soft robotics involve a number of different disciplines and sectors. They also attract the interest of an increasing number of researchers worldwide and literature in this topic is growing

22 http://smart-e-mariecurie.eu/

at a fast pace: from basically no papers until 2004, to 10 papers in 2008, and 40 papers in 2012.

While this interdisciplinary nature of soft robotics is one of its main strengths for disruptive innovation, at the same time one of the possible risks to jeopardize the full development of the potential of soft robotics is the scattered community.

In 2012 a Technical Committee of the IEEE Robotics and Automation Society has been started on Soft Robotics[23], to gather scientists in this field, at least from the robotics community, with the impressive results of collecting 379 members in less than two years.

Including other disciplines, too, the scientific community of Soft Robotics is gathering around the ICT-FET Open RoboSoft[24] Coordination Action, started in November 2013. A common forum helps soft robotics researchers to combine their efforts, to maximize the opportunities and to materialize the huge potential impact of soft robotics. RoboSoft is creating the missing framework for the soft robotics scientists, regardless of their background disciplines, and enabling the accumulation and sharing of the crucial knowledge needed for scientific progress in this field. RoboSoft is aiming not only to create and consolidate the soft robotics community, but also to establish effective links with relevant scientific communities potentially interested in exploiting soft robotics as case study.

Soft robotics is not just a new direction of technological development. The use of soft materials in robotics is going to unhinge its fundamentals. Soft robotics is going to stand as a novel approach to robotics and artificial intelligence, and it has the potential to produce a new generation of robots, in the support of humans in our natural environments.

Acknowledgments These works have been supported by the European Commission with the OCTOPUS IP (# 231608, FET Proactive Initiative ICT 2007.8.5 "Embodied Intelligence"), the OCTO-PROP grant (#269477, Marie Curie European Re-integration Grants), the STIFF-FLOP IP (#287728, ICT Challenge 2), the SMART-e ITN (#608022, Marie Curie Action) and the RoboSoft CA (#619319, ICT-FET Open) and by the Fondazione Livorno with the PoseiDRONE and PoseiDRONE II projects. The author wishes to acknowledge the work of the SSSA Soft Robotics Team and of the project partners.

21.5 References

[1] Iida F, Laschi C (2011) Soft Robotics: Challenges and Perspectives. Procedia Computer Science, 7:99-102. doi:10.1016/j.procs.2011.12.030

[2] Siciliano B, Khatib O (2008) Handbook of Robotics. Springer, Berlin

[3] Brooks RA (1991) New approaches to robotics. Science 253:1227-1232. doi: 10.2307/2879167

---

23 http://softrobotics.org/

24 http://www.robosoftca.eu/

[4] Pfeifer R, Lungarella M, Iida F (2007) Self-organization, embodiment, and biologically inspired robotics. Science 318(5853):1088-1093. doi:10.1126/science.1145803
[5] Kim S, Laschi C, Trimmer B (2013) Soft robotics: a bioinspired evolution in robotics. Trends in Biotechnology 31(5):287-294. doi:10.1016/j.tibtech.2013.03.002
[6] Pfeifer R, Bongard JC (2007) How the Body Shapes the Way We Think: A New View of Intelligence. MIT Press, Cambridge
[7] Wells MJ (1976) Octopus: Physiology and Behaviour of an Advanced Invertebrate. John Wiley & Sons, New York
[8] Laschi C, Cianchetti M (2014) Soft Robotics: New Perspectives for Robot Bodyware and Control. Front. Bioeng. Biotechnol 2(3). doi:10.3389/fbioe.2014.00003
[9] Kier WM, Smith KK (1985) Tongues, tentacles and trunks: the biomechanics of movement in muscular-hydrostats. Zool J Linn Soc 83:307-324. doi:10.1111/j.1096-3642.1985.tb01178.x
[10] Kier WM, Stella MP (2007) The arrangement and function of octopus arm musculature and connective tissue. J. Morphol. 268:831–843
[11] Mazzolai B, Margheri L, Dario P, Laschi C (2013) Measurements of Octopus Arm Elongation Movement: Evidence of Differences by Animal Sex and Size. J Exp Mar Biol Ecol 447:160–164. doi:10.1016/j.jembe.2013.02.025
[12] Margheri L, Mazzolai B, Dario P, Laschi C (2012a) A bioengineering approach for in vivo measurements of the octopus arms. J Shellfish Res 30(3):1012-1012. doi:10.2983/035.030.0342
[13] Margheri L, Ponte G, Mazzolai B, Laschi C, Fiorito G (2011) Non-invasive study of Octopus vulgaris arm morphology using ultrasound. J Exp Biol 214:3727-3731. doi:10.1242/jeb.057323
[14] Giorgio-Serchi F, Arienti A, Laschi C (2013) Biomimetic Vortex Propulsion: Toward the New Paradigm of Soft Unmanned Underwater Vehicles. IEEE/ASME Trans. Mechatronics 18(2):484-493. doi:10.1109/TMECH.2012.2220978
[15] Calisti M, Giorelli M, Levy G, Mazzolai B, Hochner B, Laschi C, Dario P (2011) An octopus-bioinspired solution to movement and manipulation for soft robots. Bioinspir Biomim 6:036002. doi:10.1088/1748-3182/6/3/036002
[16] Margheri L, Laschi C, Mazzolai B (2012b) Soft robotic arm inspired by the octopus. I. From biological functions to artificial requirements. Bioinspir Biomim 7(2):025004. doi:10.1088/1748-3182/7/2/025004
[17] Mazzolai B, Margheri L, Cianchetti M, Dario P, Laschi C (2012) Soft-robotic arm inspired by the octopus: II. From artificial requirements to innovative technological solutions. Bioinspir. Biomim. 7(2):025005. doi:10.1088/1748-3182/7/2/025005
[18] Bar-Cohen Y (2004) Electroactive Polymer (EAP) Actuators as Artificial Muscles. SPIE Press, Bellingham
[19] Carpi F, Smela E (2009) Biomedical applications of electroactive polymer actuators. Wiley, New York
[20] Cianchetti M, Mattoli V, Mazzolai B, Laschi C, Dario P A new design methodology of electrostrictive actuators for bio-inspired robotics. Sensor Actuat B-Chem 142(1):288-297. doi:10.1016/j.snb.2009.08.039
[21] Pons JL (2005) Emerging Actuator Technologies: A Micromechatronic Approach. John Wiley & Sons, Chichester
[22] Cianchetti M (2013) Fundamentals on the Use of Shape Memory Alloys in Soft Robotics, in Interdisciplinary Mechatronics: Engineering Science and Research Development. In: Habib MK and Paulo Davim J (ed) Interdisciplinary Mechatronics: Engineering Science and Research Development. Wiley-ISTE, Hoboken.
[23] Lin HT, Leisk GG, Trimmer B (2011) GoQBot: a caterpillar inspired soft-bodied rolling robot. Bioinspir. Biomim. 6:026007. doi:10.1088/1748-3182/6/2/026007

[24] Laschi C, Cianchetti M, Mazzolai B, Margheri L, Follador M, Dario P (2012) Soft Robot Arm Inspired by the Octopus. Adv Robotics 26(7):709-727. doi:10.1163/156855312X626343

[25] Chou C-P, Hannaford B (1996) Measurement and modeling of McKibben pneumatic artificial muscles. IEEE Trans Robot Autom 12:90-102. doi:10.1109/70.481753

[26] Shepherd RF, Ilievski F, Choi W, Morin SA, Stokes AA, Mazzeo AD, Chen X, Wang M, Whitesides GM (2011) Multigait soft robot. PNAS 108(51):20400-20403. doi:10.1073/pnas.1116564108

[27] Camarillo DB, Carlson CR, Salisbury JK (2009) Configuration tracking for continuum manipulators with coupled tendon drive. IEEE Trans Robot 25(4):798-808. doi:10.1109/TRO.2009.2022426

[28] Jones BA, Gray RL, Turlapati K (2007) Three dimensional statics for continuum robotics. IEEE/RSJ Int. Conf. on Intelligent Robots and Systems 11-15.

[29] Boyer F, Porez M, Khalil W (2006) Macro-Continuous Computed Torque Algorithm for a Three-Dimensional Eel-Like Robot. IEEE Trans Robot 22:763-775. doi:10.1109/TRO.2006.875492

[30] Renda F, Cianchetti M, Giorelli M, Arienti A, Laschi C (2012) A 3D Steady State Model of a Tendon-Driven Continuum Soft Manipulator Inspired by Octopus Arm. Bioinspir. Biomim. 7:025006. doi:10.1088/1748-3182/7/2/025006

[31] Chirikjian GS, Burdick JW (1994) A modal approach to hyper-redundant manipulator kinematics. IEEE Trans Robot Autom 10(3):343–354. doi:10.1109/70.294209

[32] Giorelli M, Renda F, Calisti M, Arienti A, Ferri G, Laschi C (2012) A two dimensional inverse kinetics model of a cable driven manipulator inspired by the octopus arm. IEEE Int. Conf. on Robotics and Automation 3819-3824.

[33] Giorelli M, Renda F, Ferri G, Laschi C (2013) A Feed-Forward Neural Network Learning the Inverse Kinetics of a Soft Cable-Driven Manipulator Moving in Three-Dimensional Space. IEEE/RSJ Int. Conf. on Intelligent Robots and Systems 5033-5039.

[34] Cianchetti M, Ranzani T, Gerboni G, Nanayakkara T, Althoefer K, Dasgupta P, Menciassi A (2014) Soft robotics technologies to address shortcomings in today's minimally invasive surgery: the STIFF-FLOP approach. Soft Robotics 1(2):122-131. doi:doi:10.1089/soro.2014.0001

[35] Arienti A, Calisti M, Giorgio-Serchi F, Laschi C (2013) PoseiDRONE: design of a soft-bodied ROV with crawling, swimming and manipulation ability. MTS/IEEE OCEANS

22 Flexible Robot for Laser Phonomicrosurgery

Dennis Kundrat, Andreas Schoob, Lüder A. Kahrs, Tobias Ortmaier

Institute of Mechatronic Systems, Leibniz University Hannover

Abstract In this contribution we present a customized flexible robot developed as endoscopic device for laser phonomicrosurgery. Following the idea of soft robotics we describe the conventional clinical setting and adjunct benefits of the proposed assistance device to facilitate gentle surgery and usability in the operating room. Design constraints are obtained from medical image data implementing a mechanical design comprising compliant and flexible sections, actuation unit and multifunctional tip. We present results of a proof of concept experiment using a patient phantom, demonstrating the applicability of our system for laryngeal access.

22.1 Introduction

Recent advances in surgical techniques covering minimally invasive surgery (MIS) and natural orifice transluminal endoscopic surgery (NOTES) are focused on innovative instruments minimizing trauma and optimizing dexterity in constraint anatomical cavities. However, these approaches are associated with challenging manual and cognitive surgical tasks. Introduction of robotic and mechatronic assistance in the medical environment has overcome several limitations and advanced state of the art procedures. This resulted in the success of the commercially available da Vinci system (Intuitive Surgical, CA, USA) originally developed for cardiac and now mainly used in abdominal surgery. Additionally, researchers address systems (e.g. highly dexterous manipulators or snake-like devices) for further specific applications enhancing surgeons' performance with potentially beneficial postoperative outcome.

We present our progress towards an endoscopic device for laser phonomicrosurgery. The development aims at advanced instrumentation and mechatronic assistance by use of micro mechanisms, laser control, and vision.

22.2 Phonomicrosurgery

The anatomy of the human larynx is complex due to various connections between ligaments, muscles, cartilage, and soft tissue. The larynx is located in the lower neck and is divided topographically into supraglottic, glottic, and subglottic region as shown in Fig. 22.1.

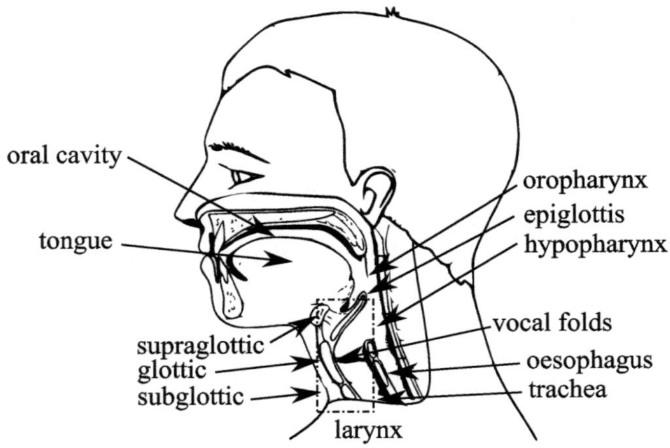

Fig. 22.1 Sagittal cut of human head showing oral cavity, pharynx, and larynx.

Besides the functionality of the epiglottic cartilage avoiding aspiration of food and liquids, the voice box in the glottic region produces voice based on vocal fold oscillations. In regard to epidemiology, laryngeal cancer is the second most common cancer in the aerodigestive system [1]. In particular, malignant or benign pathologies of the vocal folds influence the patient's life significantly because swallowing, breathing, and voice production can be abnormal. After indication of surgical treatment, phonomicrosurgery is considered as gold standard for preservation and recovery of functionality.

The conventional surgical setup is presented in Fig. 22.2a. The anaesthetized patient is positioned with extension of the neck and suspension laryngoscopy is conducted by inserting a straight rigid laryngoscope through the oral cavity to expose the surgical site in the glottic region. Intraoperatively, stereo microscopes are employed to perform surgery and examination under magnification. Tissue manipulation is performed with micro instruments or more recently with laser devices based on manually operated beam deflecting manipulators. This setup allows for control and observation of the laser ablation while removing tumor tissue [2].

Direct line-of-sight laryngoscopy is associated with disadvantages for patient and surgeon. Hyperextension of the neck can lead to high forces in the contact between laryngoscope and surrounding soft tissue resulting in tissue damages or ruptures. These complications can lead to loss of taste and sensitivity as well as cervical immobility [3]. Furthermore, current intraoperative ergonomics require intensive training to improve dexterity while operating the beam deflecting manipulator in a working distance of approx. 400 mm and simultaneous handling of micro instruments resulting in fatigue in tedious surgeries.

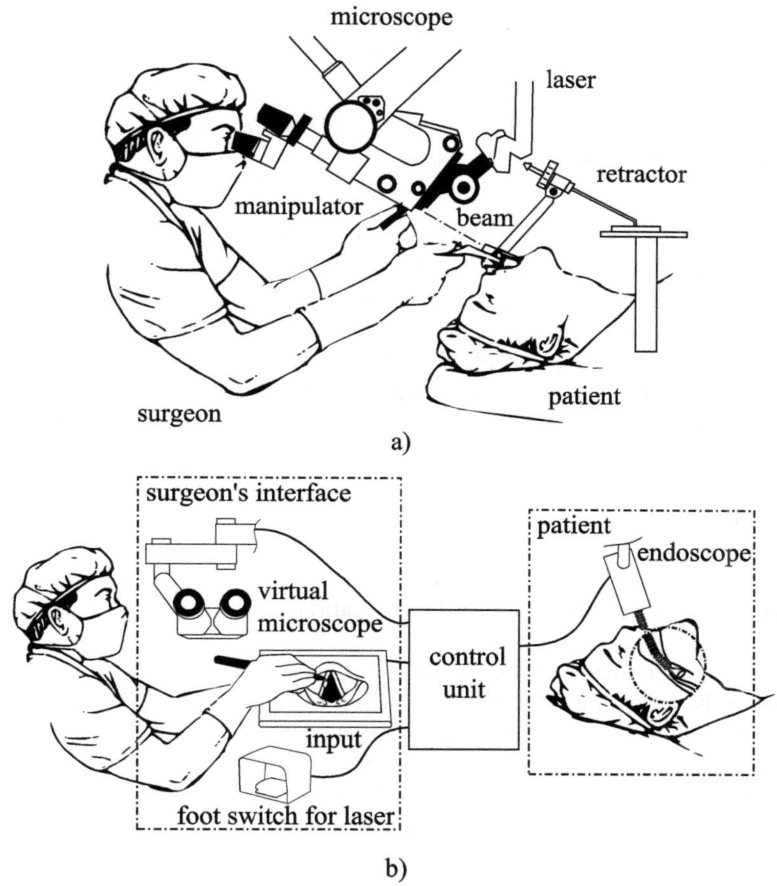

Fig. 22.2 a) Conventional setup: Microscopic observation and tissue manipulation under suspension laryngoscopy b) Proposed setup of "micro technologies for robot-assisted laser phonomicrosurgery " (μRALP) project: Novel surgeon's interface with interactive laser control, virtual microscope and endoscopic robot inserted into patient's larynx.

Contributions from researchers and companies have indicated advanced approaches compared to state of the art procedures by introducing transoral robotics surgery (TORS) addressing novel devices and instrumentation. In the following an excerpt is presented. The commercial da Vinci system has been used in TORS of laryngeal lesions [4]. Customized robots have been developed by Haifeng et al. [5] with serial architecture and Simaan et al. [6] with continuum architecture for micro manipulation and suturing in the laryngeal region. Furthermore, curved blade laryngoscopes complying with anatomical shapes have been proposed. Studies have proven total laryngeal exposure despite cervical rigidity and forces between instrument and tissue have been reduced significantly compared to straight laryngoscopy. Regarding phonomicrosurgy, Patel et al. integrated a laser beam deflecting mechanism based on Risley prisms into a rigid endoscope [7].

22.3 System Design

Taking the aforementioned developments into account we designed an adaptive curvature endoscopic robot with multifunctional tip to improve state of the art laser phonomicrosurgery. The adaptive shape can facilitate the insertion procedure and reduces forces to adjacent tissue minimizing postoperative complications. By transferring imaging from outside the patient to a chip-on-the-tip approach, direct visualization of the surgical scene with decreased extension of the neck is feasible. By incorporating a novel beam deflecting micro manipulator in the endoscopic tip [8] reduction of the working distance and workspace enlargement are achieved respectively. The µRALP setup is shown in Fig. 22.2b obviating the need for extension of the neck due to an adaptive endoscopic shaft.

22.3.1 Design Specifications and Constraints

The main requirements of the device have been specified and are summarized briefly in the following. The insertion procedure should obviate suspension laryngoscopy providing a gentle insertion without neck extension. For visualization purposes with implicit depth perception a stereo camera has to be integrated associated with appropriate illumination. Furthermore, the endoscopic tip has to incorporate the micro manipulator for beam deflection in order to provide adequate prepositioning with respect to the surgical site.

After determination of general requirements we analysed computed-tomography (CT) images of six patients in order to conduct morphological measurements and derive characteristic length and radii of specific anatomical landmarks. The minimal radius of curvature to access the larynx from the oral cavity was determined to approx. 30 mm in normal lying position.

22.3.2 Flexible Sections, Actuation Unit, and Control

In order to achieve the bending radii of the aforementioned constraints and to provide appropriate stiffness of the device we inspired our design by the work of Simaan et al. [5]. The flexible manipulator (see Fig. 22.3) consists of two sections with length of 70 mm and 15 mm, respectively. Each section has a 20 mm outer diameter and four tubular Nitinol (NiTi) backbones. These are axially aligned with an angle of 120° and 8.5 mm distance from the central backbone.

In contrast to a totally actuated system we propose hybrid control involving manual actuation of the first shaft in one degree of freedom (DoF) and automatic actuation of the distal tip in two DoF in order to facilitate pre- and intraoperative handling. For kinematic modelling we apply a model based on the assumption of constant curvature and neglecting torsion as well as backlash.

The actuation unit coordinates the movements of the endoscopic shaft and tip. It consists of a manually operated mechanism based on spindle drives and carriers for the shaft section and a motorized carrier system for distal tip deflections.
Our control architecture focuses on modular design as well as compact dimensions resulting in an embedded based architecture. The controller is hosted on a BeagleBoneBlack board running a dedicated Ubuntu system and Robot Operating System (ROS). Low level interfaces are provided for serial I/O and ethernet for distributed network operation in the overall framework. Customized hard- and software interfaces are developed to integrate kinematics and motion control into ROS.

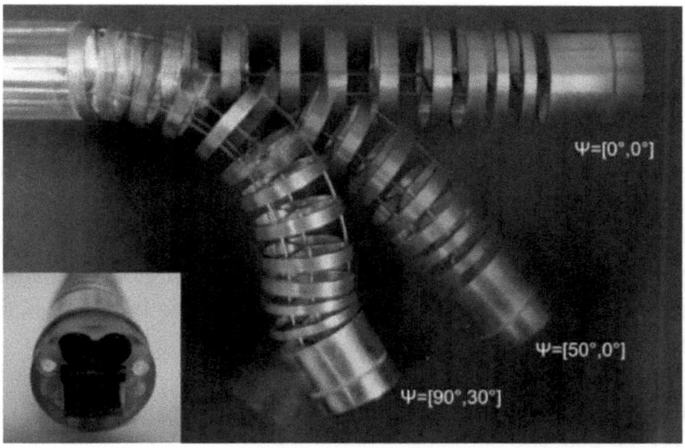

Fig. 22.3 Overlay image of flexible sections with kinematic configuration ψ defining tilt angle for first and second shaft, respectively. Lower left corner shows the tip design.

22.4 Results

Prior to a cadaver study first evaluations were conducted in a phantom (AirSim, Erler + Zimmer, Germany). The phantom comprises head, upper body, and detailed upper airway including larynx and vocal folds. The neck was slightly extended to facilitate the procedure. The main shaft was bent and the device was advanced transorally. Subsequently, a sagittal X-ray image of the phantom head was acquired to determine the compliance and fit of the manipulator with regard to anatomical constraints. As shown in Fig. 22.4, the curvature corresponds with the anatomy and enables a gentle insertion. The tip is located in the supraglottic region enabling observation and manipulation of the vocal folds in the glottic region.

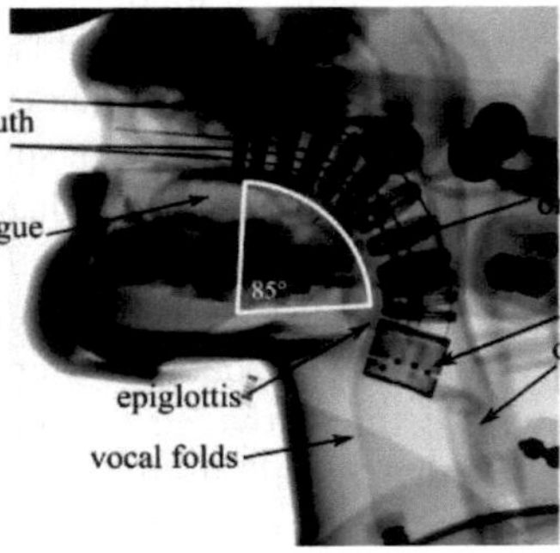

Fig. 22.4 Sagittal X-ray image of phantom head.

22.5 Conclusions

Laryngeal cancer and in particular vocal fold pathologies demand microsurgical interventions. Laser phonomicrosurgery is associated with several disadvantages for both patient and surgeon. Hybridly actuated continuum sections represent a promising approach to avoid direct laryngoscopy. Laryngeal exposure is achieved with less interaction forces as well as decreased distance between beam deflecting device, camera systems, and surgical site.

Acknowledgments This research has gratefully received funding from EU FP7 under grant agreement µRALP - n°288663.

22.6 References

[1] Rubinstein M, Armstrong WB (2011) Transoral laser microsurgery for laryngeal cancer: A primer and review of laser dosimetry. Lasers Med Sci 26:113-124.

[2] Mattos LS et al. (2013) A novel computerized surgeon-machine interface for robot-assisted laser phonomicrosurgery. Laryngoscope 123:8:1887-1894.

[3] Friedrich G, Kiesler K, Gugatschka M (2009) Curved rigid laryngoscope: missing link between direct suspension laryngoscopy and indirect techniques? Eur Arch Otorhinolaryngol 266:1583-1588.

[4] Blanco RGF et al. (2011) Transoral robotic surgery of the vocal cord. Journal of Laparoendoscopic & Advanced surgical techniques 21:2:157-159.

[5] Haifeng L et al. (2009) A master-slave robot system for minimally invasive laryngeal surgery. Proc. IEEE International Conference on Robotics & Biomimetics, pp 782-787.

[6] Simaan N et al. (2004) A dexterous system for laryngeal surgery. Proc. IEEE International Conference on Robotics and Automation, pp 351-357.

[7] Patel S et al. (2012) Endoscopic laser scalpel for head and neck cancer surgeries. Proc. SPIE 8207, Photonic Therapeutics and Diagnostics VIII, 820715. pp n/a.

[8] Lescano S et al. (2013) Kinematic analysis of a meso-scale parallel robot for laser phonomicrosurgery. Proc. Second Conference on Interdisciplinary Applications in Kinematics, IAK 2013, pp n/a.

23 Soft Components for Soft Robots

Jamie Paik

Ecole Polytechnique Federale de Lausanne

Abstract Typical robot platforms comprise rigid links with fixed degrees-of-freedom, solid blocks of transmission and actuator, and superficial positioning of sensors: they are often optimized for the given design criteria but are unable to execute instantaneous changes to the robot's initial mechanism design. The real-life incidences, however, require robots to face complex situations filled with un-programmed tasks and unforeseen environmental changes. One of the growing efforts in the field that address such juxtaposing design paradigm is *soft robotics*: augmentations of "softness" in robots to complement, adapt, and reconfigure to the contingent assignments. Although the "softness" invokes and relates to many facets of robot design in both soft and hardware, this manuscript focuses on describing some critical hardware components. I will present several on going research on actuation and sensor solutions for soft robotics application as well as novel methods and materials for sensor and actuation integration.

23.1 Introduction: What Kind of Softness?

The general concept of soft robotics stimulates interest from a wide field of scientists for its projection of autonomous systems that can safely conform to the unknown environment and un-programmed tasks. While the word "softness" attracts diverse interpretations, it connotes two aspects: intrinsic and extrinsic softness of the said mechanism. The intrinsic softness signals the compliancy achieved from the material characteristics where the level of the softness highly depends on the Young's modulus (<100MPa) of the main composing material. The extrinsic softness relates to the increased compliancy of the body and/or end-effector through mechanism design (i.e. springs, compliant joints). Evidently, a purely intrinsically soft robot suffers from reduced zero force bandwidth while a thoroughly extrinsic one has limited reconfigurability large or small. Therefore, for a truly interactive and soft system, optimal consolidation of both ends of the "softness" spectrum is crucial, followed by seamless integration of the multiple "soft" components. For any robotic system, the principal design boundaries come from the mechanical performance and capacity of the actuator, sensor and the control solutions. In the following sections, I will focus on actuator and sensor solutions that are feasible for often unconventional soft robot designs.

23.2 Actuators for Soft Robots

Most conventional actuators maintain the interface with its load as stiff as possible to maximize the work and control efficiency. For soft robots, reducing or even eliminating this interface stiffness introduces safety via lower reflected inertia, more stable force control (although the accuracy could be debatable), less inadvertent damage to the environment, the capacity for energy storage, and a high reconfigurability.

23.2.1 Actuators for Multi-DoF Designs

Mechanisms with multiple compliant joints or augmented degree-of-freedom (DoF) require specific transmissions linked with novel actuators that meet the design restrictions. There have been examples of modular connection of serial kinematic based actuators and / or foldable structures using flexible joints. The serial kinematic based actuators can have fully actuated joint; however, under-actuated systems are preferred for reducing the mechanism complexity while enhancing the flexibility. There are several ways of adding transmission or (under-) actuating these passive joints: they can either be flexures [1] or springs [2], and pro / antagonistic cables, or encapsulated geartrain that are controlled by a single actuator. Developed in the early 90's and continuing to evolve, series elastic actuators (SEAs) provide variable and controllable compliance using mechanical springs / compliant elements within the serial connection [3-5]. The elongation of the spring is used as the force measurement. When considering transmission-less actuations, smart materials like shape memory alloys (SMA) or magneto-, thermo-rheological fluids [6] that locally modulate the stiffness of joints are attractive choices. For direct actuation with restrictive body configuration, thermally activated foldable SMA actuators are effective for various types of 2D origami robots [7-11]. Depending on the application, the motion patterns are set while SMA wires [7,8], SMA sheets [9,10], and conductive polymer films [11] actuate the individual joint.

23.2.2 Pneumatic Artificial Muscles (PAMs)

PAMs, also known as Mckibben actuators, are basically pressurized air-filled rubber bladders with valves that control overall range of motion and applied force / torque: the compliance of the actuator can vary via operating pressure. The most commonly used PAMs are cylindrical balloons with rigid metal valves and shaft attachments to ensure the volume to linear displacement transmission [12, 13]. In order to increase the contraction force, PAMs with embedded fibers in the rubber air chamber are also introduced [14-16]. While the physics of the actuation principle is similar, soft pneumatic actuators (SPAs)' uniqueness comes from their air chambers that are entirely made of much softer silicone (PDMS™, Ecoflex™) instead of rubber (about 1/100 of Young's modulus); therefore, the operating pressure is a fraction of atmospheric pressure making them more compliant and safer. Owing to the new material and construction, these recent generation of pressure driven actuators are also highly customizable: their sizes and functionality (range of motion, compliance, torque / force output, actuation points) can be engineered by the geometry of the mold while the intrinsic softness can still be addressed via switching the silicone hardness [17, 18]. SPAs are effective even with fluids instead of pressurized air [19].

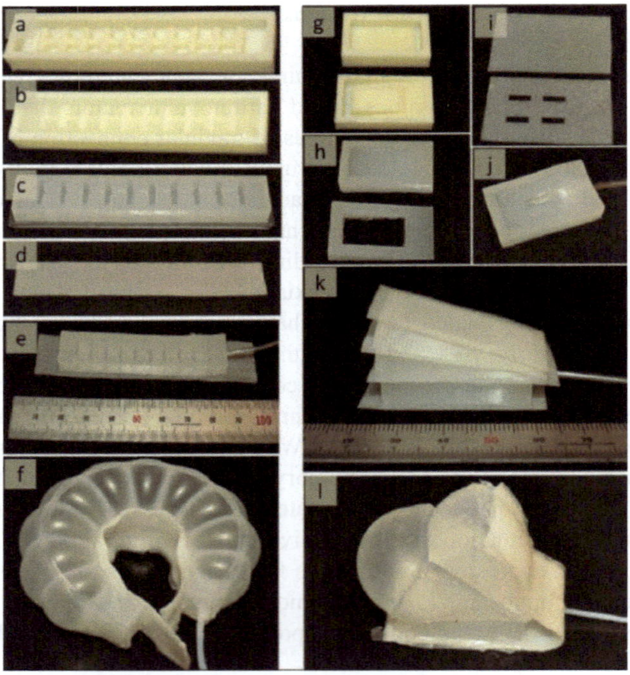

Fig. 23.1 Various SPA designs with different mouldings [17].

23.2.3 Smart Material-Based Actuators

There are continuing efforts in using smart materials to create functional actuators for soft robots: shape memory alloy (SMA), shape memory polymer (SMP), ionic polymer metallic composite (IPMC), magneto-rheological fluid, and dielectric elastomers are few of most studied smart materials. Using SMA and SMP as robot's body material has enabled researchers to achieve complicated motions and shape transformations in micro-, meso- and even macro-scale robots. Shape memory materials show different behaviour when their environmental condition changes (for most, thermally active stress-strain behaviour is of the most interest for their significant mechanical property change). When SMA passes its phase transition temperature, its phase changes from soft martensite to hard austenite crystal structure. Also while plastic deformation in martensite form is readily possible due to the material atomic structure, in austenite phase, the atomic structure pushes the bulk form of the metal back to its annealed shape. It is claimed that the secondary shape “memory” can be programmed but for practical meso-scale actuation [20], we focus mostly on a single “memory” shape. SMP shows the same behaviour and the difference is that it is softer above its glass transition than in its cold state, and that for meso-scale usage, the strain effect is still minimal relative to SMAs. The use of SMA wires and coils for making soft robots mimicking the peristaltic motion of a worm [21] or an octopus arm [22] have attained much attention already. Another application for these materials is using the elastic behaviour transformation for making adaptive body structures for robots. An example of efforts toward using these materials is a meso-scale wing structure with controllable stiffness joint that can tune its dynamic respond on demand [23]. Other methods of tuning stiffness such as exploiting phase transition in wax coated polymers [24] and phase transition in low melting point metals [25] has also been suggested. Each of these methods has their merits but this variety shows the need for an adjustable stiffness structure that can actively tune the joint stiffness, or turn on / off the existing joint’s DoF.

23.3 Soft Sensors

The ultimate sensors for soft robots are thin, stretchable and robust to reconfiguration of the moving body shapes and tasks. Due to the practical cost of adding sensors, robotic manipulators often lack thorough sensor integration and resort to a multi-axis F/T sensors at the end effector only. However, a truly conforming, safe and interactive soft robot would require sensors that are soft (stretchable), robust and small enough to be embedded, distributed directly on the robot body. The physical limitations of the sensor material contribute majorly to the difficulties in the fabrication and embed-ability of bendable sensors, let alone stretchable sensors. To overcome this challenge, some of the latest methods are: optimizing hard materials’ geometry (aspect ratio, patterns, slits), experimenting with different conductive material, and embedding discrete sensing elements within soft matrix.

23.3.1 Soft Geometry for “Hard” Conductor

Simply put, strain gauges are a standard example of sensors that have an elevated compliancy and resistive sensitivity due to its planar geometry: dense serpentine shape allows the metal layer to remain flexible. By introducing various types of serpentine patterns, the metal layer can have different sensitivities toward localized curvatures, linear strain, and pressure via resistance change in its effective conductivity. When optimizing geometric patterns, it is not limited to planar surface but also in 3D (the serpentine pattern can be pre-strained to come out of the plane [26]). Stretchable metallization prepares electrodes and conductive tracks; the gold film covered with PDMS shown in [27] stretches up to 20 % of its initial size and measures a pressure up to 160 kPa.

23.3.2 Conductive Material

A practical way of using conductive material for soft sensors is to make an effective conductive path with a cross-section that would be sensitive to different mechanical loading. There are sensors made of an elastomer with embedded micro channels filled with conductive liquid [28-30]. Upon loading, the electric resistance changes with the deformation of the cross-section areas of the micro channels. Multi-axis strain, bending curvature, normal forces and in-plane (shear forces) can be measured as well [29, 30]. Carbon nanotube composite thin films have been used as the active sensing material in [31] and these sensors can be stretched up to 2.5 %. Conductive polymer based sensors have better sensitivity than the metallic foils while being more flexible. The gauge factor (GF: relative resistance change to the mechanical strain and indicates the sensitivity) of these sensors is 50-100 and 2-5 for the metallic foil sensors. Commercial sensors such as Flexiforce®, Bend Sensor®, Bi-Flex Sensor™ also use conductive polymers. Crystalline silicone has been used in [32] which has an active area of 410 x 410 μm^2. The overall size of the sensor is 63 x 63 mm^2 x 50 μm (thick) with GF of 8.5. Piezo-resistive sensors that use intrinsic piezoelectric effect of the polymer provide the resistance change when subjected to deformations. PeDOT(poly(3,4-ethylenedioxythiophene) and conductive ink have been used in [33] which has a GF of 2.48. A micro-structured silicon with DuPont Kapton is used in [34] with a large GF of 43.

23.3.3 Discrete Sensors in Soft Matrix for Distributed Sensing

When the softness of a sensor is limited by the material property, and if the component size is relatively small, the overall softness can be augmented by embedding sensing receptor in a softer matrix. Once the hard "pixel" receptors are embedded in the soft silicone matrix, upon mechanical loading, the receptors provide an electrical output. For this application, piezoelectric materials are desirable for their high sensitivity (15 – 65 nm/V) and consistency under mechanical loading. They are micro-machinable for easy distribution in different size and shape of arrays. Also compared to conventional strain gauges, signal conditioning is easier especially in applications where there are low strains and high noise levels [35]. These are attractive characteristics for applications in wearable and embeddable structures where precise measurement with robustness is required. However, Young's modulus of piezoelectric material is high (2.5 – 63 GPa) and they are very brittle: conditions not ideal for relatively small loading. What we found, however, after embedment in the silicone membrane, the sensitivity is increased due to the large shear loading on the receptor surface [36, 37].

Polysilicon piezoresistive materials are used as strain gauges in many MEMS devices to measure the deflection of a micro-machined deformable structure [38-40]: although these materials are robust, have low cost of fabrication, and have a higher gauge factor compared to metal alloy strain gauges, they show nonlinearity with hysteresis. Piezoelectric ceramics (PZT) and piezoelectric polymers such as polyvinylidene fluoride (PVDF) [41-44] are mostly used materials for pressure / force measurements with the piezoelectric effect. To compare PZT and PVDF as sensing receptors, PZTs are less expensive, easier to fabricate, have a high dielectric constant, and provide better electromechanical transformation. However, they are highly brittle. PVDFs are very flexible but have a higher cost of fabrication (easier with nm wavelength UV laser), lower dielectric constants, lower electromechanical transformation and more signal conditioning for their voltage outputs to be used as a sensor. These sensors are supposed to provide sensing in large areas with distributed arrays.

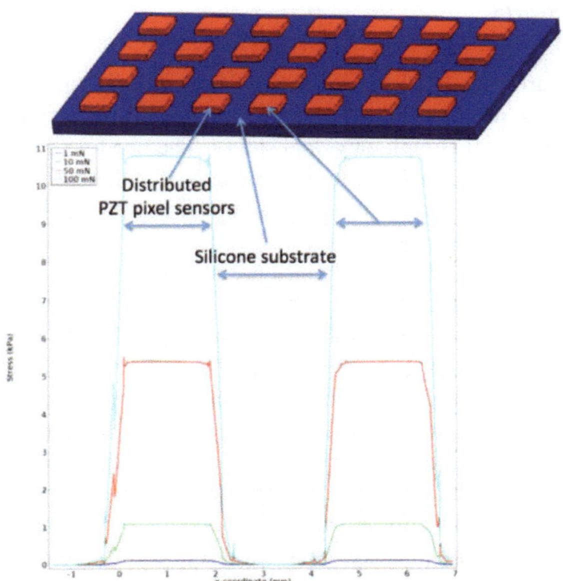

Fig. 23.2 The Sensible Skin with distributed piezoelectric elements embedded in a soft silicone matrix. The graph shows the modulation of stress depending on the input force between the PZT pixels [37].

The minimum distance between each piezo receptor depends on the stress distribution are shown in Fig. 23.2 while Fig. 23.3 displays how commercial flexible sensors would look when embedded in the silicone matrix (with PZT receptors, the measuring pixel size can go down to 1.5 x 1.5 mm^2). The matrix material property determines the stress distribution and dictates overall sensitivity and the resolution of the sensing surface.

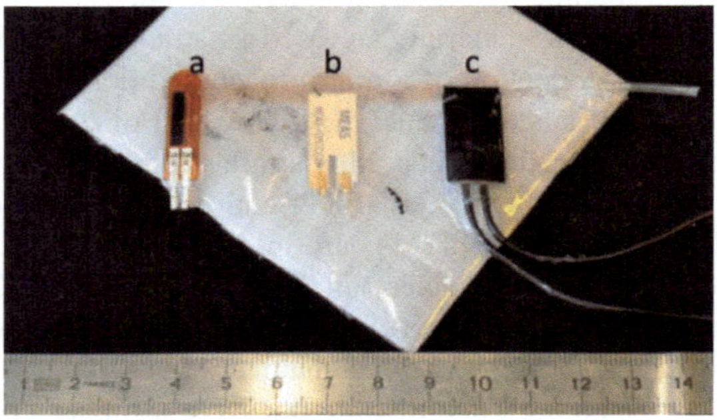

Fig. 23.3 Bending sensors embedded in 500 μm thick silicone matrix: a. Flexpoint™, b. piezo film, c. Bi-flex sensor™ [18].

23.4 Conclusions

The foremost hardware challenge of the soft robotics is exploring new solutions toward improving machine-human interaction: this can further be described as task and environmental compliancy, and safety. The need for the new components for soft robots prevail as the community still seeks for the optimal solution toward actuation, sensing, and control as well as total integration. This manuscript describes the concurrent research efforts on soft actuation and sensing with promising results, but the remaining challenges are still vast. Now, the investigation of the material and mechanical design from the conception of the robot is the key importance. The application specific functional requirements, the fabrication process of each component must be revisited and refined: this process is particularly important for investigating novel control methods and computational techniques. The design and simulation tools will greatly contribute toward improving the design-fabrication-integration iterations that are undoubtedly labour and experimentally intensive. The technology and research development toward the mentioned soft components will not only advance the robotics community but also neuroprosthetics, materials engineering, chemical engineering, computer sciences and medical studies.

Acknowledgments This work was supported by Swiss NCCR Robotics.

23.5 References

[1] Odhner, Lael U., et al. "A compliant, underactuated hand for robust manipulation." The International Journal of Robotics Research 33.5 (2014): 736-752.

[2] Potratz, Jason, et al. "A light weight compliant hand mechanism with high degrees of freedom." Journal of biomechanical engineering 127.6 (2005): 934-945.

[3] Pratt, Gill A., and Matthew M. Williamson. "Series elastic actuators." Intelligent Robots and Systems 95.'Human Robot Interaction and Cooperative Robots', Proceedings. 1995 IEEE/RSJ International Conference on. Vol. 1. IEEE, 1995.

[4] Tsagarakis, N. G., et al. "A compact soft actuator unit for small scale human friendly robots." Robotics and Automation (ICRA), 2010 IEEE International Conference on. IEEE, 2010.

[5] Sensinger, Jonathon W., and Richard Ff Weir. "Improvements to series elastic actuators." Mechatronic and Embedded Systems and Applications, Proceedings of the 2nd IEEE/ASME International Conference on. IEEE, 2006.

[6] Cheng, Nadia, et al. "Design and analysis of a soft mobile robot composed of multiple thermally activated joints driven by a single actuator." Robotics and Automation (ICRA), 2010 IEEE International Conference on. IEEE, 2010.

[7] Onal, Cagdas D., Robert J. Wood, and Daniela Rus. "An origami-inspired approach to worm robots." Mechatronics, IEEE/ASME Transactions on 18.2 (2013): 430-438.

[8] Lee, Dae-Young, et al. "Deformable wheel robot based on origami structure." Robotics and Automation (ICRA), 2013 IEEE International Conference on. IEEE, 2013.

[9] Paik, Jamie K., and Robert J. Wood. "A bidirectional shape memory alloy folding actuator." Smart Materials and Structures 21.6 (2012): 065013.

[10] Hawkes, E., et al. "Programmable matter by folding." Proceedings of the National Academy of Sciences of the United States of America 107.28 (2010): 12441-12445.

[11] Okuzaki, H., et al. "A Biomorphic Origami Actuator Fabricated by Folding a Conducting Paper." Journal of Physics: Conference Series. Vol. 127. 2008.

[12] Daerden, Frank, and Dirk Lefeber. "Pneumatic artificial muscles: actuators for robotics and automation." European journal of mechanical and environmental engineering 47.1 (2002): 11-21.

[13] Beck, Roland D. "Pneumatic actuator control system." U.S. Patent No. 3,237,529. 1 Mar. 1966.

[14] Tomori, Hiroki, and Taro Nakamura. "Theoretical Comparison of McKibben-Type Artificial Muscle and Novel Straight-Fiber-Type Artificial Muscle." Int. J. Autom. Technol 5.4 (2011): 544-550.

[15] Nakamura, Taro, and Hitomi Shinohara. "Position and force control based on mathematical models of pneumatic artificial muscles reinforced by straight glass fibers." Robotics and Automation, 2007 IEEE International Conference on. IEEE, 2007.

[16] Faudzi, A. M., Razif, M. R. M., Nordin, I. N. A. M., Suzumori, K., Wakimoto, S., & Hirooka, D. (2012, July). Development of bending soft actuator with different braided angles. In Advanced Intelligent Mechatronics (AIM), 2012 IEEE/ASME International Conference on (pp. 1093-1098). IEEE.

[17] Sun, Yi, Yun Seong Song, and Jamie Paik. "Characterization of silicone rubber based soft pneumatic actuators." Intelligent Robots and Systems (IROS), 2013 IEEE/RSJ International Conference on. Ieee, 2013.

[18] Suh, Chansu and Jamie Paik, "Soft Pneumatic Actuator Skin with Embedded Sensors." Intelligent Robots and Systems (IROS), 2014 IEEE/RSJ International Conference on. Ieee, 2014.

[19] Onal, Cagdas D., and Daniela Rus. "A modular approach to soft robots." Biomedical Robotics and Biomechatronics (BioRob), 2012 4th IEEE RAS & EMBS International Conference on. IEEE, 2012.

[20] Kohl, Manfred. Shape memory microactuators. Springer, 2004.

[21] Seok, Sangok, et al. "Meshworm: a peristaltic soft robot with antagonistic nickel titanium coil actuators." Mechatronics, IEEE/ASME Transactions on 18.5 (2013): 1485-1497.

[22] Mazzolai, B., et al. "Soft-robotic arm inspired by the octopus: II. From artificial requirements to innovative technological solutions." Bioinspiration & biomimetics 7.2 (2012): 025005.

[23] Hines, Lindsey, Veaceslav Arabagi, and Metin Sitti. "Shape memory polymer-based flexure stiffness control in a miniature flapping-wing robot." Robotics, IEEE Transactions on 28.4 (2012): 987-990.

[24] Chenel, Thomas, Jamie Paik and Rebecca Kramer. "Variable Stiffness Fabrics with Embedded Shape Memory Materials for Active Joint Stability Braces." Intelligent Robots and Systems (IROS), 2014 IEEE/RSJ International Conference on. Ieee, 2014.

[25] Schubert, Bryan E., and Dario Floreano. "Variable stiffness material based on rigid low-melting-point-alloy microstructures embedded in soft poly (dimethylsiloxane)(PDMS)." Rsc Advances 3.46 (2013): 24671-24679.

[26] Rogers, John A., Takao Someya, and Yonggang Huang. "Materials and mechanics for stretchable electronics." Science 327.5973 (2010): 1603-1607.

[27] Cotton, Darryl PJ, Ingrid M. Graz, and Stephanie P. Lacour. "A multifunctional capacitive sensor for stretchable electronic skins." Sensors Journal, IEEE 9.12 (2009): 2008-2009.

[28] Vogt, Daniel M., Yong-Lae Park, and Robert J. Wood. "Design and Characterization of a Soft Multi-Axis Force Sensor Using Embedded Microfluidic Channels." (2013): 1-1.

[29] Hammond III, Frank L., et al. "Soft Tactile Sensor Arrays for Force Feedback in Micromanipulation." IEEE SENSORS JOURNAL 14.5 (2014): 1443.

[30] Chossat, Jean-Baptiste, et al. "A Soft Strain Sensor Based on Ionic and Metal Liquids." IEEE SENSORS JOURNAL 13.9 (2013): 3405.

[31] Jung, Soyoun, et al. "Flexible strain sensors based on pentacene-carbon nanotube composite thin films." Nanotechnology, 2007. IEEE-NANO 2007. 7th IEEE Conference on. IEEE, 2007.

[32] Zhou, Lisong, et al. "Flexible substrate micro-crystalline silicon and gated amorphous silicon strain sensors." Electron Devices, IEEE Transactions on 53.2 (2006): 380-385.

[33] Correia, Vítor, et al. "Development of inkjet printed strain sensors." Smart Materials and Structures 22.10 (2013): 105028.

[34] Won, Sang Min, et al. "Piezoresistive strain sensors and multiplexed arrays using assemblies of single-crystalline silicon nanoribbons on plastic substrates." Electron Devices, IEEE Transactions on 58.11 (2011): 4074-4078.

[35] Sirohi, Jayant, and Inderjit Chopra. "Fundamental understanding of piezoelectric strain sensors." Journal of Intelligent Material Systems and Structures 11.4 (2000): 246-257.

[36] Roche, Denis, et al. "A piezoelectric sensor performing shear stress measurement in an hydrodynamic flow." Applications of Ferroelectrics, 1996. ISAF'96., Proceedings of the Tenth IEEE International Symposium on. Vol. 1. IEEE, 1996.

[37] Acer, M, Marco Salerno and Jamie Paik, "Piezo resistive sensors with Silicone Embedment." Submitted for publication.

[38] H, Yating, Katragadda, Rakesh B, TU, Hongen, et al. "Bioinspired 3-D tactile sensor for minimally invasive surgery." Microelectromechanical Systems, Journal of, 2010, vol. 19, no 6, p. 1400-1408.

[39] Ahmed, M. Chitteboyina, M.M., Butler, D.P. and Celik-Butler, Z., "MEMS Force Sensor in a Flexible Substrate Using Nichrome Piezoresistors," Sensors Journal, IEEE , vol.13, no.10, pp.4081,4089, Oct. 2013

[40] Noda, K., Hoshino, K., Matsumoto, K., & Shimoyama, I. (2006). "A shear stress sensor for tactile sensing with the piezoresistive cantilever standing in elastic material". Sensors and Actuators A: physical, 127(2), 295-301.

[41] Ottermo, Maria V., Oyvind Stavdahl, and Tor Arne Johansen. "Palpation instrument for augmented minimally invasive surgery." Intelligent Robots and Systems, 2004. (IROS 2004). Proceedings. 2004 IEEE/RSJ International Conference on. Vol. 4. IEEE, 2004.

[42] Seminara, Lucia, et al. "Piezoelectric Polymer Transducer Arrays for Flexible Tactile Sensors." IEEE Sensors Journal 13.10 (2013).

[43] Qasaimeh, M. A., S. Dargahi, and M. J Kahrizi. "pvdf-based microfabricated tactile sensor for minimally invasive surgery." Microelectromechanical Systems, Journal of (2009).

[44] Zirkla, M., et al. "PyzoFlex®: a printed piezoelectric pressure sensing foil for human machine interfaces." Proc. of SPIE Vol. Vol. 8831. 2013.

24 Soft Robotics for Bio-mimicry of Esophageal Swallowing

Steven Dirven[1], Weiliang Xu[1], Leo Cheng[2]

[1] Department of Mechanical Engineering, The University of Auckland

[2] Auckland Bioengineering Institute, The University of Auckland

Abstract The field of soft robotics is continuing to expand into exploring the possibilities for novel, non-skeletal, transport and locomotion systems inspired by biological phenomena. Application of these techniques toward development of an anthropomorphic esophageal swallowing robot requires overcoming of many soft robotic design and characterization challenges. Additionally, soft-robots require vastly different methods of specification and validation than traditional robots, as they typically exhibit less well-defined degrees of freedom. This chapter reveals a series of novel methods to: establish interdisciplinary specifications for the esophageal swallowing process, develop a soft robotic analogue in the engineering domain, and demonstrate its capability.

24.1 Introduction

Biomimetic engineering has become increasingly more popular as scientists and researchers embrace inspiration from biological forms to solve synonymous problems in the engineering domain. This has seen the emergence of new research fields which embody features observed in nature to provide elegant solutions to complex problems. The emergence of soft-robotics is a result of this trend, where the continuous actuation and sensation, as well as the high degree of freedom behaviors, of organisms and their tissues are explored.

Inspiration for soft robotics stems from the identification of useful engineering features, and their application domain. In the human body, for example, peristaltic transport is a common fluid transport technique, which occurs in the ureter, and throughout many regions of the digestive tract. The concept of peristalsis has been explored and implemented in the engineering field for a variety of applications based on transport metrics. However, these devices exhibit limitations with the continuity and uniformity of fluid deformation, which has led to empirical, physical, evaluation of these behaviors being held back. This has predominantly been caused by actuation challenges, which have prevented establishing biomimetic radial occlusion of a linear peristaltic pumping conduit. Thus, inspired by the peristaltic transport technique of esophageal swallowing, specification and development of a soft, biomimetic, swallowing robot was conducted to address these issues.

The concept of developing biomimetic swallowing robots had seen significant research interest, towards understanding swallow efficacy, modelling, and providing a new platform for medical training. The oral and pharyngeal phases of swallowing have commanded the most attention, especially by the pioneering works of Kobayashi *et al.* [1,2] and the subsequent robotic swallowing system by Noh et al. [3] which annexes their airway management research. It is observed in both instances that the actuation is undertaken by wire-driven methods which exhibit discrete surface contact, though both embody the continuity of the tissue surface, with the lumen being of a single material. Modelling of the esophageal peristaltic technique by [4] applies discrete shape-memory-alloy actuation to a series of axially arranged elements, which facilitate a semi-continuous and distributed actuation. The actuation is unidirectional in that the device occludes in a single dimension. Biomimetic exploration of the bowel region of the digestive tract by Suzuki *et al.* [5] more closely emulates the radial nature of biologic peristalsis, where pneumatic actuation offers distributed and compliant behavior. Each axially arranged element of the system exhibits promising actuation behavior. However, rigid skeletal elements that bind these elements together prevent the device from achieving a continuously propagating peristaltic occlusion.

The field of modelling peristaltic organs of the human body is becoming increasingly anthropomorphic with advances in actuation technology. The constraints, limitations and requirements of a compliant, biomimetic, peristaltic, swallowing-robot are synonymous with the emerging possibilities of soft robotic actuation. This chapter addresses the interdisciplinary communication of these concepts into the engineering domain to exploit soft robotic techniques to develop a robot which can bio-mimetically investigate the interaction between food boluses and the swallowing conduit. The resulting robot is of a continuous architecture with distributed actuation, inspired by the layout of muscles around the biological conduit. The processes of actuator specification, design, fabrication and characterization are clarified to communicate the solution architecture and address current challenges in developing and measuring continuous, soft robotic, behaviors.

24.2 Interdisciplinary Specifications

Peristaltic actuation of the esophagus results from the coordinated rostro-caudal recruitment of circular and longitudinal muscle fibers. To communicate this behavior into the engineering field interdisciplinary specifications are developed in order to facilitate arbitration of modelling assumptions, while still capturing the salient process features.

The esophagus can be viewed as a linear, continuous, and soft peristaltic pumping organ which transports food boluses from the pharynx to the stomach. It is typically 20-26 cm in length and has a comfortable maximal distension diameter of approximately 18 mm (subjects report pain for distensions on the order of 20-

22 mm) [6]. It represents the last phase of the swallowing process which actively transports boluses at an axial velocity of between 2 and 4 cm/s [7]. The conduit does not depend on a skeletal structure for its actuation effort, and instead constricts on itself via the circular muscles that surround its axis. This behavior is imperative for an engineering analogue, where the muscle system draws the conduit wall together radially.

Medically, the esophageal swallowing process is investigated by manometry, videofluorography, endoscopy, and intraluminal ultrasound [8]. The manometric and videofluorographic techniques are considered the 'gold standard' measurement techniques [9,10] and are commonly applied simultaneously as the pressure and geometrical behaviors are complementary, and both are required to fully interpret bolus transit. It is found that, for primary peristalsis, boluses travel ahead of a single peristaltic wave, which propagates in a continuous, compliant manner.

Peristaltic transport models in the mathematical field typically describe the wave profile as having trigonometric radial amplitude of the root form (24.1).

$$H(x,t) = \varepsilon + \frac{\alpha}{2}\left(1-\cos\left(2\pi\frac{x-ct}{\lambda}\right)\right) \tag{24.1}$$

This facilitates a continuous description of the surface from a series of parameters such as minimum radius (ε), wave velocity (c), amplitude (α) and wavelength (λ) with respect to axial displacement (x) and time (t). These models are then used to predict intrabolus pressure profiles for materials with differing rheology. The parameters of these models share many similarities with medical measurement and interpretation of findings in the clinical context. The mathematical models have been applied generally to peristaltic systems, and specifically to modelling esophageal swallowing. Following this method facilitates analytical expression of the wave trajectory and in future, will allow for comparison between interdisciplinary findings.

Embodiment of the esophageal swallowing process arises from knowledge and inspiration from the medical, food science and mathematical modelling fields. The quantitative and qualitative measurements and observations are used to serve as targets for biomimetic, soft robotic behavior. The exploration of these concepts in the engineering domain contributes understanding back to these fields about the peristaltic transport process, as well as generalizing the concepts for application in engineering disciplines.

24.3 Actuator Design and Manufacture

The largest obstacles to achieving biomimetic peristaltic transport of a radial nature have been in the development constraints of actuators, which prevent them from being able to orchestrate continuously propagating waves with compliant behavior. Soft robotics techniques are ideal for such applications as they exhibit distributed actuation and continuous output. The sequential rostro-caudal peristaltic technique of muscle activation in the esophagus, and study of current peristaltic device architectures inspired a regular and repeating structural architecture in the axial aspect. In order to achieve biaxial occlusion a rotationally symmetric design (about the conduit axis) was proposed.

The tissues of the esophagus undertake large deformations when they occlude the conduit. Thus, the materials and actuation technique are required to withstand similar conditions. A plausibility study of soft robotic actuation techniques was undertaken where the advantages of pneumatic actuation upon elastomer materials became evident. However, the discrete nature of pneumatic chambers in previous peristaltic devices had led to discontinuity of actuation. The axially arranged skeletal elements used to bound actuation segments both exhibit different architecture and capability than the biological system. The current swallowing robot (Fig. 24.1) exploits elastomeric division of chambers which can stretch and expand, which additionally facilitates co-actuation of neighboring chambers.

This development concept overcomes the intrinsic difficulties experienced by previous robotic devices, and significantly increases the biomimetic possibilities of the robot. The swallowing robot is cast from Ecoflex 00-30, a commercially available, two-part, silicone rubber material. It has 48 embedded pneumatic chambers which are arranged in 12 axially distributed whorls of four which inflate to occlude the conduit. This structure was favored over a single ring-shaped chamber at each level due to the increased passive luminal opening force. Additionally, it causes initiation of wall buckling in an organized manner, which is required to achieve occlusive motion of the essentially incompressible silicone rubber material.

The casting process is undertaken in five stages where the central actuator, chambers, and swallowing conduit are cast in a single stage. The remaining four stages are to seal each of the chambers and provide pneumatic inlet nipples for inflation. The sealing casts are undertaken by placing the actuator on a bath of silicone rubber, where the mating surfaces bond as the material cures. The molds and housings are each rapid prototyped from plastic by the 3D extrusion printing method. The pneumatic chambers push off the housing structure to cause the occlusive peristaltic motion which mimics muscle activation.

Each whorl of chambers is connected to a single pneumatic inlet, resulting in a 12 input device. A supporting electro-pneumatic interface consisting of 12 pressure regulating proportional valves (ITV0030) and control logic is used to assert the robot, with independent, continuously variable, pressures. This facilitates achievement of many different deformation characteristics, which can be cycled through to develop propagating peristaltic waves.

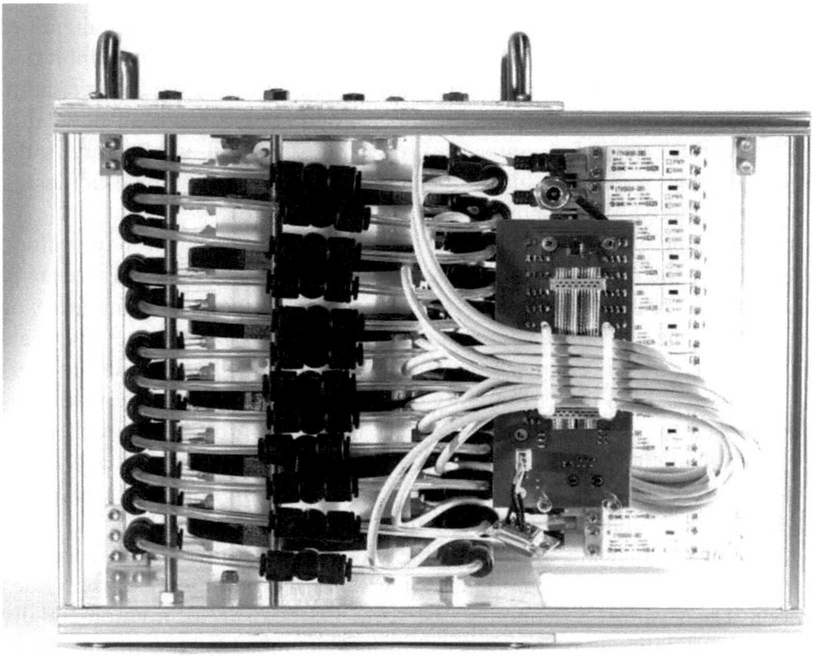

Fig. 24.1 Peristaltic swallowing robot (left) and control system (right). The internal swallowing conduit is vertical, halfway between the two handles on top. The four chambers in each of the 12 whorls are connected circumferentially (horizontal).

24.4 Experimental Characterization

In order to achieve the aforementioned specifications, and develop an open-loop model of the swallowing robot, the wave seal pressure and geometrical deformation require investigation. These parameters are synonymous with similar measurement practices in the medical field. The sensor locations and actuation concepts are demonstrated in Fig. 24.2.

In previous swallowing robotic models, the medical technique of videofluorography is used as an experimental method. However, the nature of this technique results in a silhouette view, not necessarily in the axial plane. In order to overcome this, a 3-dimensional location and orientation sensing process called articulography was used to measure the surface displacement.

Experimental characterization of the swallowing robot was conducted under dry-swallow conditions, where no bolus was present, and hence, there was no intra-bolus pressure signature. The trajectory was designed such that the pressure for three adjacent chambers either decreased or was equal to that of its neighbor in the direction of wave transit. This method guaranteed reliable single inflection waves, which is synonymous with the sinusoidal geometric specification inspired by modelling in the mathematic field.

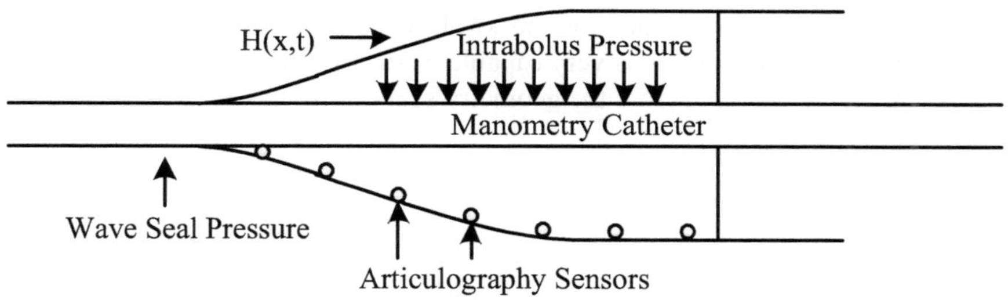

Fig. 24.2 Location of articulography sensors for radial displacement measurement and manometry catheter for wave seal and intra-bolus pressure measurement.

The characterization data represents an empirical model of the device behavior which can be exploited to generate desirable trajectories. Articulography measurements, of the surface deformation, and manometry, investigation of the wave tail seal pressure, were undertaken separately due to the incompatibility of using manometry in the resonating magnetic field of the articulograph, as well as inherent access issues with a large number of sensing devices in the small conduit.

The model undergoes a transition, from geometric to pressure output, after the conduit wall moves through its maximum displacement to contact the opposing wall at the center of the peristaltic conduit. At this point, increasing pressure in the pneumatic chambers can displace the wall no further. This results in equal and opposite occlusive forces of the conduit wall, which achieves a wave tail seal.

The methods and findings of manometry and articulography, the interdisciplinary experimental procedures, are clarified in the following sections. These techniques are inspired by medical practices surrounding investigation of the swallowing process in the human body.

24.4.1 Manometry Method and Findings

The wave tail seal is a salient feature of the esophageal swallowing process, which prevents the bolus from passing behind the wave and propagating in the wrong direction. The robotic swallowing conduit is manufactured in a distended state, which requires pneumatic pressure to deform the device into the occluded morphology. Thus, there exists a minimum pneumatic pressure to achieve wave tail seal pressure.

The wave tail seal was investigated for assertion of a single whorl of chambers with a resolution of 15 divisions over the range of 50 - 72 kPa. It was found that the wave tail pressure transition zone occurs between 60 - 64 kPa of pneumatic assertion where additional pressure results in no geometric motion of the wall, just an increase in occlusive pressure. This experiment was conducted 10 times, which demonstrated repeatable and reliable results (Fig. 24.3)

The occlusive pressure throughout human swallowing is on the order of 15 kPa [11] which presents a minimum for the swallowing robot. It is observed (Fig. 24.3) that the swallowing robot is very capable of such wave seal pressures, reaching this magnitude quickly across the geometry/seal transition zone.

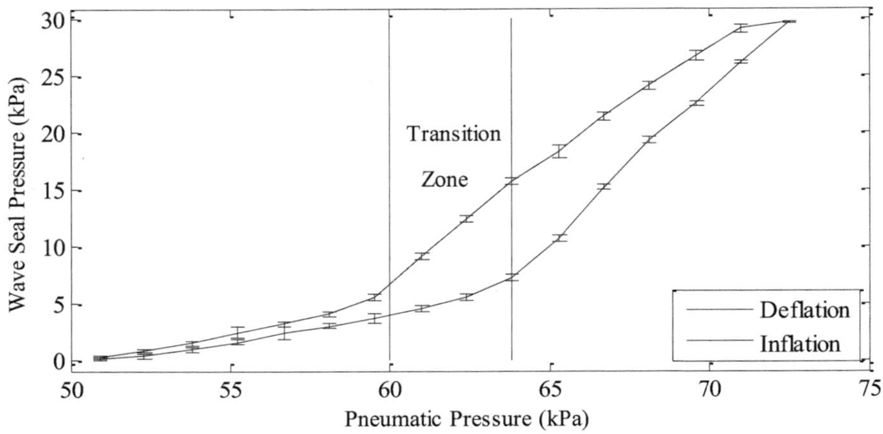

Fig. 24.3 Relationship between wave seal occlusion pressure and pneumatic pressure. The geometry/seal transition zone is identified where there is a significant change in gradient from light contact to large positive seal [12].

24.4.2 Articulography Method and Findings

The displacement behavior of soft robotic devices is challenging to characterize due to the continuous nature of their deformation, and resulting high degree of freedom. In order to investigate this behavior the technique of articulography was used, where small sensors of 4 mm diameter are placed on the luminal surface. These were spaced approximately 7.5mm apart in the axial dimension, at the center and border of 3 whorls of chambers. The dimension of the sensors is the same as the diameter of the manometry catheter, which means that their finite size can be neglected; the same datum will be used for all experimentation.

Pneumatic pressure was asserted into the three adjacent whorls of chambers to explore all combinations of $P_A \geq P_B \geq P_C$ with a resolution of 20 divisions over the range of 0 - 71.5 kPa. The sensors were supported against the surface with a pressure of 3.6 kPa. Each combination was held for one second, before moving to the next. This allowed the pneumatic system to stabilize and achieve a steady result. The displacement behavior of the sensors for constant pressure P_A = 71.5 kPa, and variable pressures P_B and P_C exhibits an intrinsically non-linear relationship between pressure and displacement (Fig. 24.4). This is due to the non-linear stress strain behavior of the silicone rubber elastomer material, and the co-operative actuation behavior of asserting adjacent whorls of chambers.

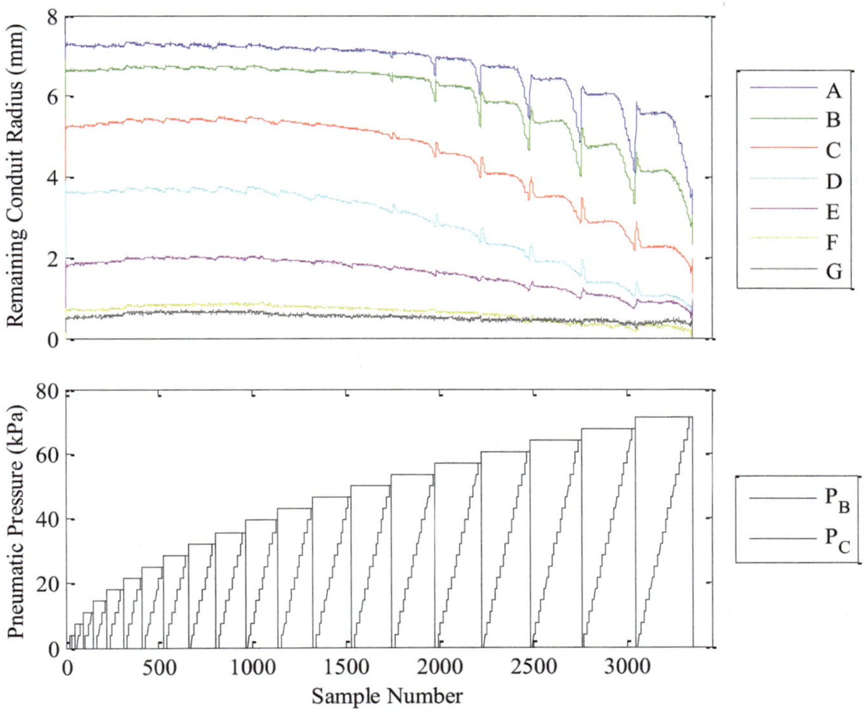

Fig. 24.4 Displacement of sensors in the radial aspect, showing the remaining conduit diameter in response to pneumatic actuation with $P_A = 71.5$ kPa, and pressures P_B and P_C as indicated. These combinations exhaustively investigate $P_A \geq P_B \geq P_C$ with a resolution of 20 divisions over 0 - 71.5 kPa

It is observed that sensors F and G remain at a completely occluded displacement as the pressure in chamber whorls B and C (P_B and P_C) are varied, which is due to the high pneumatic pressure in whorl A ($P_A = 71.5$ kPa). This is synonymous with the understanding gained from manometry where 71.5 kPa is above the pressure required at the geometry/seal transition zone (60-64 kPa). The co-actuating nature of the robotic device is also evident: where sensors A-E also exhibit displacement for actuation only in one chamber. This behavior inherently non-linear and is challenging to theoretically model, which are the motivations for empirical investigation and characterization.

24.5 Discussion and Conclusion

The specification, design, and characterization methods towards development of the soft-robotic peristaltic swallowing robot address different soft-robotic design challenges, from abstraction to application. Medical displacement and pressure measurement techniques have been found to be promising methods of investigation in the soft robotic space. This has been exacerbated by the bio-mechatronic demands of the research, as well as the motivation to develop biomimetic swallowing trajectories. This brings the model, and bio-mechanical understandings of the swallowing process closer together to facilitate interdisciplinary knowledge transfer.

The manometry and articulography methods offer complementary information as to the behavior of the novel peristaltic swallowing robot. These methods, developed for measurement of the human body, are particularly suitable in this domain as comparisons can be made between results measured in the biological and engineering systems. However, their main advantage is that they are designed to interact with soft tissues which undertake synonymous processes. This concept of using medical techniques in the soft robotic field may inspire many novel characterization methods for soft robots of differing origin. This class of robotic device requires new measurement and investigation techniques compared to their rigid robotic counterparts.

The characterization of this particular robot was undertaken to clarify the independent geometric and pressure behaviors in response to pneumatic actuation. The empirical method represents a basis for open-loop behavior description, as testing was conducted under dry-swallow conditions (without transporting a bolus).

In future, the swallowing robot is proposed to be applied as a tool to investigate swallowing of bolus materials, externally to the human body. In order to achieve this, its capabilities will be enhanced with additional self-sensing capabilities for pressure and geometry which can measure and model the device behavior online. It is anticipated to find application in the medical and food technology fields as a novel bolus investigation tool. The research initiative continues to build on the emerging concepts in the field of biomimetic, soft robotics.

24.6 References

[1] Clayton SB, Rife C, Kalbeisch JH, Castell DO (2013) Viscous impedance is an important indicator of abnormal esophageal motility. Neurogastroenterology and Motility 25(7):563-e455

[2] Dirven S, Chen F, Xu W, Bronlund J, Allen J, Cheng L (2014) Design and characterization of a peristaltic actuator inspired by esophageal swallowing. Mechatronics, IEEE/ASME Transactions on 19(4):1234-1242

[3] Gravesen FH, Gregersen H, Arendt-nielsen L, Drewes AM (2010) Reproducibility of axial force and manometric recordings in the oesophagus during wet and dry swallows. Neurogastroenterology and Motility 22(2):142-e47
[4] Jean A (2001) Brain stem control of swallowing: Neuronal network and cellular mechanisms. Physiological Reviews 81(2):929-969
[5] Kobayashi H, of Science TU (2004) Swallowing robot apparatus
[6] Kobayashi H, Minato A, Kobayashi T, Michiwaki Y (2005) Development of swallow robot for research of the mechanism for the human swallow. Nippon Kikai Gakkai Robotikusu 23:1P2N1
[7] Kuo P, Holloway RH, Nguyen NQ (2012) Current and future techniques in the evaluation of dysphagia. Journal of Gastroenterology and Hepatology 27(5):873-881
[8] Miki H, Okuyama T, Kodaira S, Luo Y, Takagi T, Yambe T, Sato T (2010) Artificial esophagus with peristaltic motion using shape memory alloy. International Journal of Applied Electromagnetics and Mechanics 33(1-2):705-711
[9] Noh Y, Segawa M, Sato K, Chunbao W, Ishii H, Solis J, Takanishi A, Katsumata A, Iida Y (2011) Development of a robot which can simulate swallowing of food boluses with various properties for the study of rehabilitation of swallowing disorders. In: Robotics and Automation (ICRA), 2011 IEEE International Conference on, pp 4676-4681
[10] Orvar KB, Gregersen H, Christensen J (1993) Biomechanical characteristics of the human esophagus. Digestive Diseases and Sciences 38(2):197-205
[11] Suzuki K, Nakamura T (2010) Development of a peristaltic pump based on bowel peristalsis using for articial rubber muscle. In: Intelligent Robots and Systems (IROS), 2010 IEEE/RSJ International Conference on, IEEE, pp 3085-3090
[12] Yang W, Fung T, Chian K, Chong C (2007) Finite element simulation of food transport through the esophageal body. World Journal of Gastroenterology 13(9):1352-1359

MIX
Papier aus verantwortungsvollen Quellen
Paper from responsible sources
FSC® C105338

If you have any concerns about our products,
you can contact us on
ProductSafety@springernature.com

In case Publisher is established outside the EU,
the EU authorized representative is:
Springer Nature Customer Service Center GmbH
Europaplatz 3, 69115 Heidelberg, Germany

Printed by Libri Plureos GmbH
in Hamburg, Germany